Measurement of Suspended Particles
by Quasi-Elastic Light Scattering

Measurement of Suspended Particles by Quasi-Elastic Light Scattering

Edited by

BARTON E. DAHNEKE

University of Rochester

A Wiley-Interscience Publication
JOHN WILEY & SONS
New York • Chichester • Brisbane • Toronto • Singapore

Library of Congress Cataloging in Publication Data

Main entry under title:

Measurement of suspended particles by quasi-elastic
  light scattering.

  "A Wiley-Interscience publication."
  Includes index.
  1. Particle size determination.   2. Optical
measurements.   3. Light—Scattering.   4. Light
beating spectroscopy.   I. Dahneke, Barton E.
TA418.8.M4   1983   620'.43'0287   82-17334
ISBN 0-471-87289-X

Printed in the United States of America

10 9 8 7 6 5 4 3 2 1

Preface

Scientists and engineers interested in aerosols, colloids, powders, and other particulate systems will, I hope, find intriguing the descriptions in this volume of several powerful measurement techniques, such as the various versions of QELS, and the complementary techniques of correlation spectroscopy, electrophoretic light scattering, fluorescence photobleaching recovery, laser Doppler velocimetry, steady and transient electric birefringence, and optical transients. These techniques have been and will yet be applied to many interesting measurements of particulate systems. Light scattering specialists will also, I hope, find this book useful because it collects together 20 contributions that focus on the theory and practice of particle measurement by light scattering and associated techniques, which are diverse in the range of application they cover.

The term "quasi-elastic light scattering" (QELS) and other terms used synonymously, namely, inelastic light scattering, dynamic light scattering, photon correlation spectroscopy, and intensity fluctuation spectroscopy, describe a powerful light-scattering technique that can provide information about random Brownian and systematic ordered motions of suspended particles and macromolecules. Because such motions frequently depend on a particle "size," such as the particle geometric diameter, hydrodynamic diameter, Stokes diameter, aerodynamic diameter, or molecular

v

weight, a QELS measurement provides information that can be used to determine the average size and the size distribution of the suspended particles or macromolecules. The QELS method thus provides a convenient, fast, nonintrusive measurement that has great promise of becoming a powerful tool in research and industrial laboratories and in industrial process development and control.

However, there are limitations on the average size and, especially, the size distribution information obtainable from QELS data by itself. These limitations result from the finite noise present in the QELS data coupled with the mathematically ill-posed nature of the inversion problem that must be solved to obtain the size distribution from the time or frequency distribution of the scattered light signals, from the extremely strong size dependence of the light-scattering cross section of small suspended particles and macromolecules, and from particle interaction and multiple scattering effects when the particle concentrations are not sufficiently dilute. All these limitations are specifically addressed in this volume, and methods are described for overcoming or reducing their impact or at least defining their consequences. Combinations of QELS with other types of measurements comprise useful tools for overcoming or reducing the inherent limitations of QELS data by itself, and several such combinations are described in detail.

The QELS method had its beginnings in the early 1960s. Today, some 20 years later, it is described by a literature of substantial extent. Most of this literature is not easily accessible because it is spread among journals addressing a wide variety of disciplines. Moreover, some important work on this method is just now emerging.

It was therefore my goal in organizing the symposium having the same title as this book, held in Santa Monica, California on February 18 and 19, 1982, to collect together a group of speakers that could authoritatively describe the current state of the art

of the measurement of suspended particles and macromolecules by QELS and complementary methods and
present and critically discuss new results and
promising directions.   The same goal has been sought
in collecting together in this volume the written
versions of the authors' presentations.

As a second goal I have sought to collect into a
single reference source, by means of the authors'
contributions, the most important previous results as
well as new results pertinent to the measurement of
suspended particles by QELS.

My third goal in organizing the symposium and
editing this volume was to expose QELS and "fine
particle" specialists to each other in the hope that
the resulting exchange of information will further
stimulate the development and application of QELS and
complementary methods in the measurement of aerosols
and colloids, especially the former.   The intensive
development and application of QELS technique to the
measurement of particles suspended in liquids has not
been accompanied by a comparable effort for aerosols.
Fortunately, much of the technique developed for
liquid suspensions can also be used for gas suspensions.   Examples of such transferability of technique
are described explicitly and implicitly throughout
this volume.   Conversely, the authors of Chapter 12
present, in their initial description of a promising
new QELS technique, a method they developed for
aerosols that can also be applied to liquid suspensions.   The strong focus of QELS development and
application to liquid suspensions has not been due to
any essential advantage of these suspensions, but
rather to the primary interest of the developers in
liquid suspensions.   This volume thus collects
together contributions on QELS theory and technique
as they apply to the measurement of suspended particles in both gases and liquids to emphasize the
generality of much of the theory and technique and
the value of such measurements in both kinds of
systems.

The response to these goals has been positive at every stage: first, by D. T. Shaw, who invited me to hold the symposium as part of the first annual meeting of the American Association for Aerosol Research (AAAR); second, by the authors who so graciously accepted their invitations to participate; third, by the publisher, who was supportive of this project from its beginning; and fourth, by the symposium audience. These consistently positive reactions have reinforced my belief in the timeliness and need for such a volume as this one that collects together in a single work both previous and new results pertaining to the measurement of suspended particles and macromolecules by QELS and, in so doing, defines the current state of the art of this method and indicates its promising directions and breadth of application.

This volume does not contain a comprehensive introductory treatment of QELS. Rather its contents are intended to focus on a specific subject range and treat it thoroughly. The excellent books of Berne and Pecora and of Chu (cited many times throughout this volume) already provide introductory treatments. Nevertheless, the two introductory chapters comprise reviews of QELS and correlation spectroscopy that are quite broad in scope.

I have organized the contents into four parts, including an introduction (two chapters), a part presenting the essentials of the measurement of particle size distribution by QELS (six chapters), a part on extended QELS and complementary methods for size distribution measurement (ten chapters), and a part treating the effects of particle interactions (two chapters).

I would like to express my thanks to the many people who assisted me in organizing the symposium and preparing this volume. These include N. C. Ford for co-chairing the symposium with me, B. Chu, N. C. Ford, I. Morrison, R. Pecora, and R. Pike for chairing the various sessions; N. C. Ford, K. Langley, I. Morrison, and R. Pecora for their help in selecting the speakers; D. T. Shaw, S. K. Friedlander, and the AAAR for a splendid background atmosphere for the symposium; the authors for their conscientious efforts in presenting their talks and preparing their chapters; B. Gray for reading the text and advising me on several technical points; B. Shube and D. Hillhouse of John Wiley and Sons for their able and compassionate editorial assistance in seeing the book through its conception and publication; and S. Beren of Computype for his competent typing of the camera-ready copy and careful attention to numerous other production details.

Barton E. Dahneke

Rochester, New York

October 1982

Contents

xi

xii   Contents

xiv   Contents

PART ONE
INTRODUCTION

CHAPTER 1

Quasi-Elastic Light Scattering of Macromolecules and Particles in Solution and Suspension

R. PECORA
Stanford University
Stanford, California

CONTENTS

I.   INTRODUCTION

Static light scattering — the measurement of time-averaged intensities of light scattered from dispersions of particles and macromolecules — is an important tool for the determination of particle structure [1]. In the early 1960s, however, it was noted that measurement of the time fluctuations of the intensity of the scattered light (or, equivalently, the spectral distribution of the scattered intensity) could be used to obtain information about the dynamics of particles and macromolecules in solution and suspension [2,3]. Since the initial work at Columbia University [2,3], this "quasi-elastic" light-scattering technique has developed into a sophisticated field with wide-ranging applications, mainly as the result of the work of a relatively few laboratories. In the past few years, however, there has been a surge of interest in the technique from a wide range of workers, including specialists in polymer chemistry, biophysics, and colloid science. This interest has been fueled by the power of the technique to yield information that is difficult, if not impossible, to obtain by other methods and by the advent of commercial instruments that allow nonspecialists to use it.

Quasi-elastic light scattering (QELS) appears in the literature under a plethora of names, some synonymous, and others referring to specific methods of analyzing the scattered light intensity. Among those names are *dynamic light scattering, intensity fluctuation spectroscopy, photon correlation spectro-*

scopy, light-beating spectroscopy, homodyne and
heterodyne spectroscopy, and *Fabry-Perot interferometry.* Dynamic light scattering is a general term
synonomous with QELS.   It includes all the other
techniques mentioned.   Photon correlation spectroscopy usually implies a "digital" technique for
measuring intensity fluctuations in which the number
of photons arriving at the detector in a given time
interval is measured and correlated.   Intensity
fluctuation spectroscopy includes all techniques in
which intensity fluctuations of scattered light are
measured (including photon correlation spectroscopy).
Fabry-Perot interferometry refers to direct measurements of the spectral distribution of the scattered
light intensity by use of Fabry-Perot interferometer
as a high-resolution monochromator.

At the foundation of the QELS technique is the
principle that the net intensity of light scattered
from a collection of scatterers is the result of
interference of the light scattered from each of the
scatterers in the illuminated region.   The phase and
polarization of the light scattered from each scatterer depends on its position and orientation.   As
these positions and orientations fluctuate as a
result of Brownian motion, the phase and polarization, and hence the net scattered intensity with a
given polarization, also fluctuate.   Thus these
fluctuations in the scattered intensity contain
information about the dynamics of the scatterers.

An alternative method of obtaining essentially
the same information is to analyze the frequency
distribution of the scattered intensity by using a
monochromator to separate the various frequency
components.   In these experiments on dispersions, the
incident light is essentially monochromatic and
exhibits a constant intensity.   The scattered light,
however, contains a distribution of frequencies very
close to that of the incident light.   The "beating"
of these components with each other gives rise to the
intensity fluctuations discussed above.   However, if
the frequencies are spread over a sufficiently broad

range, the frequency distribution could be directly measured. The decision as to which of these techniques is used in a given situation depends on the time scale of the intensity fluctuations (or, equivalently, the breadth of the scattered light frequency distribution; see below).

Some of the experimental techniques and applications of QELS to the study of liquid particle dispersions are briefly surveyed in this chapter. Velocimetry techniques, such as electrophoretic light scattering [4] and sedimentation velocity analysis [5] are not discussed at all. More complete discussions of fundamental principles and experimental techniques are given in two monographs on QELS [6,7].

II.   EXPERIMENTAL METHODS

Quasi-elastic light-scattering experiments employ a laser source, usually a continuous gas laser, and a photomultiplier detector. Argon ion lasers are the most commonly used lasers because of their combination of relatively high power, frequency and intensity stability, and general overall reliability. The choice of pre and post detection systems depends on the time scale of the dynamic quantity to be measured.

For slow processes (having time scale $\tau \gtrsim$ 1 μsec), the scattered light (after traversing some lenses, pinholes and, perhaps, a polarizer) impinges on the photomultiplier. The output of the photomultiplier (PM) is proportional to the scattered photon intensity that, as discussed in the introduction, fluctuates in time on a time scale τ characteristic of the thermal motions of the scatterers. The PM output is then passed through an amplifier discriminator system and fed into a device that computes the time autocorrelation function of the scattered intensity (or a quantity equivalent to it). This device is usually an autocorrelator, although in some cases it is a spectrum analyzer or a computer. The resulting autocorrelation function is then fed into a computer for further analysis. Most QELS experiments on particles dispersed in liquids are, in fact, of this type. These experiments are variously called *intensity fluctuation*, *photon correlation*, *optical beating*, and *homodyne* experiments.

In the case of faster fluctuations ($\tau \lesssim$ 1 μsec), a filter or monochromator is placed before the photomultiplier. This filter is usually a piezo-electrically swept Fabry-Perot interferometer, [6] although pressure-swept Fabry-Perots are sometimes used. The direct current (DC) output of the PM is measured for each filter setting as the filter is swept. On-line computers are usually available to collect and further analyze the spectrum. This technique gives the spectrum of scattered light directly.

Since for most particle dispersions, the maximum of the scattered spectrum is at the incident laser frequency and the line "width" is relatively small, this technique gave rise to the general term quasi-elastic light scattering. As mentioned in the introduction, it is usually called Fabry-Perot interferometry.

In both the intensity fluctuation and Fabry-Perot methods the incident and scattered light polarizations may be selected by appropriately placed polarizing prisms.

The plane of the incident and scattered beams define the *scattering plane*. The polarization of the incident light is usually chosen to be perpendicular to this plane. Given this choice for the incident polarization, two different polarizations for the scattered light are often chosen for study — each containing different information about the scatterer thermal motions. The first (and most common) choice is the component of scattered light perpendicular to the scattering plane (and parallel to the incident polarization). This component is usually called the *polarized component* or *vertical-vertical (VV) component*. The second choice is the scattered component in the scattering plane. This component is usually called the *depolarized component* or *vertical-horizontal* (VH) *component.*

Most QELS experiments measure the polarized component. In fact, since the intensity of the depolarized component for particles (size < 1 μm) in liquids is often several orders of magnitude less than that of the polarized component, experiments performed without any polarizers whatsoever usually yield the polarized QELS correlation function.

For small nonspherical particles (see below), the depolarized component is used to study the particle rotational diffusion coefficient. Since small particles rotate on a fast time scale, it is often necessary to use Fabry-Perot interferometry to study the rotational motions. Thus, for these small particles the polarized component will often be measured by the intensity fluctuation method whereas the depolarized scattering from the same system will often be observed by use of Fabry-Perot interferometry.

In the following, polarized and depolarized scattering are discussed separately. In this discussion the small contribution to the polarized component of terms proportional to the depolarized component is neglected since it is seldom easily detected. The last section contains an introduction to the techniques for studying particle size distributions by QELS.

III.   POLARIZED SCATTERING

A.   Translational Diffusion

Most QELS experiments on macromolecules and particles in solution and suspension measure the scatterer translational diffusion coefficient. It is easily shown that the polarized intensity time correlation function for a dilute solution of relatively small scatterers is given by [2,6,7]

$$\langle I_{VV}(0)I_{VV}(t)\rangle = A + B\exp[-2q^2Dt] \ , \tag{1}$$

where q is the length of the scattering vector, which depends on the wavelength of the incident light (in the scattering medium) λ and the scattering angle θ,

$$q = \frac{4\pi}{\lambda}\sin\frac{\theta}{2} \ . \tag{2}$$

The reciprocal of q is a characteristic length in scattering problems. For instance, Eq. (1) is "valid" for relatively small scatterers of any shape, that is, those scatterers whose characteristic size L satisfies $qL \lesssim 1$. For spheres, Eq. (1) is more general. It applies even to spheres as large as 1 μm in diameter. The decay constant of the exponential in Eq. (1)

$$\tau = \frac{1}{2q^2D}$$

is a measure of the time it takes a scatterer to travel a root-mean-square (RMS) distance q^{-1}.

The QELS technique for measuring D is rapid (for strong scatterers $\sim$ 1 min), precise (about 1-2% for clean samples of strong scatterers), requires only small volumes of scattering material (volumes as small as 0.01 ml have been used) and is nondestructive. Quasi-elastic light scattering is currently the technique of choice for measuring diffusion coefficients of relatively small particles ($\leq$ 1 μm) in dilute liquid dispersions.

The translational diffusion coefficient of a *spherical* particle is, at infinite dilution, related to the particle radius R by the Stokes-Einstein relationship,

$$D = \frac{k_B T}{6\pi\eta R} \tag{3}$$

where k_B is Boltzmann's constant, T is the absolute temperature, and η is the solvent viscosity. Thus measurements of D by QELS can be used to rapidly determine the size of spherical particles.

For ellipsoids of revolution and long thin rods, generalizations of Eq. (3) relate the particle translational diffusion coefficients to the two particle dimensions. For irregularly shaped particles and flexible particles such as random coils, Eq. (3) is often used to define a quantity called the *hydrodynamic radius* of a particle.

The Stokes-Einstein relationship and its generalizations for other shapes are valid only at infinite dilution, that is, in the regime in which the dispersed particles move independently of one another. Therefore, QELS data must usually be obtained at finite concentrations and the results must then be extrapolated to infinite dilution in order to obtain particle dimensions. Quasi-elastic light-scattering experiments at concentrations at which the particles influence each other's motions are commonly used to study particle interactions.

As a first approximation, the diffusion coefficient is usually expressed as a power series in the concentration, expressed as a particle volume fraction ϕ_S

$$D = D_0[1 + K_D\phi_S] + \cdots \tag{4}$$

where D_0 is the infinite dilution diffusion coefficient and K_D is a coefficient, which for hard spheres is predicted by various theories to be either positive or negative [8-11]. Comparisons of Eq. (4) with experiment are difficult since most of the systems studied to date contain particles that are charged, a complication not included in the theories for K_D.

If coulombic forces between charged particles are included, the theory becomes much more difficult, especially in the low-ionic-strength regime. In this regime there is little screening of the charges on the particles by the smaller counterions in solution, and the coulombic forces can extend over large distances. In fact, in this regime the apparent diffusion coefficient depends on q (or alternatively on the scattering angle) and the correlation function exhibits deviations from single exponential form.

At present there is much theoretical and experimental research being performed on these systems of concentrated, charged, colloidal particles. It is difficult to separate the effects of the hard-core interaction (excluded volume), solution nonideality, hydrodynamic interactions, and charge, even in the case of spherical particles. Excellent reviews of this field have been given by Hess [12] and Pusey and Tough [13].

Quasi-elastic light-scattering studies of the infinite dilution diffusion coefficient are now commonly used to obtain particle sizes and to examine processes in which changes in particle size or shape occur. For example, the onset of the helix coil transition in a polypeptide has been followed by examining changes in D as the transition proceeds [14]. The denaturation of proteins has been studied by similar techniques [15,16].

B. Intramolecular and Rotational Relaxation Times

Information about rotational and intramolecular relaxation times of particles and macromolecules in dilute solution can often be obtained from polarized QELS measurements.

The basis of the method is that if a particle is sufficiently large, light scattered from different parts of the same molecule interferes to produce a characteristic angular dependence of the scattered intensity. For example, therefore, spheres, rods,

and Gaussian coils exhibit a *form factor* that gives
an angular dependence to the scattering that is
characteristic of the particular particle shape.
Measurements of this angular dependence are commonly
employed in static light scattering [1,7]. In QELS,
on the other hand, the time dependence of fluctuations in orientation or shape or size is important.
The time dependence of these fluctuations cannot be
observed unless the fluctuation exhibits a sufficiently large amplitude (compared to q^{-1}) for the
different states (orientations, shapes, or sizes) to
have different scattered intensities. In this case
the fluctuation will give rise to intensity fluctuations that can be observed by QELS.

The case of a thin, rigid rod that is reorient-
ing in a fluid illustrates this principle. If the
rod is long enough, the rod, because of the different
positions of its segments relative to the scattering
vector, exhibits varying amounts of intramolecular
interference as it reorients. Thus the scattered
intensity fluctuates as the polarized QELS time
correlation function and contains information about
the rate of rotational motion.

The theory for rigid rods has been formulated by
Pecora [2,17]. If it is assumed that translation and
rotation of the rod are independent of each other,
the polarized intensity time correlation function may
be shown to be given by [17].

$$\langle I_{VV}(0)I_{VV}(t)\rangle \propto [S_0(qL)\exp[-q^2Dt]$$
$$+ S_1(qL)\exp[-(q^2D+6\Xi)t] + \cdot \cdot \cdot]^2 , \qquad (5)$$

where L is the rod length and Ξ is the rotational
diffusion coefficient of the long axis of the rod.
The amplitude factors S_0 and S_1 are *known functions*
of qL [17].

In accordance with the preceding discussion, it
is clear that as $qL \to 0$, $S_1(qL)$ must vanish and
$S_0(qL) \to 1$. Thus at low qL ($\lesssim 3$), the translational
diffusion coefficient D may be obtained and at higher

qL ($\gtrsim 3$) the *rotational term* in Eq. (5) may be "turned on" and a fit of the result to Eq. (5) with the already obtained value of D may be used to obtain the rotational diffusion coefficient. This procedure has been used to study several long, rigid, thin, rod-shaped particles, including viruses [18-21] and structural proteins [22].

Examination of the polarized time correlation function as a function of q and a comparison of the results with Eq. (5) may sometimes be used to determine whether particles are rodlike. For instance, Flamberg and Pecora [23] have concluded from QELS studies that at high salt concentrations certain micelles become long, thin rods.

Similar calculations may be done for particles of other shapes. For a rigid sphere, it is evident from the above discussion that the rotation will not affect the time correlation function and that Eq. (1) will apply over the whole range of sizes (for diameter $\lesssim 1$ μm). Tanaka [25] has numerically calculated the weight functions $S_0(qL)$ and $S_1(qL)$ for prolate ellipsoids, whereas Aragon and Pecora [26] have presented formulas for disk-shaped particles. Similarly, $\langle I_{VV}(0)I_{VV}(t) \rangle$ for rods with joints in the middle (once broken rods) have been calculated [27-29].

Calculations and polarized QELS experiments have also been performed on flexible particles such as Gaussian coil polymers and long DNAs.

The theory for Gaussian coil polymers obeying the Rouse-Zimm dynamic model shows [30,31] that the polarized intensity time correlation function is the same form as Eq. (5) when the product of q and the radius of gyration of the coil R_G is

$$qR_G \sim 1.$$

The difference from Eq. (5) is that 6Ξ is replaced by a relaxation time for the longest normal intramolecular relaxational mode of the polymer. The weight functions S_0 and S_1 also, of course, are different.

When $qR_G \ll 1$, the term in Eq. (5) containing the intramolecular relaxation time is negligible and the polarized QELS correlation function is given by Eq. (1). In this regime the fluctuations in size are small compared to q^{-1}; thus the coil appears to be a diffusing point. This method for studying intra-molecular relaxation in flexible macromolecules has been applied to high-molecular-weight polystyrenes in solution by several groups [32-38]. Recently much theoretical work [39-43] has been performed on compu-tation of polarized QELS time correlation functions of flexible polymers by use of the projection opera-tor techniques that have been introduced into polymer physics by Bixon [44] and Zwanzig [45].

If $qR_G \gg 1$, Eq. (5) is no longer adequate for treatment of experimental data since higher terms in the series become very important. DuBois-Violette and de Gennes [46] have calculated the full polarized QELS correlation functions for Rouse-Zimm chains in this limit. Adam and Delsanti [47] found good agree-ment between the shapes of their experimentally determined QELS correlation functions and the shape predicted by DuBois-Violette and de Gennes.

Many particles, especially those of biological importance, are semiflexible and require more complex dynamic theories than do those mentioned for stiff rods and flexible coils for description of their intraparticle motions. Semiflexible particles whose intraparticle motions have been studied by polarized QELS include F-Actin [48-51], plasmodium actin [52], bacterial flagella [53], DNA [54-66], and synthetic polynucleotides [67-69].

IV.   DEPOLARIZED SCATTERING

A.   Rotational Diffusion

For dilute solutions of rigid cylindrically symmetrical molecules with optical anisotropy β, the time correlation function of the depolarized inten-

sity is related to the rotational diffusion coefficient of the rod axis [70]

$$\langle I_{VH}(0)I_{VH}(t)\rangle = A + B \exp[-2(q^2D+6\Xi)t] , \qquad (6)$$

where the constant B is proportional to β^4.
Depolarized QELS experiments are sometimes performed
at zero scattering angle; that is, the light scattered into the forward direction is observed [71-73].
Since $q = 0$ for this experimental geometry, there is
no dependence of the depolarized time correlation
function on D ([$q^2D = 0$ in Eq. (6)].

If the rotational diffusion of the rod is fast
[$(6\Xi)^{-1} \lesssim 1$ µsec], depolarized intensity fluctuation
(photon correlation) experiments become very difficult to perform; however, Fabry-Perot interferometry
experiments can often be performed instead in this
regime.

In the Fabry-Perot experiment analogous to that
described above the depolarized scattered intensity
is a single Lorentzian centered at the incident laser
frequency [70]

$$I_{VH}(q,\omega) = A\beta^2 \frac{(q^2D+6\Xi)}{\omega^2 + (q^2D+6\Xi)^2}, \qquad (7)$$

where ω is the frequency difference of the scattered
light from the incident laser frequency. In Eq. (7)
q^2D is usually negligible in comparison to 6Ξ, and
the rotational diffusion coefficient is the dominant
contribution to the spectral width. For particles
dispersed in water with characteristic dimensions
$\lesssim 50$ Å, the Fabry-Perot method is most often the
method of choice for measurement of rotational diffusion coefficients.

A frequent procedure for characterizing rigid
macromolecules in solution is to combine polarized
QELS measurements of D with depolarized QELS measurements of Ξ. The rotational and translational
diffusion coefficients obtained can then be used in

conjunction with theoretical hydrodynamic relationships to obtain particle solution dimensions [74,75].

An example of such a procedure is given by
Michielsen and Pecora [75], who studied gramicidin, a
linear pentadecapeptide molecule believed to form
dimeric rod-shaped particles in alcohol solutions.
By comparing Fabry-Perot measurements of Ξ with
photon correlation measurements of D, these authors
confirmed that gramicidin exists as dimers in both
methanol and ethanol. For instance, the solution
dimensions of this rod in ethanol were found to be
29 Å for the length and 17 Å for the cross section
diameter.

B.   Intraparticle Relaxation Times

Depolarized QELS of flexible and semi flexible
particles can, in favorable circumstances, be used to
study intraparticle relaxation times [70]. The
mechanism by which these motions affect the depolarized QELS is different from that for the polarized
QELS. In polarized QELS a structural fluctuation
must be of extent $\sim q^{-1}$ to appreciably affect the
correlation function. In the depolarized case,
modulation of the scattered intensity can occur by
purely "local" reorientations of optically aniso-
tropic segments of a particle. Thus depolarized QELS
is capable of studying intraparticle motions of small
particles (or, equivalently, of large particles in
the $q = 0$ limit). The major difficulty with depolarized QELS for these purposes is the relatively low
signal level and consequent poor signal/noise ratio
of the correlation function.

Bauer et al. [76] have observed depolarized QELS
spectra of polystyrenes in dilute solution by using
both Fabry-Perot interferometry and intensity fluctuation spectroscopy. The spectra exhibit at least two
components — a slow component whose relaxation time
is proportional to molecular weight and a fast component whose relaxation time is independent of molecular weight. For dilute polystyrenes in CCl_4, the

fast relaxation time is 4.5 ± 1.0 nsec. The slow relaxation time in these systems is identified with half the relaxation time of the longest Rouse-Zimm mode for the chain, and the fast, molecular-weight-independent time is assigned to a local, correlated motion of phenyl groups about the main chain backbone. Monte Carlo simulations of the dynamics of chains on a tetrahedral lattice with optically aniso-tropic side groups give depolarized spectra that are qualitatively similar to those observed in poly-styrenes [77]. Moro and Pecora [78] have attempted to explain the dominance of the longest chain mode in the low-frequency part of the Fabry-Perot spectrum by introducing local transverse stiffness into the model for chain dynamics. Evans [79] has proposed that intramolecular collision-induced scattering is responsible for the low mode dominance.

Zero and Pecora [80,81] have constructed a model of depolarized light scattering from once broken rods with a restricted diffusion angle at the hinge. Their model has been used to interpret depolarized QELS experiments on the myosin rod. As yet very few depolarized QELS experiments have been performed on flexible particles in solution.

C. Semidilute Solutions

Rigid-rod particles in semidilute solution have been studied by both polarized [82,83] and depolar-ized [82] QELS as well as by electric birefringence [84]. Theories [85] of rotational motion of rods at these high concentrations predict a drastic slowing down of the rotational diffusion of the long axis of the rod due to steric hindrance of neighboring rods. Polarized and depolarized QELS experiments on poly-γ-benzyl-L-glutamate [82] and light meromyosin [80] have shown that there is a slowing down, although not nearly so much as that predicted by theory.

V. POLYDISPERSITY

The determination of particle size distributions is one of the most important practical problems confronting the polymer and colloid scientist. In a dispersion of particles in a liquid, each size particle has a different scattered intensity and a different translational diffusion coefficient. Thus, instead of Eq. (1), the polarized QELS correlation function consists of a weighted sum of exponentials, each with a different decay constant. The net polarized QELS correlation function observed, therefore, is nonexponential and represents an average over the distribution of sizes. Methods of analyzing measured QELS correlation functions for information about particle distributions are currently the subject of much research effort. Some of the major approaches to this problem are briefly described below.

In the following, it is assumed that the particle dispersion is sufficiently dilute that interparticle interactions are negligible and that particle rotation and intraparticle motions do not contribute significantly to the correlation functions.

Let each size particle in a dispersion have number concentrations N_i, polarizability α_i, form factor P_i [7], and diffusion coefficient D_i. Then the polarized QELS correlation function is

$$\langle I_{VV}(0)I_{VV}(t)\rangle = A + B\left(\sum_i N_i \alpha_i^2 P_i \exp[-q^2 D_i t]\right)^2, \quad (8)$$

where the sum is over the different size particles present.

It is convenient to define a continuous distribution function of sizes. Let $f(R)\, dR$ be the fraction of particles with radius R between R and $(R + dR)$. If it is noted that α_i, P_i, and D_i are all functions of R, the sum in Eq. (8) may be replaced by an integral

$<I_{VV}(0)I_{VV}(t)> - A$

$$= B\left(\int_0^\infty \alpha^2(R)f(R)P(R)\exp[-q^2D(R)t]dR\right)^2. \qquad (9)$$

The problem is then to obtain information about $f(R)$ from measurement of $<I_{VV}(0)I_{VV}(t)>$. There are three main approaches to this problem, each of which is briefly discussed in turn.

A. Cumulants

The first approach, known as the method of *cumulants* [86], is the simplest method. It yields the first two moments of the distribution.

Take the logarithm of the normalized QELS correlation function and expand it as a power series in t

$$- \frac{1}{2} \ln \left\{ \frac{[<I_{VV}(0)I_{VV}(t)> - A]}{B'} \right\}$$

$$= K_1 t - K_2 \frac{t^2}{2} + \cdot \cdot \cdot , \qquad (10)$$

where K_1, K_2, $\cdot \cdot \cdot$ are known as the first, second, $\cdot \cdot \cdot$ cumulants. If the correlation function were a single exponential (monodisperse dispersion), it is evident that the first cumulant is

$$K_1 = q^2 D$$

and all higher cumulants are zero. In general, the first two cumulants are given by

$$K_1 = q^2 <D> \qquad (11)$$

and

$$K_2 = q^4 <(D - <D>)^2> ,$$ (12)

where the $<\cdot \cdot \cdot>$ indicate a rather complicated average. For instance,

$$<D> \equiv \frac{\int_0^\infty D(R)\alpha^2(R)f(R)P(R)dR}{\int_0^\infty \alpha^2(R)f(R)P(R)dR} .$$ (13)

Thus the first cumulant yields an "average" diffusion coefficient, and the second cumulant yields a mean-squared deviation from this average.

If we consider the distribution of molecular weight instead of size it is necessary only to replace R in Eqs. (9) and (13) with molecular weight. Usually α for a given type of polymer is proportional to the molecular weight. Then, if the P_i are *negligible*, it is easy to show that the first and second cumulants in Eqs. (11) and (12) become the "Z average diffusion coefficient" and the mean-square fluctuation of the Z average diffusion coefficient

$$<D>_z \equiv \frac{\int_0^\infty D(M)M^2 f(M)dM}{\int_0^\infty M^2 f(M)dM} .$$ (14)

In summary, from measurements of the polarized QELS correlation function at small times, an "average" diffusion coefficient and a mean-square deviation from the average may be obtained. In practice it is difficult to measure cumulants higher than the second. Thus the cumulants method gives no information on the form of the distribution function. It cannot, for instance, distinguish between a monomodal and a bimodal distribution of sizes. It can, however, be quite useful in showing whether there is some sample polydispersity and giving a rough measure of it.

B. Fit to a Known Distribution Function

If the *form* of the distribution function [87-89] and the dependences of α, P, and D on size (or molecular weight) are known for a particular particle dispersion, the integral in Eq. (9) may be calculated in terms of the parameters characterizing the distribution function.

For instance, there are many cases in which the particle dispersion might be expected to obey a Schulz distribution [90,91]

$$ f(R) = \frac{1}{Z!} \left[\frac{Z + 1}{\langle R \rangle} \right]^{z+1} R^{z} \exp \left[- \frac{(Z+1)R}{\langle R \rangle} \right], \quad (15) $$

where Z and $\langle R \rangle$ are parameters characterizing the distribution. The Schulz distribution is unimodal and is skewed toward larger sizes. As the parameter Z becomes larger, the distribution becomes more narrow and Gaussianlike and as Z approaches infinity, it becomes monodisperse.

If Eq. (15) is assumed to be the form of the distribution function, numerical calculations of the integral in Eq. (9) can be performed in terms of the parameters Z and $\langle R \rangle$ for given particle shapes [which determine $\alpha(R)$, $P(R)$, and $D(R)$]. Measurement of $\langle I_{VV}(0)I_{VV}(t) \rangle$ may then be used to extract values of Z and $\langle R \rangle$ [87-89].

This method is rather cumbersome and difficult to apply. In addition, a major difficulty arises if the distribution deviates from Eq. (15). For instance, even dispersions that initially obey a

Schulz distribution may undergo aggregation in time to form a multimodal distribution.

C. Inversion of the Laplace Transform

The basis for this method is to note that Eq. (9) may be written in terms of a Laplace transform,

$$\langle I_{VV}(0)I_{VV}(t)\rangle - A = B'\left(\int_{0}^{\infty} \exp[-\Gamma t]G(\Gamma)d\Gamma \right)^2 \quad , \quad (16)$$

where $\Gamma \equiv q^2 D$ and $G(\Gamma)$ is easily found from Eq. (9),

$$G(\Gamma) = \frac{dR}{d\Gamma} \alpha^2(R)P(R)f(R) \quad . \quad (17)$$

The quantity $dR/d\Gamma$ may be evaluated from a knowledge of the dependence of the particle translational diffusion coefficient on R. For spheres, Eq. (3) is used.

Numerical techniques may then be used to invert the Laplace transform in Eq. (16) to find the function $G(\Gamma)$ from the measured $\langle I_{VV}(0)I_{VV}(t)\rangle$. Note that $G(\Gamma)$ gives the fraction of the correlation function that relaxes with time constant Γ^{-1}; however this itself is not the distribution $f(R)$ [which must be found from Eq. (17)].

The numerical techniques that are used to invert the transform in Eq. (16) depend on choosing only "smoothed" $G(\Gamma)$. Various approaches are described in this volume. The "histogram" method is described by Chu and DiNapoli [92], Bedwell et al. [93], and Tang et al. [94], the "spline" method is described by Goll and Stock [95], and more complex numerical techniques are described by Pike et al. [96] and Bott [97]. The reader is referred to these chapters for further information on these inversion techniques. They are currently being applied to QELS experiments on a wide variety of systems and are providing rapid and reliable $G(\Gamma)$.

Most work on particle distribution analysis by QELS has been performed by use of the polarized component. Since depolarized QELS correlation functions usually exhibit more unfavorable signal/noise ratios than do the polarized ones, it is more difficult to extract particle distribution information from them. Crosby et al. [98] have demonstrated that it is feasible to obtain distribution function information from depolarized QELS. They used a histogram analysis of a forward-angle depolarized correlation function to find the molecular-weight distribution function of a rigid macromolecule in solution.

REFERENCES

1. M. Kerker, *The Scattering of Light and Other Electromagnetic Radiation*, Academic Press, New York, 1969.

2. R. Pecora, J. Chem. Phys. *40*, 1604 (1964).

3. H. Z. Cummins, N. Knable, and Y. Yeh, Phys. Rev. Lett. *12*, 150 (1964).

4. See, for instance, B. A. Smith and B. R. Ware, Contemp. Top. Anal. Clinical Chem. *2*, 29 (1978).

5. I. Chabay, Chapter 13, this volume.

6. B. Chu, *Laser Light Scattering*, Academic Press, New York, 1974.

7. B. J. Berne and R. Pecora, *Dynamic Light Scattering*, Wiley-Interscience, New York, 1976.

8. J. M. Burgers, Proc. Acad. Sci. Amst. *44*, 1045, 1177 (1941); *45*, 9, 126 (1942).

24   R. Pecora

9. C. W. Pyun and M. Fixman, J. Chem. Phys. *41*, 937 (1964).

10. A. R. Altenburger and J. M. Deutch, J. Chem. Phys. *59*, 894 (1973).

11. G. K. Batchelor, J. Fluid Mech. *52*, 243 (1972); *74*, 1 (1976).

12. W. Hess, in *Light Scattering in Fluids and Macromolecular Solutions*, V. Digiorgio, M. Corti, and M. Giglio, Eds., Plenum Press, New York, 1980.

13. P. N. Pusey and R. J. A. Tough, in *Dynamic Light Scattering: Applications of Photon Correlation Spectroscopy*, R. Pecora, Ed., Plenum Press, New York, in press. QC45+L6 3 D96

14. N. C. Ford, W. Lee, and F. E. Karasz, J. Chem. Phys. *50*, 3098 (1969).

15. D. F. Nicoli and G. B. Benedek, Biopolymers *15*, 2421 (1976).

16. C.-C. Wang and R. Pecora, Biophys. Chem. *11*, 439 (1980).

17. R. Pecora, J. Chem. Phys. *48*, 4126 (1968).

18. H. Z. Cummins, F. D. Carlson, T. J. Herbert, and G. Woods, Biophys. J. *9*, 518 (1969).

19. D. W. Schaefer, G. B. Benedek, P. Schofield, and E. Bradford, J. Chem. Phys. *55*, 3884 (1971).

20. E. Loh, E. Ralston, and V. N. Schumaker, Biopolymers *18*, 2549 (1979).

21. E. Loh, Biopolymers *18*, 2569 (1979).

22. J. S. Hwang and H. Z. Cummins, J. Chem. Phys.
     77, 616 (1982).

23. A. Flamberg and R. Pecora, in *Scattering Tech-
     niques Applied to Supramolecular and Nonequil-
     ibrium Systems*, S.-H. Chen, B. Chu, and R.
     Nossal, Eds., Plenum Press, New York, 1981.

24. R. Pecora and S. R. Aragon, Chem. Phys. Lipids
     13, 1 (1974).

25. T. Tanaka, J. Colloid Interface Sci. *64*, 171
     (1976).

26. S. R. Aragon and R. Pecora, J. Chem. Phys. *66*,
     2506 (1977).

27. R. Pecora, Macromolecules *2*, 31 (1969).

28. M. Fujiwara and N. Saito, Polym. J. *11*, 249
     (1979).

29. M. Fujiwara, N. Numasawa, and N. Saito, Rep.
     Progr. Polym. Phys. Jpn. *23*, 531 (1980).

30. R. Pecora, J. Chem. Phys. *43*, 1562 (1965); 49,
     1032 (1968).

31. A. Perico, P. Piaggio, and C. Cuniberti, J.
     Chem. Phys. *62*, 2690 (1975).

32. T. F. Reed and J. F. Frederick, Macromolecules
     4, 72 (1971).

33. O. Kramer and J. E. Frederick, Macromolecules 5,
     69 (1972).

34. W. Huang and J. E. Frederick, Macromolecules 7,
     34 (1974).

35. T. A. King, A. Knox, and J. D. G. McAdam, Chem. Phys. Lett. *19*, 351 (1973).

36. T. A. King, A. Knox, and J. D. G. McAdam, J. Polym. Sci. Polym. Symp. *44*, 195 (1974).

37. J. D. G. McAdam and T. A. King, Chem. Phys. *6*, 109 (1974).

38. B. E. A. Saleh and J. Hendrix, Chem. Phys. *12*, 25 (1976).

39. Z. Akcasu and H. Gurol, J. Polym. Sci., Polym. Phys. *14*, 1 (1976).

40. A. Z. Akcasu, M. Benmouna, and C. C. Han, Polymer *21*, 866 (1980).

41. W. Burchard, Polymer *20*, 577 (1979).

42. W. Burchard, Macromolecules *11*, 455 (1978).

43. W. Burchard, M. Schmidt, and W. H. Stockmayer, Macromolecules *13*, 580 (1980).

44. M. Bixon, J. Chem. Phys. *58*, 1459 (1973).

45. R. Zwanzig, J. Chem. Phys. *60*, 2717 (1974).

46. E. DuBois-Violette and P. G. deGennes, Physics *3*, 181 (1967).

47. M. Adam and M. Delsanti, J. Phys. Lett. *38*, L271 (1977).

48. S. Ishiwata and S. Fujime, J. Phys. Soc. Jpn. *30*, 302 (1970); *31*, 1601 (1971).

49. S. Ishiwata and S. Fujime, J. Molec. Biol. *68*, 511 (1972).

50. F. D. Carlson and A. B. Fraser, in *Photon Corre-lation and Light Beating Spectroscopy*, H. Z. Cummins and E. R. Pike, Eds., Plenum Press, New York, 1974.

51. F. D. Carlson and A. B. Fraser, J. Molec. Biol. *89*, 283 (1974).

52. S. Fujime and S. Hatano, J. Mech. Chem. Cell Motil. *1*, 81 (1972).

53. S. Fujime, M. Maruyama, and S. Asakura, J. Molec. Biol. *68*, 347 (1972).

54. S. B. Dubin, J. H. Lunacek, and G. B. Benedek, Proc. Natl. Acad. Sci. (USA) *57*, 1164 (1967).

55. K. S. Schmitz and J. M. Schurr, Biopolymers *12*, 1543 (1973).

56. R. L. Schmidt, Biopolymers *14*, 521 (1973).

57. K. S. Schmitz and R. Pecora, Biopolymers *14*, 521 (1975).

58. K. L. Wun and W. Prins, Biopolymers *14*, 111 (1975).

59. D. Jolly and H. Eisenberg, Biopolymers *15*, 61 (1976).

60. J. M. Schurr, Quart. Rev. Biophys. *9*, 109 (1976).

61. F. C.-Chen, A. Yeh, and B. Chu, J. Chem. Phys. *66*, 1290 (1977).

62. R. L. Schmidt, J. A. Boyle, and J. A. Mayo, Biopolymers *16*, 317 (1977).

63. R. L. Schmidt, M. A. Whitehorn, and J. A. Mayo, Biopolymers *16*, 327 (1977).

64. M. Caloin, B. Wilhelm, and M. Daune, Biopolymers *16*, 2091 (1977).

65. S. C. Lin and J. M. Schurr, Biopolymers *17*, 425 (1978).

66. N. Parthasarathy, K. S. Schmitz, and M. K. Cowman, Biopolymers *19*, 1137 (1980).

67. P. Mathiez, G. Weisbuch, and C. Mouttet, J. Phys. Lett. *39*, L139 (1978).

68. P. Mathiez, C. Mouttet, and G. Weisbuch, Biopolymers *18*, 1465 (1979).

69. P. Mathiez, C. Mouttet, and G. Weisbuch, J. Phys. *41*, 519 (1980).

70. R. Pecora, J. Chem. Phys. *49*, 1036 (1968).

71. A. Wada, N. Suda, T. Tsuda, and K. Soda, J. Chem. Phys. *50*, 31 (1969).

72. C. Han and H. Yu, J. Chem. Phys. *61*, 2650 (1974).

73. J. C. Thomas and G. C. Fletcher, Biopolymers *18*, 1333 (1979).

74. S. B. Dubin, N. A. Clark, and G. B. Benedek, J. Chem. Phys. *54*, 5158 (1971).

75. S. Michielsen and R. Pecora, Biochemistry *20*, 6994 (1981).

76. D. R. Bauer, J. I. Brauman, and R. Pecora, Macromolecules *8*, 443 (1975).

77. C. Cornelius, Ph.D. dissertation, Stanford University, 1979.

78. K. Moro and R. Pecora, J. Chem. Phys. *69*, 3254 (1978); *72*, 4958 (1980).

79. G. T. Evans, J. Chem. Phys. *71*, 2263 (1979).

80. K. Zero and R. Pecora, Macromolecules *15*, 1023 (1982).

81. K. M. Zero, Chapter 14, this volume.

82. K. M. Zero and R. Pecora, Macromolecules *15*, 87 (1982).

83. J. F. Maguire, J. Chem. Soc. Faraday II 77, 513 (1981).

84. J. F. Maguire, J. P. McTague, and F. Rondelez, Phys. Rev. Lett. *45*, 1891 (1980); *47*, 148 (1981).

85. M. Doi and S. F. Edwards, J. Chem. Soc. Faraday II *74*, 560 (1978).

86. D. E. Koppel, J. Chem. Phys. *57*, 4814 (1972).

87. Y. Tagami and R. Pecora, J. Chem. Phys. *51*, 3293, 3298 (1969).

88. T. F. Reed, Macromolecules 5, 771 (1972).

89. S. R. Aragon and R. Pecora, J. Chem. Phys. *64*, 2395 (1976).

90. G. V. Schulz, Z. Phys. Chem. 43, 25 (1939).

91. F. W. Billmeyer, Jr. and W. H. Stockmayer, J. Polymer Sci. 5, 122 (1950).

92. B. Chu, and A. DiNapoli, Chapter 3, this volume.

93. B. Bedwell, E. Gulari, and D. Melik, Chapter 8, this volume.

94. H. I. Tang, P. L. Johnson, and E. Gulari, Chapter 17, this volume.

95. J. H. Goll, and G. B. Stock, Chapter 6, this volume.

96. E. R. Pike, D. Watson, and F. McNeil Watson, Chapter 4, this volume.

97. S. E. Bott, Chapter 5, this volume.

98. C. R. Crosby, N. C. Ford, F. E. Karasz, and K. H. Langley, J. Chem. Phys. 75, 4298 (1981).

CHAPTER 2

Theory and Practice of Correlation Spectroscopy

N. C. FORD
Langley-Ford Instruments
Amherst, Massachusetts

CONTENTS

I.   INTRODUCTION

In 1965 two groups demonstrated independently that fluctuations in laser light-scattering intensity could be used to measure properties of the medium responsible for the scattering [1,2]. In both cases analysis of the low-frequency components of the fluctuations was used to deduce the thermal diffusivity of a fluid near its critical point. Two years later it was demonstrated [3] that similar techniques could be used to measure the diffusion constant of small particles or macromolecules in solution. Since then the use of correlation spectroscopy has expanded rapidly, and today several hundred laboratories are equipped to make the necessary measurements. Indeed, several manufacturers are now making integrated systems that utilize light scattering to measure the size of small particles. As the technique becomes more widely used, the need for technical descriptions that are accessible to the nonphysicist increases.

It is the purpose of this chapter to provide such a description, bearing in mind the needs of the researcher studying aerosols.

We begin with a brief description of the scattering process with emphasis on the time dependence of the scattered intensity. In particular, we show how information such as the translational diffusion constant can be obtained by making measurements of the spectrum of the scattered light.

With this material as background, we then describe the typical light-scattering apparatus in considerable detail. Each major component of the overall apparatus is the subject of a separate section. We conclude the chapter with a brief mention of several other applications of the correlator.

II. LIGHT SCATTERING

A. Background

Light will be scattered by a molecule in solution if the molecule has a polarizability different from its surroundings. In this case the oscillating dipole moment induced by the electric field of the incident light beam will radiate light in all directions. The intensity of the scattered light will be related to the direction of polarization of the incident light, scattering angle, and solution parameters. If it is assumed that incident light is linearly polarized and if the angles are defined as in Fig. 1, the scattered intensity due to a single molecule with dimensions much smaller than λ is given by [4]

$$I_s(1) = \frac{4\pi^2 M^2 \sin^2\phi (dn/dc)^2 I_0}{N_A^2 \lambda^4 R^2} = \frac{S I_0}{R^2} . \qquad (1)$$

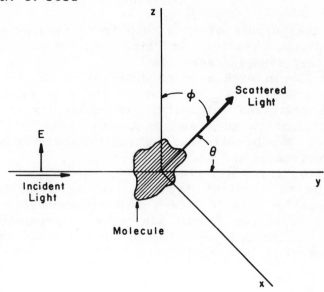

FIGURE 1.   The angles required to give a mathematical description of light scattering are defined here. The electric field in the incident beam is along the z direction, and the incident beam propagates in the y direction; ϕ and θ are the angles between the propagation direction of the scattered beam and the z and y axes, respectively.

In this equation M is the molecular weight of the molecule in daltons; dn/dc is rate of change of index of refraction of the solution as the concentration of the solute is changed, I_0 is the incident light intensity, N_A is Avogadro's number, λ is the wavelength of light in the solution, and R is the distance from the scattering point to the observation point. The scattered light will be linearly polarized if the incident light is polarized. The direction of polarization will lie in the plane determined by the direction of polarization of the incident light beam and the scattering direction. Notice that if these two directions are parallel, the angle ϕ is zero and the scattered intensity vanishes according to Eq. (1). Thus there is no ambiguity in determining the direction of polarization of the scattered light.

If a molecule is nonspherical or if it has unequal polarizabilities along two directions, there will be a component of the scattered light with polarization perpendicular to the direction defined above. The intensity of this depolarized light (frequently given the symbol I_{VH}) is usually orders of magnitude lower than that of the polarized component (I_{VV}); however, useful information concerning the rotational diffusion constant of the scattering molecule can be obtained from measurements of I_{VH}.

B. Fluctuations

When more than one molecule is responsible for the light scattering (as is usually the case), we must first calculate the scattered electric field $\mathbf{E}_{si}$ for each scattering molecule and then sum the electric fields and square the result to obtain the scattered light intensity [5]. If all the molecules are identical, the electric field contributions due to the individual molecules will have the same magnitude but the phases will be different. The summation of the electric fields is most conveniently visualized by use of a two-dimensional plot in which each scattered electric field is represented by a line segment with length proportional to the magnitude of $\mathbf{E}_s$ and making an angle with the abscissa equal to the phase angle of $\mathbf{E}_s$ with respect to light scattered from the center of the scattering region. The vector sum of the line segments will then give the resultant total scattered electric field as shown in Fig. 2.

The sum illustrated graphically in Fig. 2 may be done explicitly. For the component of $\mathbf{E}_s$ in phase with the reference beam, we find

$$E_s^i = E_s \sum_i \cos \delta_i \tag{2}$$

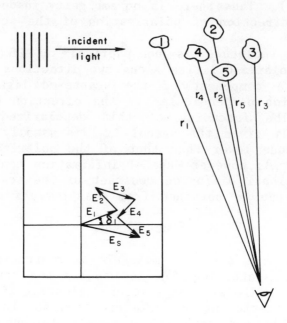

Figure 2.   For calculation of the intensity of light
            scattered from a collection of molecules,
            the contribution to the electric field of
            the scattered light E_s due to the indi-
            vidual molecules must be added, taking
            into account the different phase angles.
            This process may be visualized by repre-
            senting each contribution to E_s as a
            vector whose length gives the magnitude
            and whose angle with respect to the x axis
            gives the phase angle of that contribution
            to E_s.   Calculation of E_s is then equiva-
            lent to performing vector addition of the
            individual contributions as shown in the
            inset.

and for the component with a phase angle of 90°

$$E_s^0 = E_s \sum_i \sin \delta_i \tag{3}$$

where E_s represents the magnitude of $\mathbf{E}_s$.
According to the theory of light, the resulting intensity will be given by

$$I_s = \frac{(E_s^0)^2 + (E_s^i)^2}{377}$$

$$= \frac{1}{377} E_s^2 \left(\sum_{i=1}^{N} \cos \delta_i \right)^2 + \left(\sum_{i=1}^{N} \sin \delta_i \right)^2 \tag{4}$$

where I_s is in W/m^2 and N is the number of molecules in the scattering volume. Equation (4) provides the starting point for a discussion of photon correlation spectroscopy. By proper study of this equation, we can learn not only the principles of photon correlation spectroscopy, but the factors that must be understood for application of the technique.
We can write Eq. (4) in a more meaningful form by first separating the sums into terms involving only one molecule and terms involving two molecules. The result is

$$\left(\sum_{i=1}^{N} \cos \delta_i \right)^2 + \left(\sum_{i=1}^{N} \sin \delta_i \right)^2$$

$$= \sum_{i=1}^{N} (\cos^2 \delta_i + \sin^2 \delta_i)$$

$$+ 2 \sum_{j>i=1}^{N} (\cos \delta_i \cos \delta_j + \sin \delta_i \sin \delta_j)$$

$$= N + 2 \sum_{j>i=1}^{N} \cos (\delta_i - \delta_j) , \qquad (5)$$

where the second step is the result of the use of the identities

$$\sin^2 \theta + \cos^2 \theta = 1$$

$$\cos(\theta_1 - \theta_2) = \cos \theta_1 \cos \theta_2 + \sin \theta_1 \sin \theta_2 .$$

We can then rewrite the total scattered intensity as

$$I_s = I_s(1)[N + 2 \sum_{j>i=1}^{N} \cos(\delta_i - \delta_j)] , \qquad (6)$$

where $I_s(1)$ is the intensity of light scattered by one molecule.

Now recall that the phases of the various components of the electric field are changing with time because the molecules are moving about in the solution. Consequently, the δ_i and δ_j values depend on time with the result that the entire last term of Eq. (6) is time dependent. On the average, this term is zero with excursions in both positive and negative directions. Therefore, the time average of the scattered intensity is

$$\langle I_s \rangle = N I_s(1) , \qquad (7)$$

which, taken together with Eq. (1), is the basis of the traditional molecular weight determination by use of light-scattering intensities.

The instantaneous value of the last term in Eq. (6) is not zero. In fact, this term may be found more or less anywhere in the range from $-N$ to $+N$, leading to the result that the scattered intensity may be observed to fluctuate between zero and about twice its average value. The time required for a fluctuation between extremes is roughly equal to the time required for two molecules to move with respect to each other far enough to change the relative phase of the light scattered from each from 0 to π radians. This time will depend on the scattering angle as well as the size of the molecule; typical values range from a few microseconds for molecules with a molecular weight of 25,000 to many milliseconds for objects a few microns in diameter.

C. Properties of Fluctuations

The intensity fluctuations described here are seldom observed for two reasons: (1) they take place on a time scale faster than many photometers (and the human eye) will respond; and (2) the fluctuations as described here take place only at a point. This fact is of great importance in properly designing an experiment and thus is examined in detail.

When we say that the fluctuations take place only at a point, we mean that they may be observed at any point in the scattered field, but that the fluctuations at two different points will not be coincident unless the two points are very close together. To see why this is true, consider scattering from two molecules a distance d apart. As shown in Fig. 3, the scattering pattern will have maxima at any angle for which $d \sin \theta$ is a multiple of λ. Let us define the angular separation between adjacent maxima as $\Delta\theta$. A simple calculation shows that if $\Delta\theta$ is small, then to good approximation $\Delta\theta = \lambda/(d \cos \theta)$. For example, if $\lambda = 500$ nm and $d = 0.1$ mm, we find an angular separation of $\Delta\theta = 5 \times 10^{-3}/\cos \theta$ radians or $\Delta\theta = 0.3/\cos \theta$ degrees between bright spots. Thus a detector that subtends an angle significantly greater than this value would detect several maxima and several minima simultaneously. As the two molecules move about, the fluctuations in the detected signal will consequently be a smaller fraction of the average intensity than would be the case for a smaller detector.

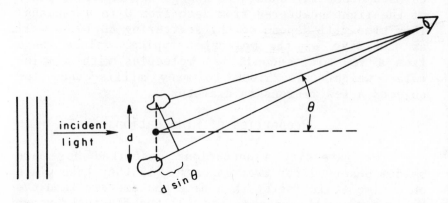

Figure 3.   Geometry for calculating the intensity of
             light scattered from two molecules. The
             scattered intensity will depend on θ and d
             and thus the exact positions at which
             observations are made as well as the
             relative positions of the molecules.

    The situation for the more realistic case of
light scattered from many molecules distributed
randomly throughout a small volume is more difficult
to analyze, although the basic ideas are similar. If
the scattering volume has radius a and the detector
radius is b at a distance R from the scattering
volume, the fluctuations will be substantially the
same as they would be at a point if the detector area
is one coherence area or less, where the coherence
area A_{coh} is defined as [6]

$$A_{coh} = \pi b^2 = \frac{\lambda^2 R^2}{\pi a^2} . \tag{8}$$

The effect of the number of coherence areas on the
character of the fluctuations is illustrated in
Fig. 4, where photographs of oscilloscope traces are
reproduced for several values of N_{coh}.

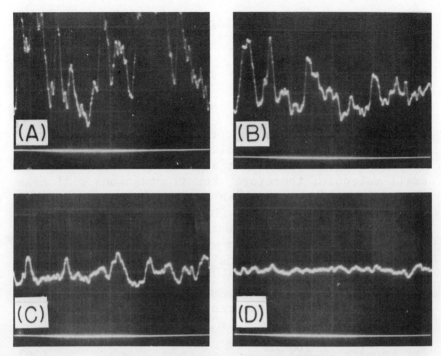

FIGURE 4.   Effect of number of coherence areas on
fluctuations in the scattered light intensity. All signals were recorded for the
same sample of 0.085 μm polystyrene latex
spheres. The laser intensity was adjusted
so that the average scattered intensity
was the same in all cases. The number of
coherence areas was 0.064 in part a, 2 in
part b, 8 in part c, and 100 in part d.

D.   Time Dependence

The time required for the fluctuations to take
place is the most important characteristic of the
signal because that time contains information about
the size of the solute molecules. The simplest
information obtained is the translational diffusion
constant, which, for a spherical molecule, is related

to the radius r according to the Stokes-Einstein relationship [7]

$$D = \frac{k_B T}{6\pi\eta r} ,$$

(9)

where k_B is Boltzmann's constant, T is the temperature, and η is the viscosity. Thus a large molecule will have a diffusion constant smaller than that of a small molecule; therefore, the fluctuations will take place more slowly, as shown in Fig. 5.

The analysis of the fluctuations may be carried out by using either a spectrum analyzer or a correlator. The more familiar of these instruments, the spectrum analyzer, obtains the power spectrum of a signal such as that shown in Fig. 5. The spectrum for the case of translational diffusion of a monodisperse molecule is a Lorentzian

$$P(\omega) = \frac{2Dq^2/\omega}{\omega^2 + (2Dq^2)^2} ,$$

(10)

where $q = (4\pi/\lambda) \sin(\theta/2)$ is the scattering vector.

The use of correlation techniques are relatively new in signal analysis, although the mathematical form of the correlation function has been employed in theoretical treatments for many years. The correlation function is defined for a signal $I(t)$ as

$$G(\tau) = \lim_{\tau \to \infty} \frac{1}{2T} \int_{-T}^{T} I(t)I(t + \tau)dt$$

(11)

The correlation function has two features that are of particular interest to us here: (1) it is easily measured, with the use of modern digital techniques, for light signals of very low level; and (2) it can be shown to be the Fourier transform of the power

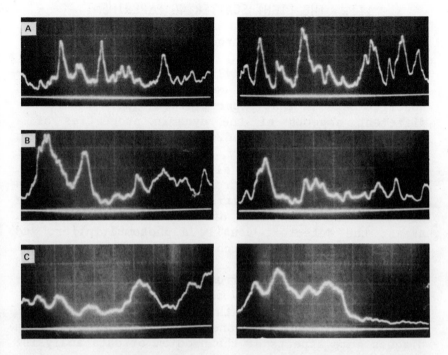

Figure 5.   Effect of size of PLS spheres on the
             fluctuations in the scattered light inten-
             sity.   All signals were recorded at the
             same average light intensity and for 0.32
             coherence areas.   The diameter of the
             spheres was 0.085 µm (a), 0.220 µm (b),
             and 1.011 µm (c).

spectrum. Thus it is both experimentally accessible
and theoretically interpretable in terms of the
system parameters of interest. Indeed, the correlation function corresponding to Eq. (10) is

$$G(\tau) = 1 + \exp[-2Dq^2\tau] \, , \tag{12}$$

a wave form that is usually easier to analyze than is
Eq. (10).

III.   THE LIGHT SCATTERING EXPERIMENT

A.   Introduction

In general terms, all light-scattering experiments contain the same components; it is only in specific details that experiments designed to measure different aspects of the dynamic properties of a solution are distinguished. The major components, as illustrated schematically in Fig. 6, are the light source (which is almost always a laser, but doesn't need to be); the spectrometer, which contains an optical system for definition of the scattering angle and also for limitation of the number of coherence areas; the detector, usually a photomultiplier; and the signal analyzer, which may be either a spectrum analyzer or a correlator. Often a computer is used to abstract information from the spectrum or correlation function obtained by the signal analyzer.

In the remainder of this chapter we discuss each of the four major components of Fig. 6. We then discuss some aspects of the problem of data analysis, thus covering the function of the dotted connection in Fig. 6.

FIGURE 6.   A block diagram of the complete light-scattering apparatus. Each block is discussed in detail in the text.

B.   The Light Source

It is commonly believed that the light source employed in photon correlation spectroscopy must be a laser in order to meet spectral purity requirements. The argument is based on the fact that changes in

frequency of the incident light of only a few kilo-
Hertz (or even a few tens of Hertz) are detected and
thus that the frequency width of the incident beam
must be only a few Hertz to obtain the maximum resolution. However, this argument is fallacious. As
has been shown by Jakeman et al. [8] intensity fluctuation spectroscopy can be performed by use of a
conventional light source.

It is not likely, however, that conventional
light sources will replace lasers in photon correlation applications because of the relatively low
intensities obtained when the light is focused to a
point. In the experiment mentioned above the scattered light detected by use of a mercury arc source
was a factor of 10 lower than was detected when a
10^{-4} W laser was substituted for the arc source.
Since the cost of a mercury arc lamp source exceeds
that of a 10^{-3} W helium-neon (He-Ne) laser, it
is clear that the laser is by far the most cost-
effective source. Therefore, we restrict the
remainder of this discussion to lasers.

The spectrum of light emitted by a laser is as
shown in Fig. 7. The laser itself consists of a
cavity formed by two spherical mirrors separated by
distance L. With the intervening space filled with
an active medium providing gain at optical wavelengths, the laser will oscillate (lase) at wavelengths satisfying the relationship

$$n \frac{\lambda}{2} = L \tag{13}$$

and for which the gain of the medium-mirror combination exceeds unity. The laser will produce light in
1 to 20 or more peaks, depending on its length and
construction.

The intensity patterns of the spectral lines
represented by Eq. (13) are circularly symmetrical
and are called TEM_{00n} modes. Modern laser design
attempts to eliminate other modes of oscillation of
the laser called *off-axis modes*, but occasionally

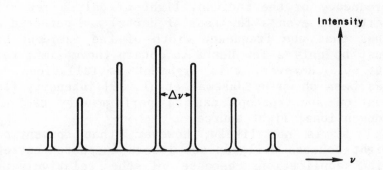

FIGURE 7. Representation of the spectrum of light
emitted from a laser.

they will be found. Unfortunately, off-axis modes
can cause undesired contributions to the analyzed
signal and hence should be eliminated.

The spectral lines in Fig. 7 can beat against
each other and will, in principle, give a contribution to the correlation function having a minimum
frequency equal to the separation of adjacent lines,

$$\Delta \nu = \frac{c}{2L} = \frac{1.5 \times 10^8}{L} \ \sec^{-1} \ , \qquad (14)$$

where L is in meters. The maximum frequency of any
importance in real experiments is 5×10^6 Hz, so that
beating between TEM_{00n} lines will be important only
if $L > 30$ m. Laboratory lasers are in fact more than
an order of magnitude smaller than this so no effects
may be expected due to beating between axial modes.
The situation with regard to off-axis modes is
entirely different. Here the spacing may be only a
few kiloHertz and beating between the various TEM_{lmn}
modes, all having the same n, can produce spurious
components in the correlation function. Consequently, it is always desirable to eliminate the
off-axis modes. Some lasers, particularly medium and
high-power Ar^+ lasers, have an intercavity iris that

may be adjusted to eliminate such modes. Most low-power modern lasers will not support off-axis modes. Nevertheless, the laser beam should be examined periodically to check its mode purity. This is especially important when the correlation function has an unexpected shape.

In selecting a laser there are four factors that must be taken into consideration. These are (1) wavelength, (2) power, (3) polarization, and (4) fluctuations in power intrinsic to the laser. The first two factors are coupled because the power required to achieve a given signal/noise ratio depends on the wavelength. The fourth factor enters in a negative sense: any fluctuations in laser power should be avoided if at all possible.

The wavelength to be selected is governed to a great extent by the nature of the experiments to be performed. The first requirement is that any absorption bands in the sample must be avoided. Otherwise, substantial local heating of the sample will take place, leading to convective flow of the solution, a thermal lens effect, and difficulty in interpreting the data. Similarly, it is undesirable to excite fluorescence lines in the sample, although unwanted light from this source could be removed with filters. Beyond these requirements there is relatively little scientific reason to choose one wavelength over another. As we see in Section V, the signal/noise ratio is, except for photomultiplier efficiency, independent of wavelength despite the fact that the scattering power is proportional to $1/\lambda^4$; therefore, it is preferable to work at the wavelength giving the best photomultiplier efficiency if low light levels are anticipated. If there is an adequate amount of scattered light, the cost of both purchase price and operating expenses is the dominant consideration. At the present time the best choices are as follows:

1. *For low light-scattering levels*: up to 200-mW AR$^+$ laser. Any greater power will

not greatly increase the range of possible
experiments. The principle lines are at
488.0 and 514.5 nm.

2. *For high light-scattering levels*:   1-50-mW
He-Ne laser. The principal line is at
632.8 nm.

Because of the relationship between the polarization of the incident light and intensity of the
scattered light [see Eq. (1)], it is essential that
the laser be capable of producing plane-polarized
light. The light with "random" polarization produced
by some inexpensive lasers has in fact two components
that are linearly polarized in perpendicular directions. Alternate lines in Fig. 7 will have one
polarization, the remainder will have the perpendicular polarization. The relative intensities of the
two components as well as the directions of polarization will change with time, particularly during the
first half-hour of operation. Consequently, there
will be fluctuations in the intensity of the scattered light that are related to the laser rather than
the sample under study.

Another source of intensity fluctuation intrinsic to the laser is the plasma oscillation found in
some large (50-mW) He-Ne lasers. The laser intensity
will be modulated at the plasma frequency, about
100 kHz in a typical laser. Lasers subject to this
instability usually have an optional RF excitation
that will reduce the plasma oscillations to an insignificant level. This option is an absolute necessity
if offered. Any large He-Ne laser without the option
is suspect and should be carefully tested before use.

There is a more mundane source of intensity
fluctuations not associated with the sample under
study. Under some circumstances vibrations in various parts of the apparatus or of one part with
respect to another can cause severe intensity fluctuations at the detector. Experiments that are prone
to interference from vibration are those in which
light scattering from a surface (either liquid or

membrane) is studied or any experiment employing a local oscillator. In these cases relative motion of two parts of the experiment by a quarter wavelength of light or less can cause large, entirely erroneous signals. A successful experiment will frequently require use of a vibration isolation table, elimination of mechanical pumps, enclosing the experiment in a sound deadening environment, and construction of experimental apparatus using massive materials.

Simple light-scattering experiments from solutions of macromolecules in which no local oscillator is employed are almost immune to this type of problem. The reason is that all the sources of light that will ultimately interfere at the detector are located within the relatively small sample volume (these sources are the molecules themselves, of course), and because of the relatively incompressible nature of the solvent, typical laboratory sources of vibration are incapable of causing relative motion of the various macromolecules by sufficient amounts to be important. Note, however, that if the laser beam and total sample move with respect to each other by a distance comparable to the diameter of the laser beam (typically 0.1 mm), and if this happens in a time on the order of or less than the duration of the correlation function, the true sample will consist of different sets of molecules at different times and one can then expect to observe effects in the correlation function. To avoid this type of effect, it is a good idea to attach the laser and light-scattering spectrometer to a single plate.

A laser with an unstable cavity will be affected by vibration. The problem is most severe in lasers having water-cooled plasma tubes. The flow of water, particularly if there are trapped bubbles, will cause vibrations that lead to objectionable laser intensity fluctuations. Reduction of the flow rate to the minimum consistent with the cooling needs, elimination of trapped bubbles, and careful alignment of the cavity mirrors for maximum power constitute the only recourses.

C.   The Spectrometer

The design and construction of a spectrometer suitable for use in correlation spectroscopy applications is not difficult, as is evidenced by the large number of "home-built" instruments to be found in research laboratories. Indeed, the wide range of experiments currently being performed make it almost impossible to design a single commercial spectrometer capable of performing all the experiments. Those currently on the market are suitable only for studies of solutions and have no provisions for the introduction of local oscillators, electric fields, or other variations of interest.

The design of a spectrometer is not complex; however, several principles must be employed if optimum results are to be obtained. We begin our discussion with a diagram (Fig. 8) of a spectrometer showing the components necessary for correlation spectroscopy. The important decisions to be made concern the focal length of the entrance lens, the design of the collection apertures, and the sample cell. An additional difficulty is introduced if a local oscillator is to be used.

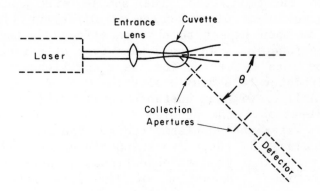

FIGURE 8.   The elements of the spectrometer.

The selection of the entrance lens is governed by the need, when weakly scattering samples are to be investigated, to scatter as much light as possible into a single coherence area. By combining Eq. (1) for the scattered intensity of a single molecule, Eq. (7) to account for N molecules in the scattering volume, and Eq. (8) defining the coherence area, we find that the detected power is

$$P_s = NS \, \frac{\lambda^2}{\pi a^2} \, I_0 \, . \tag{15}$$

If we assume that the entrance lens focuses the laser beam to a diameter 2a, then

$$I_0 = \frac{P}{\pi a^2} \tag{16}$$

where P is the laser power. The number of molecules in the scattering volume is

$$N = C_n V \, , \tag{17}$$

where C_n is the concentration of molecules in molecules per unit volume and V is the volume of the scattering region. Assuming that the scattering volume is spherical with radius a (a slight underestimate), $V = 4/3(\pi a^3)$ and we find

$$P_s = \frac{4}{3} \, C_n \, \frac{S\lambda^2 P}{\pi a} \, . \tag{18}$$

The conclusion is immediate: to maximize the detected signal, we must make a as small as possible.

For a simple lens of focal length f, the radius of the beam at the focal point will be determined by the intrinsic divergence of the original laser beam α, which is usually about 10^{-3} radians. The radius of the laser beam at the focal point is $f\alpha$, and we

see that the power scattered into a single coherence
area may be increased by decreasing the focal length
of the entrance lens.

The sample cell may assume a wide variety of
shapes and sizes depending on the specific application. Square cells with path length as small as 1 mm
have been successfully used, as have round cells 1 cm
or more in diameter. It is common to use ordinary
spectrophotometer cuvettes with all four sides
polished. Even the inexpensive plastic cuvettes may
be used in a number of cases.

The chief problem arising in the use of square
cells is that stray light may be generated by
scratches or dirt on the outside of the cell and
eventually find its way into the detector, where it
forms a weak local oscillator. The resulting change
in the measured correlation function leads to a loss
of accuracy of the measurements. Otherwise, as long
as one makes measurements only at a scattering angle
of 90° or corrects carefully for refraction at the
cell surface, square cells may be easily used.

For measurements at a number of different
angles, a cylindrical cell would seem to offer many
advantages. It is, however, difficult to achieve the
alignment necessary to realize the potential for
accuracy inherent in the light-scattering method of
measuring diffusion constants. The problem arises
because of refraction at the cell surface and is
particularly severe at small scattering angles. It
is possible to show that the fractional error in the
measured time constant ($\Delta\Gamma/\Gamma$) due to an error in
alignment ΔX of either the laser beam or the detector
is

$$\frac{\Delta\Gamma}{\Gamma} = \frac{\Delta X}{r} \left(1 - \frac{1}{n}\right) \cot \frac{\theta}{2} , \qquad (19)$$

where r is the cell radius, n is the index of refraction of the solution, and θ is the scattering angle.
It is assumed that the cell wall is very thin. For
example, to achieve a 1% error at a scattering angle

of 20° with use of a 1-cm radius cell would require $\Delta X \sim 0.007$ cm or less. Alignment accuracy of this quality is difficult to achieve.

Convection currents in a cell can lead to a substantial contribution to the signal for the same reason as mentioned earlier in connection with the effects of vibration. They are particularly prevalent when a thermal control bath is poorly designed so that the bottom of the cell is warmer than the top. However, proper design of the temperature control system, together with the use of cells with the smallest practical interior dimensions, virtually elimintes the problem. Note, however, that this makes scattering angles other than 90° (using a square cell) almost impossible.

The collection apertures should be designed to limit the detector to one or two coherence areas. The first aperture can be regarded as the pinhole of a pinhole camera. The second aperture should then have a diameter equal to the diameter of the image of the laser beam at that point. If this is done, the laser beam size, the first aperture size, and the distance from the laser beam to the first aperture will determine the number of coherence areas according to Eq. (8).

D.   The Detector†

1.   The Photomultiplier. The detector or, more properly, the detection system usually consists of a photomultiplier tube and a pulse amplifier-discriminator. The photomultiplier contains a cathode usually made of one or more alkali metals that will absorb a photon and immediately emit an electron. The electron is accelerated by an electric field and then collides with a sheet of metal (a

---

†This section assumes a rudimentary knowledge of the operation of a correlator. A brief reading of Section III.D.2 will provide the necessary information.

dynode), knocking out several electrons. The group of electrons is again accelerated, collides with a second dynode, and so forth until after 9 to 14 dynodes the single electron has been "multiplied" into 10^5-10^7 or more electrons. This group of electrons originating from the capture of a single photon forms the output of the photomultiplier.

Some light-scattering experiments provide sufficient light to permit the direct use of the photomultiplier output as an analog signal with either a spectrum analyzer or an analog correlator. However, in most cases single photon events are processed by a digital correlator. The first step is to convert the relatively small single photon signal (rarely more than 50 mV high and 50 nsec in duration) into a pulse of the proper amplitude and duration for employment by the correlator. At the same time, very small photomultiplier pulses are rejected as most likely arising from sources other than the detection of a photon. These two functions are normally combined in a single unit called a *pulse amplifier-discriminator* (PAD).

The properties of a photomultiplier of interest to us here are the quantum efficiency (probability of detecting a photon) at the wavelength to be employed, the gain of the tube, the width of the anode pulse, the degree of afterpulsing, and the dark current. In addition, there are two major types of construction, the side window and end window tubes. Since the construction has a major influence on the properties of a tube, we discuss it first.

The side window tube has a totally enclosed photocathode. The photoelectrons leave the cathode from the front and are usually (but, unfortunately, not always) attracted to the first dynode. The construction of this tube leads to a nonuniform quantum efficiency over the active area of the photocathode. This is not a serious drawback in photon correlation applications because the light to be detected falls on a small area (a few coherence areas at most), and as long as this is selected to be the most sensitive area of the photomultiplier, the nonuniform quantum efficiency will not be a serious disadvantage. The advantages of the side window tube are low cost, high gain at modest accelerating voltage, and fast pulse rise time.

The end window photomultiplier was developed for applications in which a uniform quantum efficiency over a large area is required. The photocathode is deposited on the inside of the flat end of the tube and is less than 100 nm thick. Photons are absorbed on the front side, and photo electrons are ejected from the back side of the cathode. These tubes have been widely employed in correlation spectroscopy at least partly because some manufacturers offer tubes specially designed or selected for low afterpulsing effects. However, there seems to be no reason for rejecting the side window tube, particularly in less critical applications.

Returning to the list of properties mentioned above, the tube should be chosen to have a quantum efficiency that is as large as possible at the wavelength, or range of wavelengths, to be employed. The photomultiplier companies publish curves giving the necessary information for all the tubes they manufacture.

The gain of the tube and width of the anode pulse are important in that they must be matched to the PAD. Most tubes have maximum gains somewhere in the range of 10^5-10^7, and the gain of an individual tube can be varied by a factor of 5 or more simply by changing the cathode potential. The width of the output pulse should be less than 20 nsec for applications involving correlator sample times in excess of 100 nsec. A tube with a correspondingly reduced output pulse width should be chosen if shorter correlator sample times are anticipated.

Afterpulsing is the term used for the phenomenon in which a positive ion generated at some time during an electron cascade returns to the cathode or an early dynode and initiates a second electron pulse. The second pulse follows the first by several tenths

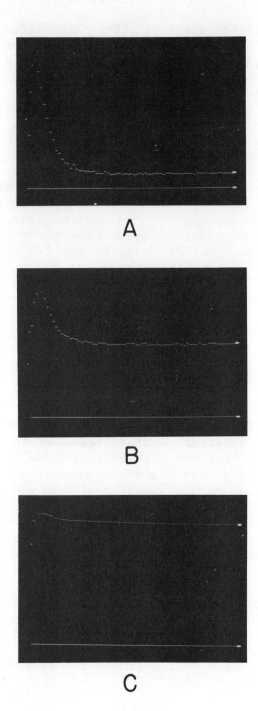

A

B

C

FIGURE 9   (facing).   The contribution of afterpuls-
            ing to the correlation function at average
            photon rates of (a) 800 counts/sec, (b)
            6000 counts/sec, and (c) 80,000 counts/
            sec.   The correlator sample time in all
            three cases is 1.0×10^{-7} sec.   The total
            probability of afterpulsing is calculated
            from Eq. (35) to be 0.2%. Note that the
            first channel appears high at the low
            count rate and low at the higher rates.
            This is caused by competition between
            instrumental dead-time effects depressing
            the first channel and pulse amplifier-
            discriminator double pulsing increasing
            the first channel. The photomultiplier is
            a   Hamamatsu   R928P,   the   correlator   a
            Langley-Ford Instruments Model 1096, and
            the PAD a Langley-Ford Instruments Model
            PAD-1.

of a microsecond and results in a correlation function with a peak such as that shown in Fig. 9. The
probability of having an afterpulse P_A is independent
of the pulse rate. If a total of N photons are
detected during time T, the absolute area ΔA of the
afterpulse peak in the correlation function will be

$$\Delta A = NP_A \text{ counts.} \tag{20}$$

These counts will be distributed over about 10 channels of the correlation function if the sample time
ΔT is 0.1 μsec. The baseline count rate is, assuming
the photons are uncorrelated,

$$G(\tau) = (\frac{N}{T} \Delta T)^2 \frac{T}{\Delta T} = \frac{N^2 \Delta T}{T} , \tag{21}$$

where the term in brackets is the average number of
photons detected in a single sample time and $T/\Delta T$ is
the number of samples taken. Dividing Eq. (20) by
Eq. (21), we find for the ratio of the area of the
afterpulse peak to the DC portion of the baseline

$$\frac{\Delta A}{G(\tau)} = \frac{P_A T}{N \Delta T} , \tag{22}$$

which shows that the afterpulse peak is reduced in
significance as the photon detection rate (N/T) or
sample time increases. At any rate, if ΔT exceeds
1 μsec, the effect can be eliminated by ignoring the
first channel. At faster times it is possible to use
careful measurements of the shape of the afterpulse
peak, and Eq. (22) to correct measured correlation
functions.

Of course, it would be best to eliminate the
afterpulse problem at its source. This may be done
to an extent by purchasing photomultipliers selected
for and/or designed for a low afterpulse rate. The
extra cost is warranted, however, only if corre-

lator sample times less than 1 μsec are to be used. Finally, it should be pointed out that not all white-light correlations can be traced to the photomultiplier. As we discuss below, the PAD can also generate a correlation function that is completely unrelated to the detection of photons.

For a variety of reasons, photomultipliers produce pulses indistinguishable from photon pulses, even in the dark. This contribution to the anode current, known as *dark current*, is often reduced by one or two orders of magnitude by cooling the tube to 0 to -30°C. This is done only if the dark current forms a significant part of the overall signal. The dark current for tubes sensitive to light from a He-Ne laser (632.8 nm) corresponds to about 300 photons/sec·cm^2 of cathode area. Therefore, a tube with a 2 cm^2 or less photocathode can be expected to produce at most 600 dark pulses per second. This pulse rate is smaller than those obtained in virtually all photon correlation experiments. Consequently, there is no need to cool a photomultiplier for these applications as long as the tube is properly selected.

After selecting a tube, it is necessary to obtain a housing and power supply. Excellent housings containing the tube socket and a resistor divider to obtain the proper dynode voltages from the cathode voltage are readily available. It is necessary to use the type containing RF shielding and, if a front window tube is selected, magnetic shielding as well.

2.   Photomultiplier Power Supply. A power supply capable of delivering the maximum voltage and current requirements must also be acquired. It is extremely important that the ripple and short-term voltage fluctuations be reduced to an absolute minimum because of the very strong dependence of the photomultiplier gain on voltage (the gain is proportional to V^6 to V^{10}, depending on the number of dynodes). Since the probability of detecting a photon depends

on the gain, any fluctuation in cathode voltage on a time scale comparable with the total correlation time will show up in the measured correlation function.

3.   Pulse Amplifier-Discriminator.   The interaction between the photomultiplier and the PAD presents the most difficult electronics problem in the light-scattering system.   The photomultiplier output consists of small, short, negative-going pulses, whereas the PAD must produce pulses conforming to the logic levels used by the correlator.   The overall gain of the PAD is frequently of the order of 1000 or more, resulting in the possibility that a coupling from output to input can generate a spurious pulse and thus spurious correlations.   Consequently, it is very important that a high-quality PAD be used.

E.   Signal Analyzers

The signals at the output of the photomultiplier are similar to those shown in Figs. 4 and 5.   At first sight the signals appear to be random noise, but careful analysis shows that the spectrum of the signal contains information about the system responsible for the light scattering.

Early experiments were performed with the use of a single-channel wave analyzer capable of obtaining a spectrum in about an hour, provided the photomultiplier signal was large.   In this system the photomultiplier output was used as an analog signal — no attempt was made to detect single-photon events.

Technical advances soon provided two new instruments, both capable of obtaining the spectrum of the scattered light in a time as short as a few seconds. Both instruments have continued to be developed and are widely used in correlation spectroscopy today. They are the real-time spectrum analyzer and the correlator.

1.   The Spectrum Analyzer.   The real-time spectrum
analyzer obtains the spectrum of an analog input
signal by use of digital techniques.   In a typical
instrument the input signal will be digitized at 512
sequential equally spaced points in time.   During the
time that the second 512 samples are being taken, 256
Fourier components of the first set are obtained and
a 200-point spectrum is presented.   The results from
a number of samplings are summed to provide a smooth
spectrum.   If the instrument can perform the Fourier
transform in a time less than that required to obtain
the second set of 512 samples, the instrument will
operate in real time and no information will be lost.
The maximum frequency at which real-time operation is
obtained is usually in the range of 1-5 kHz.

The spectrum analyzer is useful in experiments
in which there is a high detected light level, such
as when an optical local oscillator is used.   A
typical application is in electrophoretic light-
scattering experiments.   Here the fact that the
spectrum is obtained directly facilitates the interpretation of data as each species having a different
electrophoretic mobility results in a distinct spectral peak.

2.   The Correlator.   We turn now to correlators,
instruments capable of obtaining the correlation
function of an electrical signal.   Equation (11) has
partially anticipated the general mathematical definition of a correlation function,

$$G(\tau) = \lim_{T\to\infty} \frac{1}{2T} \int_{-T}^{T} I(t)J(t + \tau)dt , \qquad (23)$$

where $I(t)$ and $J(t)$ are signals that depend on time
and $G(\tau)$ is called the *autocorrelation function* or
cross correlation function depending on whether $I(t)$
and $J(t)$ are the same or different signals.   Instruments designed to obtain an approximation to
Eq. (23) for a number of values of τ are available

from several manufacturers. Some are designed primarily to accept analog input signals; others accept
pulse trains similar to those obtained from the
output of a single photon detector. We describe the
operation of the latter correlator.

A block diagram of a typical digital correlator
is shown in Fig. 10. The timing and operation of the
correlator is controlled by the sample time generator, which divides time into intervals of equal
duration $\Delta\tau$. The number of pulses at input A occurring during each sample time is counted by the shift
register counter. This situation is illustrated in
Fig. 11, where n_0, n_1, . . . are the number of pulses
appearing at input A and counted by the shift register counter.

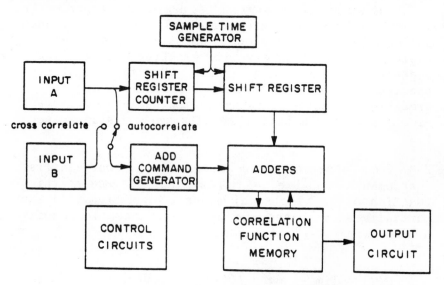

FIGURE 10. Block diagram of a correlator. The
principles of operation are discussed in
the text.

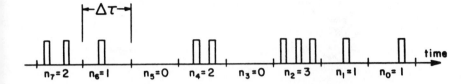

FIGURE 11.   The train of pulses processed by a correlator. The small marks represent the
sample time clock which divides time into
increments of $\Delta\tau$.

At the end of each sample time the number in the
shift register counter is entered into the first
stage of the shift register, the number that was in
the first stage is shifted to the second, the number
that was in the second stage is shifted to the third,
and so on. As a consequence, after the correlator
has been in operation for a brief period of time, the
first stage contains $I(t - \Delta\tau)$, the second $I(t - 2\Delta\tau)$, the third $I(t - 3\Delta\tau)$, and the kth stage $I(t - k\Delta\tau)$.

During the "present" sample time each pulse
appearing at input A (when in the autocorrelate mode)
or input B (when in the cross-correlate mode) is
processed by the add command generator and instructs
all adders in the correlator to add each of the
numbers stored in the shift register to the number
stored in the associated channel of the correlation
function memory. As an example, consider the pulse
sequence in Fig. 11. During the sample time interval
2, the product $n_2 n_3 = 0$ is added to correlation
function memory channel 1, $n_2 n_4 = 6$ is added to
channel 2, $n_2 n_5 = 0$ is added to channel 3, and so on.

Thus the correlator will accumulate in the first
channel:

$$G(\Delta\tau) = n_0\tilde{n}_1 + n_1\tilde{n}_2 + n_2\tilde{n}_3 + \cdots$$

$$= \sum_{i=0}^{N-1} n_i\tilde{n}_{i+1} \tag{24}$$

and in the second channel

$$G(2\Delta\tau) = n_0\tilde{n}_2 + n_1\tilde{n}_3 + n_2\tilde{n}_4 + \cdots$$

$$= \sum_{i=0}^{N-1} n_i\tilde{n}_{i+2} \,. \tag{25}$$

In general, the kth channel will contain

$$G(k\,\Delta\tau) = n_0\tilde{n}_k + n_1\tilde{n}_{k+1} + n_2\tilde{n}_{k+2}$$

$$= \sum_{i=0}^{N-1} n_i\tilde{n}_{i+k} \,, \tag{26}$$

which is a good approximation to the true correlation function whenever the change in the value of the correlation function during the time $\Delta\tau$ is small. In each of these expressions the numbers n_i represent the number of times the content of each stage of the shift register is added to its respective correlation function memory channel and the numbers $\tilde{n}_{i+k}$ are the numbers stored in the shift register.

The characteristics of a correlator that are important in light-scattering experiments are the efficiency of operation; the capacity of the shift register counter and, therefore, of the shift register; the range of sample times available; and the number of channels. Typical characteristics of correlators used in fluctuation spectroscopy are as follows:

Minimum sample time	100 nsec
Shift register capacity (each step)	4 bits
Conditions for batch mode operation	none
Number of channels	64

IV. SIGNAL/NOISE RATIO

A. Introduction

Consideration of the signal/noise ratio is complicated by the large number of factors that enter into the final answer. Some factors (e.g., dust) can be dealt with only in the most general terms while others (such as photon shot noise) are susceptible to a precise theoretical description. Because of the complexity of the subject and the presumed interest of the reader in proceeding with the experiments, the present discussion is limited to "factor of 2" calculations, thus avoiding some of the complex details required by an exact analysis. Given the nature of some of the imponderables, the results will probably be as close to reality as would be possible even with a more exact theory.

We recognize at the outset three sources of noise that limit our ability to measure the properties of the scattered light with arbitrarily high precision: (1) effects due to the finite intensity of the scattered light; (2) effects due to a finite duration of the experiment; and (3) effects due to light scattered by unwanted effects (e.g., dust). We discuss each of these contributions.

B. Effects Due to Finite Intensity

A contribution to fluctuations in the scattered light intensity is caused by the fact that the number of photons detected during each sample time is finite. If the instantaneous intensity of the scattered light corresponds to N photons per sample time, we expect most of the time to detect from $N - \sqrt{N}$ to $N + \sqrt{N}$ photons during a sample time. The number detected will obey a Poisson distribution law. The expected uncertainty in the correlation function due to this effect is given by $\sqrt{G}$ where G is the number given by Eq. (26). This contribution to the noise may be reduced by increasing the laser intensity or solute concentration, by scattering at a smaller angle, or by focusing the laser beam to a smaller diameter.

To understand the reasons for these actions, we calculate the number of photons scattered into a single coherence area each time constant of the exponentially decaying correlation function. If this number is much less than 1, the measures just mentioned will improve the signal/noise ratio. If, however, it is much greater than 1, the signal/noise ratio is limited by an entirely different process, which is described below.

The geometry of the scattering region is illustrated in Fig. 12. The laser beam is focused to a beam of radius r by lens L. In most cases r is determined by the intrinsic divergence of the laser beam and the focal length f of lens L. For our purposes, we can take $r = 10^{-3}f$. A system of apertures, A_1 and A_2, defines a scattering region that we will assume is a cylinder of radius r and length $2r/(\sin \theta)$. In this way, the scattering region as viewed from the photomultiplier will be a square with sides $2r$. The volume of the scattering region will be $(\pi r)^2 \, 2r/(\sin \theta)$, and the number of solute molecules in that volume is

$$N = \frac{2\pi r^3}{\sin \theta} \, C_n \, , \tag{27}$$

where C_n is the number density of solute molecules.

To find the intensity of the scattered light I_s at the detector, we use Eq. (1), $I_s = I_0 \, S/R^2$. For a

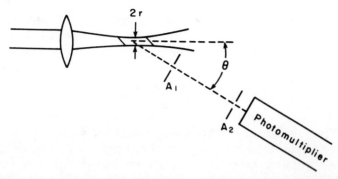

FIGURE 12.   Detailed figure showing the scattering
geometry.

total laser power P_0, we have $I_0 = P_0/2\pi r^2$. The
total power incident on a single coherence area at
the detector is $P_d = I_s A_{coh}$. Combining all these
factors and using Eq. (8) for A_{coh}, we finally find
for the power incident on the detector

$$P_d = P_0 \frac{c_n s\lambda^2}{\pi r \sin\theta} .$$ (28)

We must now multiply P_d by the exponential time
constant obtained from Eq. (12), $(2Dq^2)^{-1}$, where $q =$
$(4\pi/\lambda)\sin(\theta/2)$, to obtain the total energy incident
on the detector each time constant

$$E = P_0 \frac{c_n s\lambda^4}{32\pi^3 Dr \sin\theta \sin^2(\theta/2)} .$$ (29)

For small scattering centers, $S \propto \lambda^{-4}$. Thus Eq. (29)
shows directly that an increase in P_0 or C_n or
decrease in r or $\sin\theta \sin^2(\theta/2)$ will all result in
increase of the light energy detected for each exponential decay constant.
    It might appear from Eq. (29) that arbitrarily
small values of the solute concentration C_n could be

studied by increasing the detected light energy through the other three factors. There is, however, a practical limit achieved when the light scattered by the solvent is roughly equal in intensity to that scattered by the solute molecules. For a solute molecule with molecular weight 25,000, this will occur at a concentration of about 0.5 mg/ml.

The importance of focusing the laser to a small diameter must also be emphasized. A 50 mW laser focused into a cell with a 10 cm focal length lens is no more effective than a 5 mW laser focused with a 1 cm focal length lens. The cost of the second option is a small fraction of the cost of the first.

C. Effects Due to Finite Experiment Duration

A second limitation in accuracy is due to the fact that data are collected for a finite number of decay times of the correlation function. If the correlation function decays as

$$G(\tau) = 1 + \exp[-\Gamma\tau] , \tag{30}$$

and the total duration of an experiment is T, the number of decay times during the experiment is ΓT. The corresponding signal/noise ratio, even if the detected light level is infinite, is

$$\frac{\text{Signal}}{\text{Noise}} = \sqrt{\Gamma T} . \tag{31}$$

The only way to improve this contribution to the signal/noise ratio (other than increasing the duration of the experiment) is to *increase* Γ, which requires that light be scattered at a larger angle.

D. Effects Due to Unwanted Scattered Light

The presence of unwanted signals in the scattered light provides the third major limitation to the quality of the light-scattering results. This topic is frequently neglected in discussions of signal/noise ratios, perhaps because of its very complexity, and yet it provides the ultimate limit in signal/noise ratio for the vast majority of real experiments. Included in this category are such effects as

Fluctuations in laser intensity

Unwanted laser light due to reflections or flare that has not been scattered but acts as a local oscillator

Convection currents in the scattering cell

Dust, air bubbles, glass particles, bacteria, and other foreign matter in the solution

Light scattered at the wrong angle present because of reflections in the cell

Molecules or other artifacts resulting from improper or inadequate sample preparation

Light scattered by the solvent

The principle difficulty with many of these contributions to the noise signal is that systematic effects take place so that the measurements are consistent from experiment to experiment but unfortunately give the wrong answer. The clearest example of this phenomenon is provided by the second effect. When a small amount of light (small in comparison to the real scattered light) is unscattered and is able to act as a local oscillator, the correlation function will contain two exponentials, one with a decay rate of $2Dq^2$ and another at Dq^2 proportional in amplitude to the intensity of the local oscillator. If the resulting correlation function is fit to a single exponential, the caculated decay rate will differ from the correct result according to [9]

$$\frac{\Delta\Gamma}{\Gamma} = -\frac{16}{9}\frac{I_{LO}}{I_s}.$$ (32)

Equation (32) shows that extreme caution must be taken to avoid small amounts of local oscillator if the greatest possible accuracy is to be obtained. Of the remaining effects, much the same philosophy must be adopted; whatever is possible ought to be done to eliminate the cause.

V.   DATA ANALYSIS

A.   Introduction

The correlation functions we have discussed here have been simple exponentials resulting from the translational diffusion of a monodisperse molecule or particle. Many other processes also lead to exponential contributions to the correlation function. There are, in addition, contributions that may be damped cosine functions or Gaussian functions. Given the diversity of wave forms observed, it is clearly impossible to give a single data analysis scheme that will encompass all experiments. We can, however, discuss the steps that must be taken.

B.   Selecting the Theoretical Form

The first step in data analysis is thus the selection of the form (or combination of forms) expected in the correlation function. This selection depends on the purpose of the experiment under consideration and, to a certain extent, prejudices the outcome of the experiment. Fortunately, if judicious use of the well-known χ^2 test is made, it is possible to determine whether the choice of the theoretical expression used to fit the correlation function is adequate for the job. On the other hand, it is not always possible to prove that the suspected physical process is responsible for the details of the correlation function. This is especially true of the many physical processes known to lead to exponential terms in the correlation function.

Following determination of the form of the expected correlation function, the usual procedure is to write a general expression containing one or more parameters that may be adjusted to make the theoretical expression have as nearly as possible the same values as the experimental correlation function. For example, if the correlation function is expected to have the forms of a single exponential, we would write as the theoretical expression

$$G_{th}(\tau) = A + B \exp[-C\tau] \qquad (33)$$

and vary A, B, and C to make $G_{th}(\tau)$ resemble the measured correlation function $G_{ex}(\tau)$ as closely as possible at all values of τ.

The theory of determining the "best" values of the parameters has been extensively developed [10]. Without discussing the details of the arguments, we simply state that the "best" values are those that minimize χ^2 defined as

$$\chi^2 = \sum_{i=1}^{N} \frac{[G_{th}(\tau_i) - G_{ex}(\tau_i)]^2}{\sigma_i^2} , \qquad (34)$$

where the sum is to be taken over all values of τ for which $G_{ex}(\tau)$ has been measured and σ_i is the expected error in $G_{ex}(\tau_i)$. The required minimization is almost always performed by a computer by use of one of several standard techniques.

The quantity σ_i is the standard deviation for the quantity $G_{ex}(\tau_i)$. If $G_{ex}(\tau_i)$ were measured many times and a plot made showing the number of times $G_{ex}(\tau_i)$ takes on each value, a bell-shaped curve should be obtained with a peak at the true value of $G_{ex}(\tau_i)$ and a half-width at half height of σ_i.

Roughly 68% of all points would lie in the range
between $G_{i,peak} - \sigma_i$ and $G_{i,peak} + \sigma_i$. With the use
of an ideal correlator, the value of σ_i is $\sqrt{(G_{i,peak})}$.
This value will hold for a 4-bit correlator operated
in a mode in which the shift register is rarely
saturated.

C. Use of the χ^2 Test

Having obtained the desired values of the parameters, it is appropriate for us to return to the
question as to whether the initial form chosen for
$G_{th}(\tau)$ is consistent with the experimental data.
(Notice that we have not asked if the chosen form is
correct. Even in the best of circumstances we can
determine only that the form is consistent because
there are always an infinite number of expressions
that will fit the data to an arbitrary accuracy.
Presumably, only one of them is "correct.") The
answer is obtained with the aid of χ^2.

The quantity χ^2 defined in Eq. (34) may be
normalized by dividing by $N - p$, where p is the
number of parameters in the expression to be fit to
the data. The resulting quantity is expected to be
close to 1, particularly when $N - p$ is large (i.e.,
there are many more experimental points than parameters). The probability $P(x)$ that the quantity $x =
\chi^2/(N - p)$ will be in the range $x \pm \Delta x$ diminishes
rapidly as x increases beyond 1. The theoretical
prediction for $N - p = 60$ is shown in Fig. 13. If an
experiment is performed a large number of times and
x calculated for each run, a plot of the number of
times x has values between 0 and 0.1, 0.1 and 0.2,
and so on versus x should resemble Fig 13. If the
resulting graph is significantly different from this
figure, it may be presumed that the chosen form for
the correlation function is inadequate for representation of the data. The presumption becomes stronger
as the value of $p(x)$ at large values of x increases
and also as the number of runs increases.

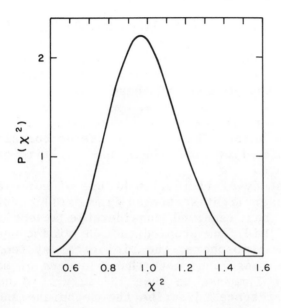

FIGURE 13. $P(\chi^2)d\chi^2$ gives the probability that χ^2 lies between χ^2 and $\chi^2 + d\chi^2$ if the form of the equation chosen is correct. This curve is calculated according to the assumption that there are 64 channels in the correlation function and three free parameters in the equation to be fit. A histogram showing the number of times that each value of χ^2 is obtained in a real experiment should reproduce this curve. An experimental curve peaking at a higher value of χ^2 indicates that the chosen function is not correct.

   We have discussed, admittedly in rather general terms, the concepts involved in obtaining information of interest from the measured correlation function. The process is in fact rather mechanical in nature, and researchers unfamiliar with the appropriate computer technique can almost always find help from their local computer experts.

VI.   OTHER APPLICATIONS OF CORRELATORS

Many additional applications of correlation technology have been developed over the last few years. Although some of these have no relevance to the study of aerosols (e.g., applications to studies of vibration in mechanical systems), two warrant discussion here. They are the use of correlators to study fluid flow and their use in time-of-flight measurements.

The study of fluid flow by use of correlators to analyze laser light scattered from small particles in the fluid has received considerable attention. The types of fluid flow studied include wind tunnels, tow tanks, the atmosphere, and biological systems. Fluid velocities range over more than 6 orders of magnitude from under 1 mm/sec to over 10^3 m/sec. A number of general references describe the technique and application to specific problems [11-13].

Three different techniques have been employed to measure local velocities in a moving fluid: (1) measurements of the Doppler shift of laser light scattered by particles entrained in the fluid; (2) measurements of periodic fluctuations in scattered light as a particle passes through the interference pattern produced when two beams from a single laser cross at a small angle; (3) the two-beam method in which the passage of particles between two points is timed by observing light scattered from the particles. The third method is an example of a time-of-flight measurement.

A number of advantages favor the laser velocimetry method over other methods in common use. They include a resolution of less than 0.1 mm, the ability to measure direction as well as magnitude of flow, the noninvasive nature of the measurement, and a rapid measurement time that allows variations in velocity with time to be followed.

Digital correlators are frequently used to analyze the signals produced in laser velocimetry measurements. The principal advantage of the corre-

lator over other instruments is the ability to produce signals with very low detected light levels.
Consequently, measurements may frequently be made
without the need to add particles to the fluid; those
present naturally are sufficient to yield an excellent signal.

A detailed description of the use of correlators
in laser velocimetry is beyond the scope of this
chapter. We describe only the time-of-flight measurement useful not only in laser velocimetry, but
for many types of velocity measurements. For example, a form of particle sizing can be accomplished by
measuring the terminal velocity with which particles
fall through still air.

In any time-of-flight measurement a cross-
correlation function is obtained with the two signals
representing the detection of the flowing particles
at the beginning and end of the path over which the
flight time is to be measured. The correlation
function will then have a peak at the time required
for the particles to traverse the flight path. If
the particles have a distribution of velocities, the
distribution will be represented by a broadening of
the peak.

One advantage of the cross-correlation technique
in time-of-flight measurements is that many particles
may be in the measurement path at any given time and
yet the velocity may still be obtained. To understand this fact, we calculate the signal/noise ratio
to be expected, assuming all particles to have the
same velocity. We further assume that $<P>$ particles
per second pass through the apparatus and that a
correlator sample time ΔT is chosen such that $<P> \Delta T$
< 1. No assumption is made about the magnitude of
the velocity nor the size of the apparatus; thus the
number of particles within the apparatus at any one
time may be indefinitely large. The experiment is
run for a time T that must be much longer than the
transit time of the particles.

The cross-correlation function will consist of
two parts: a peak containing one count for each

particle passing through the apparatus, or $T<P>$ counts; and a background containing the square of the average number of counts per sample time multiplied by the number of sample times during the experiment, or $<P>^2 T \Delta T$ counts.

Every channel (including the channel with the peak) will have the second contribution and it thus, constitutes a DC background except for variations that can be expected to be roughly $\sqrt{(<P>^2 T \Delta T)}$ in amplitude. It is these variations that are the noise – not the full DC component. The signal/noise ratio is then

$$\frac{\text{Signal}}{\text{Noise}} = \sqrt{\frac{T}{\Delta T}} \qquad (35)$$

and is independent of both $<P>$ and the total number of particles in the apparatus. It would appear that the signal/noise ratio could be improved without limit by reducing ΔT. However, when the particles have a distribution of velocities, ΔT will have a minimum sensible value and the peak will be spread out over a number of channels so that a limit to the ratio will be set. If the spread in transit times is δT and we spread the peak out over m channels, we find both signal and DC background are multiplied by $1/m$ so that, because signal/noise is proportional to signal/$\sqrt{\text{(background)}}$,

$$\frac{\text{Signal}}{\text{Noise}} = \sqrt{\frac{T}{m \, \delta T}} \, . \qquad (36)$$

This ratio decreases both as the peak becomes broader and as the resolution demanded becomes higher. We have assumed here that there are a sufficient number of particles during time T so that the noise in the peak itself is negligible.

VII. CONCLUSIONS

In this discussion we have attempted to give the novice to light scattering a firm understanding of the tools of the trade. At the end we have described briefly several additional applications of the correlator developed in recent years. In contrast to what we hope was a thorough treatment of the basic instrumentation, the description of advanced techniques can serve only to whet the appetite for new methods. There are many to choose from; only a few were mentioned here. The field offers many opportunities to the researcher interested in developing new techniques. It is the belief of the author that many useful and exciting advances remain to be realized.

REFERENCES

1. S. S. Alpert, Y. Yeh, and E. Lipworth, Phys. Rev. Lett. *14*, 486 (1965).

2. N. C. Ford and G. B. Benedek, Phys. Rev. Lett. *15*, 649 (1965).

3. S. B. Dubin, J. H. Lunacek, and G. B. Benedek, Proc. Natl. Acad. Sci. (USA) *57*, 1164 (1967).

4. C. Tanford, *Physical Chemistry of Macromolecules*, Wiley, New York, 1961, Chapter 5.

5. B. Chu, *Laser Light Scattering*, Academic Press, New York, 1974; B. Berne and R. Pecora, Dynamic Light Scattering, Wiley, New York, 1976.

6. E. Jakeman, C. J. Oliver, and E. R. Pike, The Effects of Spatial Coherence on Intensity Fluctuation Distributions of Gaussian Light, J. Phys. A *3*, L45-L48 (1970).

7. F. Reif, *Fundamentals of Statistical and Thermal Physics*, McGraw-Hill, New York, 1965, Section 15.6.

8. E. Jakeman, P. N. Pusey, and J. M. Vaughan, Intensity Fluctuation Light-Scattering Spectroscopy Using a Conventional Light Source, Opt. Commun. *17*, 305-308 (1976).

9. N. C. Ford, Jr., Biochemical Applications of Laser Rayleigh Scattering, Chem. Scripta *2*, 193-206 (1972).

10. P. R. Bevington, *Data Reduction and Error Analysis for the Physical Sciences*, McGraw-Hill, New York, 1969.

11. L. Danielsson, Ed., Photon Correlation Techniques in Fluid Mechanics: *Proceedings from 2nd International Conference*, June 14-16, 1978.

12. W. F. Mayo, Jr. and A. E. Smart, Eds., *Photon Correlation Techniques: Proceedings from 4th International Conference*, August 24-27, 1980.

13. L. E. Drain, *The Laser Doppler Technique*, Wiley, Chichester, England, 1980.

PART TWO
ESSENTIALS OF SIZE DISTRIBUTION MEASUREMENT

CHAPTER 3

Extraction of Distributions of Decay Times in Photon Correlation of Polydisperse Macromolecular Solutions

B. CHU and A. DINAPOLI
State University of New York at Stony Brook

CONTENTS

I. INTRODUCTION

Photon correlation spectroscopy allows us to measure the translational diffusion coefficient D, of structureless, monodisperse particles with great accuracy and ease. For a solution of polydisperse particles, we determine the electric field correlation function as

$$g^{(1)}(\tau) = \int_0^\infty G^*(\Gamma)\exp[-\Gamma\tau]d\Gamma ,$$ (1)

where $\Gamma = DK^2$, the magnitude of the momentum transfer vector is K [$=4\pi n \sin(\theta/2)/\lambda_0$, with n, θ, and λ_0 being the refractive index, the scattering angle, and the wavelength of light in vacuo, respectively], τ is the delay time, and $G^*(\Gamma)$ represents the normalized linewidth (Γ) distribution. The Laplace transform represented by Eq. (1) can be performed approximately depending on the finite ranges accessible by experiments and the precision of $g^{(1)}(\tau)$. In addition to particle interactions, complications may arise when modes of scattering from motions other than translation, such as rotational diffusion of rod particles or internal motions of large coiling polymers, contribute to the time correlation function. In some cases the decay time component related to translational motions may be separated from other types of motion by examining $G^*(\Gamma)$ as a function of scattering angle and concentration. In this chapter we are concerned mainly in trying to extract estimates of $G^*(\Gamma)$ due solely to translational motions by use of a variety of approximation techniques and on comparing the different $G^*(\Gamma)$ estimates in a semiquantitative manner.

Most distributions can be described in terms of the mean and moments about the mean. The cumulants method [1] has been used for some time and is the standard technique for analysis of dynamic light scattering data for polydisperse polymer solutions or colloidal suspensions. The mean linewidth $\bar\Gamma$ is defined as

$$\bar\Gamma = \int_0^\infty \Gamma G^*(\Gamma)d\Gamma$$ (2)

and the normalized variance,

$$\text{VAR} = \frac{\mu^{(2)}}{\bar{\Gamma}^2} = \frac{\int_0^\infty (\Gamma - \bar{\Gamma})^2 G^*(\Gamma)d\Gamma}{\bar{\Gamma}^2} \qquad (3)$$

measures the width of the distribution. Usually higher-order moments cannot be determined accurately and extrapolation to short delay times for determination of $\bar{\Gamma}$ and VAR is required for moderately broad linewidth distributions (VAR $\overset{\sim}{>}$ 0.2). We may compute $\bar{\Gamma}$ and VAR by using a double exponential fit to $g^{(1)}$ (τ), even when the normalized linewidth distribution function $G^*(\Gamma)$ is not bimodal in form; we return to this point in the discussion.

Although the mean and variance give important information characterizing the linewidth distribution, we are, at times, interested in the skewness of the distribution or want to know whether $G^*(\Gamma)$ has two peaks. For this reason, we compute $\bar{\Gamma}$ and VAR by use of the cumulants method and the double exponential form as a preliminary first step in our correlation function profile analysis. Then we test three different expressions to represent $G^*(\Gamma)$ in a more precise manner. They are a simple distribution that allows the integral of Eq. (1) to be computed analytically, a Pearson's type I distribution, and a histogram representation. Details of these three approaches together with the cumulants and the double exponential methods are presented, and three goodness-of-fit measures are used to assess our ability to extract $G^*(\Gamma)$ from generated data based on molecular parameters for polystyrene in cyclohexane at the theta temperature, experimental data from the same system, and measured data from two polyacrylamide samples in water and in 0.5 N NaNO$_3$ solutions.

II.   METHODS OF DATA ANALYSIS

The net, measured autocorrelation function, after baseline subtraction, has the form

$$G^{(2)}(\tau) = A\beta \; |g^{(1)}(\tau)|^2 \; , \qquad (4)$$

where β is an instrumental constant and A is the baseline. The factor A can be handled in a number of different ways [4,5] The correlation coefficients are taken as the square root of $G^{(2)}(\tau)$ and are proportional to $g^{(1)}(\tau)$, for which we have written the five expressions to be used to obtain $G^*(\Gamma)$. The fit is accomplished by use of a standard, nonlinear least-squares algorithm due to Marquadt [2] for all cases. To monitor the quality of the fit, we compute the normalized sum of squared errors

$$\chi^2 = \frac{\sum_{i=1}^{n} (y_i - y_i')^2}{n - L + 1} \qquad (5)$$

where y_i is the measured value of the correlation function, y_i' is the value computed from the fitting function to be tested, n is the number of correlator channels (92 in our case), and L is the number of fitting parameters. The χ^2 parameter, which measures the standard deviation of the fit, is compared to the statistical error of the measurement, approximated as the square root of the baseline. We define the ratio

$$R = \sqrt{\frac{A}{\chi^2}} \; , \qquad (6)$$

which approaches unity for a good fit. It is also necessary to examine the behavior of $(y_i - y_i')$ at each delay time. Systematic trends indicate that our fit may overemphasize a particular frequency component of $G^*(\Gamma)$, whereas a random distribution of

$(y_i - y_i')/y_i$ about zero indicates a good fit.   This is quantified by the correlation parameter Q:

$$Q = 1 - \frac{\Sigma \, (y_i - y_i')(y_{i+1} - y_{i+1}')}{\Sigma \, (y_i - y_i')^2} \, , \qquad (7)$$

as suggested by Mathiez et al. [3].   A random distribution of errors gives a value of Q near unity.

We have somewhat arbitrarily set the lower limits of both R and Q to be 0.7 for an acceptable fit.   The actual values of RSQ (defined below), R, and Q depend on the signal/noise ratio, the signal/background ratio, and the range of correlation delay times used.   Mathematical justifications to obtain proper fits have been discussed for multiple exponentials even though their detailed use has not yet been made readily available [6].   It has been our experience that poor fits will give R and $Q < 0.1$, whereas good fits result in values usually greater than 0.8 or 0.9.   We have assumed that the measured correlation function is of sufficient precision to allow for a sophisticated analysis.   The empirical criteria for this assumption to hold are that the ratio of the net correlation coefficient in the first channel to the background should have a value of at least 0.2 and that the correlation coefficient of the first channel should be about 8×10^6 above baseline.   In such cases RSQ, as defined by

$$RSQ = 1 - \frac{\Sigma \, (y_i - y_i')^2}{\Sigma \, (y_i)^2 - (\Sigma \, y_i^2)/n} \, , \qquad (8)$$

will have values greater than 0.9999.   Whereas R and Q vary over a range of 0 to 1, RSQ is greater than 0.9 for even very noisy data.   The scattered intensity from all the samples in this study is reasonably large; therefore, $RSQ \gtrsim 0.9999$ in all cases except where noted.   For values of RSQ less than 0.99 or

0.999, great care should be exercised in data interpretation.

The cumulants technique utilizes the expansion
of the logarithm of $g^{(1)}(\tau)$ about the mean linewidth. The resulting expression can be approximated
as

$$g^{(1)}(\tau) \cong \exp\left[-\bar{\Gamma}\tau + \sum_{i=2}^{L} \frac{(-1)^i \mu^{(i)} \tau^i}{i!}\right]. \qquad (9)$$

Equation (9) is valid only as τ approaches zero,
necessitating measurements of the correlation function at about four or five delay increments and a
graphical extrapolation for both $\bar{\Gamma}$ and VAR. Delay
increments are selected such that $\bar{\Gamma}\tau_{max}$, where τ_{max}
is the maximum delay time, ranges from about $1/2$ to 4
or 5. The value of L must be increased as $\bar{\Gamma}\tau_{max}$
increases. The extrapolation makes use of lower L
values at small delay times and higher L values at
larger delay times. The graphical procedure is
illustrated in Fig. 1 for a polyacrylamide (PAM-A)
solution. Figure 1 demonstrates that the extrapolation is ambiguous and can lead to significant uncertainties in $\bar{\Gamma}$ and VAR. The convergence criteria, R
and Q, are not easily applied in this case, and we
can only compare the cumulants results to those from
other procedures to assess their reliability.

The double exponential fit represents $G^{\star}(\Gamma)$ as

$$G^{\star}(\Gamma) = A_1 \delta(\Gamma - \Gamma^{(1)}) + A_2 \delta(\Gamma - \Gamma^{(2)}) , \qquad (10)$$

from which we obtain

$$g^{(1)}(\tau) = A_1 \exp[-\Gamma^{(1)}\tau] + A_2 \exp[-\Gamma^{(2)}\tau] . \qquad (11)$$

Equation (10) approximates the distribution as two
narrow spikes at $\Gamma^{(1)}$ and $\Gamma^{(2)}$ with amplitudes A_1 and
A_2. $\delta(\Gamma-\Gamma^{(1)})$ is simply the Dirac delta function.
Whereas Equations (10) and (11) are normally used to
fit a bimodal linewidth distribution function com-

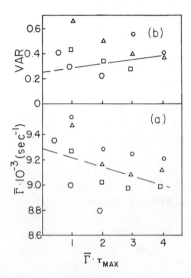

FIGURE 1.   Cumulants analysis for PAM-A in 0.5 N NaNO$_3$; concentration is 0.098 wt.%; θ = 90°.³ Symbols represent the following: hexagon, L = 2; square, L = 3; triangle, L = 4; circle, L = 5.

posed of both self-beating and homodyne components or to characterize a monomer-dimer type equilibrium, we have found the technique useful for computation of $\bar{\Gamma}$ and VAR for unimodal polydisperse, polymeric systems. The moments are computed directly by means of Eqs. (10), (2), and (3).

If we assume that the linewidth distribution can be represented as

$$G^*(\Gamma) = \Gamma^{\nu-1} \exp[-\alpha\Gamma] , \qquad (12)$$

the electric field correlation function takes the simple form

$$g^{(1)}(\tau) = \underset{\sim}{\Gamma}(\nu) \cdot (\tau + \alpha)^{-\nu} \qquad (13)$$

where $\underset{\sim}{\Gamma}(\nu)$ is the gamma function of ν. The mean and variance are then

$$\bar{\Gamma} = \frac{\underset{\sim}{\Gamma}(\nu + 1)/\underset{\sim}{\Gamma}(\nu)}{\alpha} \tag{14a}$$

and

$$VAR = \left[\frac{\underset{\sim}{\Gamma}(\nu + 2)\underset{\sim}{\Gamma}(\nu)}{\underset{\sim}{\Gamma}(\nu + 1)} \right]^2 - 1 . \tag{14b}$$

Equation (13) can be fitted to the measured time correlation function by using a desk top calculator, rather than a larger computer. Unfortunately, at low values of VAR, the exponent, ν becomes large and may cause calculator overflows. When VAR is higher, the distribution function defined by Eq. (12) remains finite even at very low frequencies. Results of this technique are reported only for completeness. Its applications are not pursued.

A Pearson's type I distribution of the form given by Eq. (15) can also be used to approximate unimodal linewidth distributions:

$$G^*(\Gamma) = A_N \left[\frac{\Gamma}{B} - 1 \right]^C - \left[1 - \frac{\Gamma}{D} \right]^E . \tag{15}$$

In this approach we fit Eq. (15) to $[G^{(2)}(\tau)/A\beta]^{1/2}$ where $A\beta$ has been estimated by the cumulants technique. The factor A_N compensates for any error in $A\beta$ and also normalizes the distribution without affecting its shape. The values of B and D are the lower and upper limits of Γ, respectively. whereas C and E govern the shape of the curve. The integrations called for by Eqs. (1), (2), and (3) are performed numerically by using Simpson's rule with 100 or 150 steps. The step size below the peak of the distribution $G^*(\Gamma)_{max}$ is smaller than the step size above $G^*(\Gamma)_{max}$ to improve the resolution at the low-frequency end of $G^*(\Gamma)$. To obtain good values of the four important parameters of Eq. (15), the fitting procedure requires on the order of 10,000 numerical integrations. For this reason, we only use a high-speed Univac computer for the Pearson's fit.

The histogram fit has been described in detail elsewhere [4,5]. We assume that $G^*(\Gamma)$ is constant over a certain range and approximate $g^{(1)}(\tau)$ as

$$g^{(1)}(\tau) = \sum_{i=1}^{L} G^*(\bar{\Gamma}_i) \int_{\Gamma_a}^{\Gamma_b} \exp[-\Gamma_i \tau] \, d\Gamma \, , \qquad (16)$$

where Γ_a and Γ_b are $\bar{\Gamma}_i - \Delta\Gamma/2$ and $\bar{\Gamma}_i + \Delta\Gamma/2$, respectively; $\bar{\Gamma}_i$ is the mean linewidth for the ith step, and $\Delta\Gamma$ is a constant step width. In principle, the step sizes should increase exponentially with $\bar{\Gamma}_i$, but we have found that equal step sizes are more convenient and are sufficient for our analysis. The number of steps L is normally 4 to 10. To avoid negative coefficients, the fitting equation uses a_i^2, which is proportional to $G^*(\bar{\Gamma}_i)$. Again, we use only a high-speed Univac computer to execute the Fortran histogram program. The use of slower microprocessors necessitates employment of different programming techniques to allow for convenient usage of the histogram method.

III. EXPERIMENTAL TECHNIQUES

Simulated data were based on the parameters of polystyrene in cyclohexane at 35°C, the theta temperature, as given by Han and McCracken [7]. A weighted average molecular-weight distribution $f_w(M)$ was generated by using the Schulz distribution [8], and the correlation function was computed as

$$g^{(1)}(\tau) = \frac{\sum f_w(M_i) M_i P(X, M_i) \exp[-D_{0,i} K^2 \tau]}{\sum f_w(M_i) M_i P(X, M_i)} \, , \qquad (17)$$

The Debye form for the particle scattering factor of Gaussian coils was used [9]:

$$P(X, M_i) = \frac{2}{X^2} [\exp(-X) + X - 1] , \qquad (18)$$

where $X = r_g^2 K^2$ and r_g is the radius of gyration. The molecular weight enters the expression since we must use a relationship between r_g and M_i, as well as D_0, the diffusion coefficient in the limit of infinite diluation, and M. For both cases, we used the values given by Han and McCrackin [7]:

$$D_0 = 1.37 \times 10^{-4} M^{-0.5} \text{ cm}^2/\text{sec} \qquad (19a)$$

$$r_g^2 = 9 \times 10^{-18} M \text{ cm}^2 . \qquad (19b)$$

The net, measured correlation function, after baseline subtraction, was then generated as

$$G^{(2)}(\tau) = A\beta|g^{(1)}(\tau)|^2 + RND \cdot ER \cdot \sqrt{\bar{A}} , \qquad (20)$$

where we took $\beta = 1/2$ and let $A = 50 \times 10^6$. The term RND is a random number between -1 and 1. If $ER = 1$, we obtained a correlation function that was unrealistically smooth. We used $ER = 5$ as a more realistic simulation of experimental statistical errors. Data generated in this manner were designated as sets theta I → theta VIII with VAR varying from 0.005 to 0.17. A molecular-weight distribution was also obtained graphically by using digitized results from liquid chromatography. The digitized form was used in Eq. (17) and the resulting correlation function was designated as LT-I. Finally, two bimodal linewidth distributions were generated as

$$G^*(\Gamma) = C_1 G^{*(1)}(\Gamma) + C_2 G^{*(2)}(\Gamma) , \qquad (21)$$

where $G^{*(i)}$ represents the Pearson distributions, with C_1 and E being 0.5 and 1.0, respectively. Values of $B^{(1)}$ and $B^{(2)}$ were respectively 1000 and 32,000

sec^{-1} and $D^{(1)}$ and $D^{(2)}$, 11,000 and 36,000 sec^{-1}. Data set BM-I used $C_1 = C_2 = 0.5$, whereas BM-II had $C_1 = 0.8$ and $C_2 = 0.2$. The values of $g^{(1)}(\tau)$ were computed numerically by use of Eq. (1).

A light-scattering spectrometer described in detail elsewhere [10] was used to obtain dynamic measurements at a variety of scattering angles. The light source was a Spectra-Physics Model 165 argon-ion laser operating at $\lambda_0 = 488.0$ nm with a typical output below 150 mW. We used a Malvern K7023 96-channel single-clipped correlator modified to allow direct measurement of the baseline.

The polystyrene used was a National Bureau of Standards sample, NBS-705A, with $M_w = 179,300$ g/mol and $M_w/M_n \sim 1.07$. The solvent, cyclohexane, was purified and filtered by use of a Millipore filter (Millipore Corp., Bedford, Mass.) with a nominal pore diameter of 0.05 μm. The stock solution was filtered by using filters of 0.22 μm nominal pore diameter. The polyacrylamide sample (PAM-A) with $M_w \cong 2.5 \times 10^5$ g/mol, was purchased from Polyscience (catalogue no. 8248, lot no. 93-3, Polyscience, Inc., Warrington, Pa.). A higher molecular-weight sample (PAM-B) with $M_w \sim 5 \times 10^5$ g/mol was kindly provided by Dr. C. Y. Cha of Calgon Corporation, Pittsburgh, Pa. The solvent, pure water or 0.5 N $NaNO_3$, was filtered by use of a 0.05 μm pore-diameter filter, but in this case the polymer solution itself was never filtered. The solutions, however, were quite dust free.

IV.   RESULTS AND DISCUSSION

The expected value of $\bar{\Gamma}$ for simulated data was computed from Eq. (19) and the generated molecular-weight distributions $f(M)$. It is tabulated for each set in Table 1 along with the results obtained by fitting the generated time correlation function by use of a double exponential fit, a Pearson type I distribution, and the histogram method. We see that values of $\bar{\Gamma}$ can be extracted to within 1-2% for the unimodal distributions in all cases. We can only measure the variance to within 25% for some of the data sets. In the case of a bimodal distribution, the Pearson method fails as it should, whereas the other two approaches give quite reasonable values. The parameters R and Q are both above 0.7, except for the Pearson fits of BM-I and BM-II. The fact that the expected value of VAR and the determined value agree more closely for BM-I and BM-II may be fortuitous.

Results from the NBS $\underline{7}$05 standard polystyrene in cylcohexane at 35°C show $\bar{D}$ = 3.27 × 10^{-7} cm^2/sec a$\underline{t}$ θ = 60° and 90° and C = 0.2 wt.%. This value of $\bar{D}$ was computed from a Pearson fit of the linewidth distribution, not the second-orde$\underline{r}$ cumulants fit. Han and McCrackin [7] reported a $\bar{D}$° of 3.25 × 10^{-7} cm^2/sec at infinite dilution. If we were to$\underline{\ }$ take into account the concentration dependence of $\bar{D}$ for polystyrene in cyclohexane, agreement with the literature value would still be within about 5%. As the NBS 705 standard polystyrene in cyclohexane has been analyzed by use of the cumulants method, the Pearson distribution, and the histogram method, the agreement reconfirms the feasibility of obtaining the linewidth distribution function using the above-mentioned approximation techniques.

Table 2 lists the measured results for the PAM-B sample in pure water at 0.21 and 0.04 wt.% concentrations. In addition to the standard techniques of cumulants, Pearson, and histogram, we have used Eqs. (12)-(14) to analyze the measured correlation functions. We could find a range of values for α and ν that satisfied the convergence criteria. The values of α and ν presented in Table 2 were the averages of all the acceptable fitting parameters with the spread of values given$\underline{\ }$by the percentages in parentheses. The values of $\bar{\Gamma}$ and VAR listed in Table 2 are comparable to those obtained by othe$\underline{r}$ methods. For the 0.04 wt.% concentration, the $\bar{\Gamma}$ value by means of Eqs. (12)-(14) appears to deviate from the other mean by more than 10%. Therefore,

TABLE 1. Correlation Function Profile Analysis of Simulated Data Based on Polystyrene in Cyclohexane

Designation	Computed from $f(M)$ directly [$10^{-4}\bar{\Gamma}$ (sec^{-1}), VAR]	From the Correlation Function[a]		
		Double exponential [$10\,\bar{\Gamma}$ (sec^{-1}), VAR]	Pearson's [$10^{-4}\bar{\Gamma}$ (sec^{-1}), VAR]	Histogram [$10^{-4}\bar{\Gamma}$ (sec^{-1}), VAR]
Theta-VIII	2.25, 0.005	2.26, 0.01	2.28, 0.015	2.26, 0.015
Theta-I	2.23, 0.042	2.21, 0.04	2.21, 0.03	2.21, 0.04
Theta-III	2.12, 0.171	2.10, 0.13	2.07, 0.08	2.10, 0.13
Theta-IV	1.21, 0.040	1.20, 0.03	1.23, 0.03	1.26, 0.03
Theta-VI	1.22, 0.158	1.21, 0.11	1.22, 0.11	1.21, 0.13
LT-1	0.64, 0.161	0.64, 0.13	0.63, 0.11	0.64, 0.15
BM-I	2.00, 0.547[d]	1.98, 0.53	1.76, 0.57[b]	1.99, 0.54
BM-II	1.12, 1.15[d]	1.06, 0.97	0.81, 0.53[c]	1.13, 1.17

[a] $R > 0.7$ and $Q > 0.7$ except where noted.

[b] $R = 0.03$ and $Q = 0.05$. In any case, the Pearson type I distribution is not applicable for a bimodal distribution.

[c] $R = 0.03$ and $Q = 0.13$. In any case, the Pearson type I distribution is not applicable for a bimodal distribution.

[d] Computed from $G^{*}(\Gamma)$ directly.

TABLE 2. Correlation Function Profile Analysis of PAM-B ($M_w \sim 5 \beta 10^5$) in Water at 25°C

A. Equations (12) - (14)[a]

Conc. (wt.%)	θ	α (sec)	ν	$\bar{\Gamma}$ (sec^{-1})	VAR
0.21	60	7.6×10^{-4} (1%)	3.06 (2%)	4.03×10^3 (2%)	0.33 (2%)
0.04	90	6.2×10^{-4} (6%)	4.3 (5%)	7.00×10^3 (6%)	0.23 (5%)

B. Double Exponential Fit

Conc. (wt.%)	θ	$\bar{\Gamma}$ (sec^{-1})	VAR	ρ	R	$\bar{D} \cdot 10^7$ (cm^2/sec)
0.21	60	3.92×10^3	0.23	0.82	0.87	1.34
0.04	90	8.07×10^3	0.22	0.70	0.49	1.38

C. Histogram Method

Conc. (wt.%)	θ	$\bar{\Gamma}$ (sec^{-1})	VAR	ρ	R	$\bar{D} \cdot 10^7$ (cm^2/sec)
0.21	60	4.00×10^3	0.31	0.83	0.78	1.36
0.04	90	8.33×10^3	0.32	0.68	0.74	1.42

D. Pearson Type I Distribution

Conc. (wt.%)	θ	$\bar{\Gamma}$ (sec^{-1})	VAR	ρ	R	$\bar{D} \cdot 10^7$ (cm^2/sec)
0.21	60	3.96×10^3	0.24	0.88	1.1	1.35
0.04	90	8.17×10^3	0.23	0.78	0.61	1.39

[a]$\bar{\Gamma}$ and VAR were computed by means of Eq. (14) using mean values of α and ν which satisfied the convergence criteria. Percentage in parentheses represent the spread of values covered by acceptable fits.

Eqs. (12)-(14) can be used only for making estimates, whereas the double exponential _fit represents a reliable approach for obtaining $\bar{\Gamma}$ and VAR values, even for broad linewidth distributions. Both the histogram method and the Pearson type I distribution are applicable for fairly broad distributions (VAR ~ 0.3). However, it should be emphasized that the Pearson type I distribution requires extensive computer time and is valid only for unimodal distributions.

Thus far we have considered the cumulants method only as a reference technique. Table 3 shows a comparison of $\bar{\Gamma}$ and VAR values computed from histogram fits and from cumulants fits with appropriate extrapolation to short delay times for PAM-A in 0.5 N NaNO$_3$ solution. As expected, the $\bar{\Gamma}$ values are consistent when the cumulants and the histogram method are used.

The variance differs by up to 30%. In our analysis the cumulants technique consistently yields a lower estimate of the variance than does the histogram approach because, for large variances, higher-order terms must be taken into account. It is much more difficult to extract a bimodal linewidth distribution from the correlation function. Although a value of the variance exceeding about 0.6 is a strong indication that $G^*(\Gamma)$ is bimodal, the evidence is not conclusive, nor does the variance offer any quantitative information concerning the form of the distribution function, that is, whether it is unimodal or multimodal. The cumulants technique, which yields mainly the average linewidth and the variance, is thus of more limited value for bimodal linewidth distributions.

As shown by the analysis of generated data sets (see BM-I and BM-II in Table 1), we have found that the histogram approach can be used to extract a bimodal linewidth distribution from the correlation function. The following two examples show that suitable care must be exercised. Figure 2 shows a series of histograms for Polyscience PAM-A in pure water at different scattering angles. The abcissa is scaled by $\sin^2(\theta/2)$, indicating that the histograms should be similar if the decay of $g^{(1)}(2)$ is due solely to translational motion. The segments in parts b and d, and to an extent e and c, which seem to indicate some bimodality, are more reasonably attributed to uncertainties in the fit at higher frequencies.

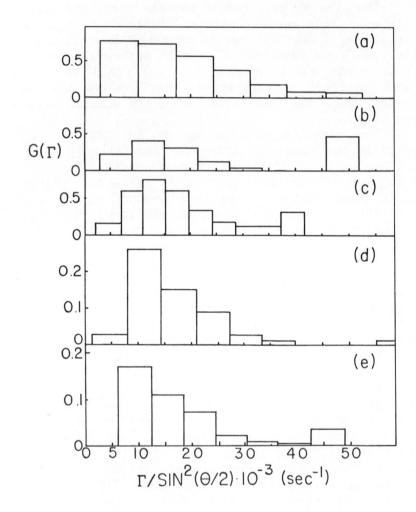

FIGURE 2.   Histograms for PAM-A in pure water; concentration = 0.1 wt.%:   (a) θ = 50°; (b) θ = 70°; (c) θ = 85°; (d) θ = 105°; (e) θ = 130°.   Note that the Γ axis has been divided by $\sin^2(\theta/2)$; therefore, the histograms are proportional to $G^*(D)$ rather than $G^*(\Gamma)$.

TABLE 3. Comparison of $\bar{\Gamma}$ and VAR Values Computed from Histogram Fits and from Cumulants Fits with Appropriate Extrapolation to Short Delay Times for PAM-A in 0.5 N NaNO$_3$ Solution

Concentration (wt.%)	θ (deg)	Cumulants[a]		Histograms	
		$10^{-3} \cdot \bar{\Gamma}(sec^{-1})$	VAR	$10^{-3} \cdot \bar{\Gamma}(sec^{-1})$	VAR
0.4	60	5.00	0.35	5.05	0.53
0.198	90	9.80	0.25	9.30	0.32
0.198	60	4.30	-	4.40	0.35
0.098	60	4.30	0.36	4.30	0.44
0.05	60	4.20	0.30	4.20	0.37

[a]Values of VAR are too low because of large variances. Higher-order terms must be taken into account.

97

A similar bimodal histogram is shown in Fig. 3a.
In this instance, however, the convergence criteria
are not met, as they are in the histograms in Fig. 2.
In Table 4, we begin with a reasonable fit (a) that
does not meet our criteria for R and Q. As the area
of the high-frequency peak is decreased and the
high-frequency component reaches even higher frequencies, fits c, d, e, g, and h increase in goodness-of-
fit properties. Above fit i, R and Q decrease. In
fits b and f, the high-frequency peak is arbitrarily
omitted and the convergence criteria computed from
the resulting unimodal distribution. All fits with
acceptable criteria are then averaged, as given in
the bottom four lines in Table 4.

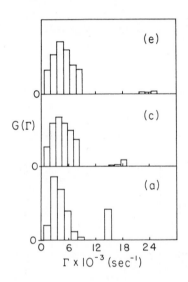

FIGURE 3.   Histograms of PAM-A in 0.5 N NaNO$_3$; concentration = 0.4 wt.%; θ = 60°. Letter
designations refer to the same fits as
presented in Table 4. The quality of the
fits increases from part a to part c to
part e.

TABLE 4. Bimodal Character of 0.4 wt.% PAM-A in 0.5 N NaNO$_3$ at $\theta = 60°$

FIT	$\bar{\Gamma}$	VAR	$\bar{\Gamma}^{(L)}$ [a]	$\bar{\Gamma}^{(H)}$	R	Q	RSQ	A_L/A_H [b]
a	5610[c]	0.57	3900	14750	0.60	0.38	0.9999942	5.3
b	4930	0.36	Unimodal[d]		0.68	0.49	0.9999956	–
c	5000	0.44	4440	17500	0.76	0.61	0.9999968	22
d	5020	0.48	4530	21100	0.91	0.77	0.9999976	33
e	5040	0.52	4580	23900	0.97	0.85	0.9999978	40
f	4790	0.24	Unimodal[e]		0.32	0.23	0.9999980	–
g	5070	0.58	4610	27100	1.00	0.93	0.9999980	47
h	5110	0.65	4620	30800	1.04	0.97	0.9999981	53
i	5290	0.95	4610	37900	0.65	0.46	0.9999951	48

From the fits that meet the convergence criteria d, e, g, and h, we conclude:

$$\bar{\Gamma} = 5040 \pm 25^{c} \qquad \text{VAR} = 0.53 \pm 0.05 \qquad \bar{\Gamma}^{(L)} = 4750 \pm 40$$

$$\bar{\Gamma}^{(H)} = 24{,}000 \pm 3000 \qquad A_L/A_H = 40 \pm 10$$

From a Pearson's fit ($Q = 0.59$, $R = 0.30$, $RSQ = 0.9999915$)

$$\bar{\Gamma} = 4800 \qquad \text{VAR} = 0.24$$

[a]$\bar{\Gamma}^{(L)}$ and $\bar{\Gamma}^{(H)}$ are the mean linewidths of the low- and high-frequency peaks, respectively.

[b]A_L/A_H is the ratio of the area under the low-frequency peak to the area of the highfrequency peak.

[c]The accuracy of any linewidth presented in this work can be no better than 2% since that is the deviation we find for the measured value of $\bar{D}$ of polystyrene latex spheres as a function of $\sin^2(\theta/2)$ in the calibration of the spectrometer. A greater number of significant figures is presented here to indicate the precision (reproducibility) of the fits.

[d]Range of Γ 500 → 13900 (sec^{-1}).

[e]Range of Γ 500 → 9600 (sec^{-1}).

We have studied the time correlation function generated from a broader molecular-weight distribution that obeys a Pearson type I distribution function. It was also necessary to use $ER = 2$, rather than 5, to obtain reasonable fits. The characteristics of the generated distributions are given in Table 5, with P-2 and P-3 plotted in Fig. 4. The

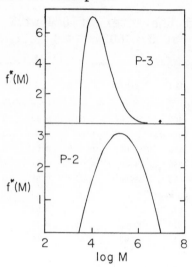

FIGURE 4.   Generated, broad distributions of $f(m)$ versus $\log_{10}M$ (see Table 5 for details). Note that P-2 was generated from a symmetrical distribution in $\ln M$ space but is normalized such that $\Sigma f(M_i) \, \Delta M_i = 1$, rather than $\Sigma f(M) \, \Delta \ln M_i = 1$.

molecular-weight distribution was transformed to a time correlation by use of Eq. (17), where the parameters again were those of polystyrene in cyclohexane at the theta temperature. Note that the skewness of the distributions presented in Table 5 resembles that of real data encountered for polyethylene.

The correlation functions were again analyzed; the results are presented in Table 6. We observe that the fits are in general not very good, with deviations in $\bar{\Gamma}$ on the order to 10% and those for VAR, larger. The double exponential fit yields reasonable values of R and Q for all data analyzed, and the Pearson method of approximating $G^*(\Gamma)$ is satisfactory only for P-4. Whereas the histogram fits give the best values of Q, indicating the closest fit to the data of all three methods, the distributions obtained are qualitative in the sense

TABLE 5. Characteristics of Generated, Broad $f(m)$ Distributions

Designation	Pearson's exponents[a] C and E in Eq. (15)	M_W	$M_N:M_W:M_Z$	$\bar{\Gamma}$ (sec^{-1}; $\theta = 90°$)	VAR
P-2	(1;1)	2,590,000	1:4.1:6.6	9100	0.22
P-3	(1;5)	295,400	1:5.9:18.6	24300	0.46
P-4	(1;10)	46,300	1:3:11.5	62900	0.39

[a]The distributions were computed as ln M on the x axis; therefore, a generated symmetric distribution (P-2) is skewed in $f(m)$ versus M space. For the generation, the limits B and D in Eq. (15) were 8 and 16, respectively.

101

that the high-frequency components are represented as a second peak rather than a continuous distribution.

V. CONCLUSIONS

We have examined five approaches for extracting information from polydisperse polymer solutions or colloidal suspensions. For broad size distributions and bimodal distributions, the traditional cumulants expansion technique should be used with caution. For absolute determinations of $\bar{\Gamma}$ and VAR, we find that appropriate cumulants fit gives values that are consistent with the histogram approach. Thus, in principle, we need only a single measurement of $G^{(2)}(\Gamma)$ for determination of an approximate $G^*(\Gamma)$.

The double exponential fit can be used to obtain the moments of distributions ranging from narrow to bimodal. Therefore, the fact that a correlation function can be fitted well to a double exponential form is not sufficient evidence to prove a dual relaxation process, such as monomer-dimer equilibria. We often use this technique initially to obtain reliable estimates of the moments to assist us in further analysis.

Equation (12) describes a clearly unimodal distribution for which the correlation function can be expressed in a closed and simple form. The approach may be applicable to distributions that are not too broad, but we feel that the four-parameter Pearson's fit is more versailte than this two-parameter equation. The Pearson type I distribution function has been found to adequately describe any unimodal distribution. Additionally, the endpoints of the linewidth distribution are fitted explicitly. In the transformation from gamma space to molecular weight space we find that the precise determination of the low-frequency limit of $G^*(\Gamma)$ is of utmost importance. Although the Pearson's fit requires longer computational time (on the order of eight times more than for a histogram fit), the programmer

TABLE 6. Results of Analysis of Correlation Functions Generated for Table 5

Designation	Double exponential $\bar{\Gamma}$; VAR; R; Q	Pearson $\bar{\Gamma}$; VAR; Q	Histogram $\bar{\Gamma}$; VAR; Q
P-2	7,900; 0.28; 0.86; 0.73	7,590; 0.20; 0.42	8,600; 0.59; 0.92
P-3	26,800; 0.28; 0.20; 0.64	23,900; 0.25; 0.15	28,900; 0.44; 0.77
P-4	57,800; 0.17; 0.46; 0.94	59,600; 0.22; 0.87	—

must input only the necessary experimental conditions and crude estimates of the fitting parameters.

The histogram technique is computer interactive. We fit only the amplitudes of each segment while manipulating the positions of the segments along the Γ axis manually in a systematic way. Computational time is decreased, but the programmer must acquire more expertise. With this approach we feel confident in fitting even bimodal distributions that could not be studied conclusively by use of most other techniques described here. We have tried the delta function approach due to McWhirter and Pike [11] with some success to fit our generated bimodal distributions, BM-I and BM-II. On the basis of our limited experience, it seems that a fair degree of expertise is also required [12]. The evidence given in Table 4 is impressive in that we can extract a high-frequency component that contributes only 2% of the total scattered intensity with reasonable accuracy. We cannot distinguish between a small peak and a long tail by means of correlation function profile analysis alone. However, if we have means to determine the modality of the linewidth (or size) distribution, the proposed method of data analysis permits us a semiquantitative determination on the skewness of the distribution or the magnitude of the second component, even when its contribution to the total scattered intensity represents only a few percent when compared with the dominant component.

ACKNOWLEDGMENTS

We gratefully acknowledge the support of this research by the Polymer Program, Division of Materials Research, the National Science Foundation (DMR 8016521), the Petroleum Research Fund administered by the American Chemical Society, and the U.S. Army Research Office.

REFERENCES

1. D. E. Koppel, J. Chem. Phys. 57, 4814 (1972).

2. P. R. Bevington, *Data Reduction and Error Analysis for the Physical Sciences*, McGraw-Hill, New York, 1969.

3. P. Mathiez, G. Weisbuch, and C. Moutlet, Biopolymers 18, 1465 (1979).

4. E. Gulari, E. Gulari, Y. Tsunashima, and B. Chu, Polymer 20, 347 (1979).

5. E. Gulari, E. Gulari, Y. Tsunashima and B. Chu, J. Chem. Phys. 70, 3965 (1979).

6. E. R. Pike, in Proceedings of NATO ASI on Scattering Techniques, Plenum Press, New York, 1981.

7. C. C. Han and F. L. McCrackin, Polymer 20, 427 (1979).

8. L. H. Peebles, *Molecular Weight Distributions in Polymers*, Interscience, New York, 1971.

9. M. Kerker, *The Scattering of Light and Other Electromagnetic Radiation*, Academic Press, New York, 1972.

10. A. DiNapoli, Ph. D. dissertation, SUNY at Stony Brook, New York, 1981.

11. J. G. McWhirter, Optica Acta 27, 83 (1980); J. G. McWhirter and E. R. Pike, J. Phys. A 11, 1729 (1978).

12. D. Soronette and N. Ostronsky, "Kinetics of Phospholipid Vessicle Growth Around the Phase Transition," presented at the *NATO ASI on Scattering Techniques*, Plenum Press, New York, 1981.

CHAPTER 4

Analysis of Polydisperse Scattering Data II

E. R. PIKE
Royal Signals and Radar Establishment
Malvern, Worcestershire, England

D. WATSON and F. MCNEIL WATSON
Malvern Instruments, Ltd.
Malvern, Worcestershire, England

CONTENTS

I. INTRODUCTION

This chapter is intended as a sequel to an earlier paper of the same title [1] in which the

107

basic problem of the inversion of quasi-elastic light scattering (QELS) data and the reasons for its ill-conditioned nature were addressed. The first requirement in interpreting such data is to perform an inverse Laplace transform on the first-order photon correlation function $g^{(1)}(\tau)$, after which the effects of Mie factors and scattering strengths must be considered to arrive at size distributions. We ignore these latter effects in the present work to concentrate on the Laplace inversion.

We showed in the previous paper that a procedure known as *numerical filtering*, using the eigenfunctions and eigenvalues of the Laplace transform discovered in 1978 by McWhirter and Pike [2], led to the concepts of "exponential sampling" and associated interpolation that are generalizations of the well-known Nyquist sampling and interpolation of standard information theory.

According to this generalization of information theory, the "resolution" with which structure in the particle radius (diffusion constant) distribution can be obtained is a function of the noise in $g^{(1)}(\tau)$ that, in turn, is a function of the duration of the scattering experiment, and this resolution is to be defined on a geometric progression of values of radii between give upper and lower values. The theory has been developed and implemented experimentally by Ostrowski et al. [3] by use of an on-line desk computer.

It was noted in both references above that an a *priori* knowledge of the support of the distribution would improve the resolution possible in the inversion. In more recent work by Bertero et al. [4] this fact has been quantified by a numerical computation of the singular value spectrum of the Laplace transform when the object is of bounded support. It is indeed found that the common ratio of the geometric progression, the so-called resolution ratio or dilation factor δ, reduces, as the ratio $\gamma = b/a$ of the known upper and lower limits b and a, respectively, of the distribution decreases, and that the resolution obtainable is a function only of this ratio γ, and not of the individual values of a and b. In a further contribution Bertero et al. [5] calculated the trace of the singular value spectrum of the inversion onto a finite support from a fixed number of geometrically or linearly spaced points of the correlation function between given limits and hence further refined the theoretical analysis of Laplace inversion toward the actual experimental situation.

In the present chapter we wish, first, to show how the generalization of Nyquist sampling theory adumbrated above implies not only that the radius distribution should be given geometrically, but also (as quantified numerically in Bertero et al. [5]) that the information content of the correlation function at sufficiently high photon rates is also so distributed. We also confirm recent work by Ostrowski and Sornette (private communication) along these lines in which successful inversions have been performed with exponentially sampled data. These findings have led to the design and construction of a new photon correlator, which is described elsewhere [6], in which the correlation coefficients (or structure function coefficients in another implementation) are computed at points of a quasi-geometric progression.

Second, we wish to make some theoretical observations about the Laplace inversion problem from the practical end, as it were, along the lines taken some years ago by Pusey et al. [7] and hope to show that these also lead to the same general conclusions as the above more elaborate analysis. We also comment from the same point of view on the question of positivity and regularization constraints and, contrary to our previous conjectures, show that our simple linear exponential sampling analysis is adequate for achievement of the required results. Finally, the resolution that may be achieved in QELS in practice is quantified and verified by extensive numerical simulation and comparison with experiment, using a prototype of the new Malvern Instruments "information-theoretic" correlator mentioned above.

II.   EXPONENTIAL SAMPLING OF DATA

By a simple extension of the eigenvalue analysis of McWhirter and Pike [2], we will show that the correlation function at a series of exponentially spaced samples of delay time

$$\tau, \ \exp[\pi/\omega_m]\tau, \ \exp[2\pi/\omega_m]\tau \ \cdot \ \cdot \ \cdot$$

or

$$\tau, \ \delta\tau, \ \delta^2\tau \ \cdot \ \cdot \ \cdot$$

together with a given interpolation procedure will allow reconstruction of the correlation function at all values of τ within the band limit ω_m, defined by the experimental noise.

In the above-mentioned eigenvalue analysis a set of functions $\phi_\omega(\tau)$ were found which transformed into themselves, apart from a scalar factor λ_ω, which can be shown to decrease to zero as $\omega \to \infty$ under the Laplace transformation, that is

$$\lambda_\omega\phi_\omega(\tau) = \int_0^\infty \exp[-\Gamma\tau]\phi_\omega(\Gamma)d\Gamma \ , \tag{1}$$

where Γ is the light-scattering linewidth variable. These "eigenfunctions" from a complete orthonormal basis for expression of the following formal solution to the inversion problem:

$$g^{(1)}(\tau) = \int_0^\infty \exp[-\Gamma\tau]p(\Gamma)d\Gamma \ . \tag{2}$$

Let b_ω be the components of $g^{(1)}(\tau)$ in this basis

$$g^{(1)}(\tau) = \int_{-\infty}^\infty b_\omega\phi_\omega(\tau)d\omega \ . \tag{3}$$

Then, using Eq. (1) and the orthonormality of the basis, we have

$$p(\Gamma) = \int_{-\infty}^{\infty} a_\omega \phi_\omega(\Gamma) d\omega , \qquad (4)$$

where

$$a_\omega = \frac{b_\omega}{\lambda_\omega} . \qquad (5)$$

In the numerical filtering technique we recognize that for values of ω above a maximum called ω_m, the value of λ_ω becomes so small that any noise contribution to b_ω is multiplied so much by the $1/\lambda_\omega$ factor that it makes the recovery of the true b_ω impossible. The integrals over ω must therefore be truncated and the correlation function is then expressed, using the explicit results for ϕ_ω given by McWhirter and Pike,

$$g(\tau) = \int_{-\omega_m}^{\omega_m} \lambda_\omega \beta(\omega) \tau^{-1/2} \exp[i\omega \log \tau] d\omega , \qquad (6)$$

where $\beta(\omega)$ is a given function of the a_ω above. If the substitutions

$$\tau = \exp[y] \qquad (7)$$

and

$$A(\log \tau) d(\log \tau) = g(\tau) d\tau \qquad (8)$$

are made, there results

$$\exp[-y/2] A(y) = \int_{-\omega_m}^{\omega_m} \lambda_\omega \beta(\omega) \exp[i\omega y] d\omega . \qquad (9)$$

The function on the left-hand side of this equation is, therefore, a band-limited Fourier transform for which the Nyquist sampling theorem applies. That is to say that the function may be fully reconstructed from its values at the points τ_n listed at the beginning of this section and that it may be interpolated between these points by use of the function

$$I(\tau,\tau_n) = \left(\frac{\tau}{\tau_n}\right)^{-1/2} \frac{\sin \omega_m(\log \tau - \log \tau_n)}{\omega_m(\log \tau - \log \tau_n)} \qquad (10)$$

as was shown in a similar way for the reconstruction of $p(\Gamma)$ in Pike [1]. It is clear, therefore, that the inversion may be carried out without loss of accuracy by knowledge of $g(\tau)$ at this set of exponentially sampled points.

In a first series of experiments the well-known cumulants analysis [8,9] was used to measure polydispersity values from data gathered from the same sample and over the same experimental times with both linearly and geometrically spaced correlation functions. Although extensive comparisons have not been made as yet, our results show that similar accuracies may be achieved by use of both methods.

III. SOME SIMPLE CONSIDERATIONS
ON THE RESOLUTION OF EXPONENTIALS

The eigenvalue analysis discussed above has indicated a resolution criterion that may be considered from a rather simpler point of view. We propose the problem of finding the resolution ratio for three exponentials such that the calculated difference between the correlation function due to the sum of the two outer ones with equal 50% weights is equal to that of the centre one alone, within the root-mean-square (RMS) noise level of the experiment. This difference may be written as follows,

$$\Delta g = -\frac{1}{2} \exp[-\tau\delta/\tau_0] + \exp[-\tau/\tau_0]$$

$$-\frac{1}{2} \exp[-\tau/\tau_0\delta] \tag{11}$$

where the exponentials are placed on the geometrically spaced points τ_0/δ, τ_0, and $\tau_0\delta$, respectively. We write the dilation factor as

$$\delta = 1 + \varepsilon \tag{12}$$

and expand the expression (11) in powers of ε. After some straightforward manipulations and writing

$$t = \frac{\tau}{\tau_0} , \tag{13}$$

we find that

$$\Delta g = -\frac{\varepsilon^2}{2} \exp[-t](t^2 - t) . \tag{14}$$

By equating the derivative with respect to t to zero, we find the turning points at

$$t = \frac{3 \pm \sqrt{5}}{2} , \text{ that is, } 0.382 \text{ and } 2.618, \tag{15a}$$

that is ,

$$\tau = 0.382\tau_0 \text{ and } 2.618\tau_0 . \tag{15b}$$

The excursions of the function from zero at these points are

$$\Delta g = 0.080\varepsilon^2 \text{ and } -0.155\varepsilon^2 . \tag{16}$$

The function in Eq. (14), of course, goes to zero at $\tau = 0$ and $\tau = \infty$. We note the vanishing of the linear

term, which explains the significance of the geometric spacing in the resolution criterion; even with
no noise, to a first approximation in ε the two
exponentials cannot be distinguished from a single
one at their geometric mean position.

The question arises as to whether the separation
should be increased until the value of Δg exceeds the
noise, at which point we might place a criterion for
their resolution. To implement this, we first need
to quantify the noise value, and we have used the
same procedure as in our previous work referred to
above of assuming that the known and experimentally
checked variance of correlation coefficients for a
single exponential photon correlation function of
decay time τ_0 as calculated by Jakeman et al. [10]
will be a sufficiently good approximation for the
situation under consideration here. These authors
give the expression,† when the number of photons per
coherence time $\gg 1$,

$$\text{var } g^{(2)}(\tau)$$

$$= \frac{1}{N_c} \left[\frac{1}{2} (1+8\exp[-x] - \exp[-2x](5 + 2x)) \right] , \quad (17)$$

where N_c is the number of coherence times in the
duration of the experiment and

$$x = 2 \frac{\tau}{\tau_0} . \quad (18)$$

It is, unfortunately, not possible easily to obtain
an analytical formula for the RMS error on $g^{(1)}(\tau)$
over a suitable range of τ using this formula but
some approximate values for the delays of the new
correlator were found by numerical means. These are

---

†In Equation (50) in Pike [1], $2/\Gamma$ should read $2/r$,
and in Eq. (52), τ should be deleted.

given in Table 1, together with the corresponding
value of ε for a τ_0 corresponding to 256 channels.
It is found that the error weighting up to 1000
channels is close to uniform.

TABLE 1.   RMS Error in $g^{(1)}(\tau)$ over 1000 Channels

N_c	RMS Error	ε	$\delta_{1.0}$	$\omega_m/2$	$\delta_{0.1}$
10^4	$8.2\ 10^{-3}$	0.23	1.51	7.5	2.98
10^5	$2.6\ 10^{-3}$	0.13	1.28	13	1.99
10^6	$8.2\ 10^{-4}$	0.073	1.15	22	1.51

We also give in this table values of dilation factor
$\delta_{1.0}$ and band limit that correspond to the separation
of the two components.

These simple calculations may be extended to
solve the following problem. Suppose that we compare
a correlation function due to a single exponential
function at τ_0 with one due to two exponential func-
tions at τ_0/δ and $\tau_0\delta$ with weights $\alpha/2$, plus a single
exponential function at τ_0 with weight $1 - \alpha$. We may
then ask at what values of α for a given ε these
cases are indistinguishable within the experimental
noise. It is easy to show that this will be the case
for the values of ε of the previous case divided by
$\sqrt{\alpha}$. The last column of Table 1 shows this effect for
$\alpha = 0.1$. We see, for example, that for $N_c = 10^5$, 10%
of a single exponential component can be removed and
redistributed at equal weights on the adjacent geo-
metrical spaced points, the higher one at twice the
ordinate value of the lower one without disturbing
the correlation function within the limits of the
noise.

These elementary results clarify from another point of view the ill-conditioned nature of the problem, showing explicitly how various quite different distributions may be quite indistinguishable from each other within the experimental noise on the correlation function and that no method of inversion can claim to produce the input distribution to any more accuracy than is imposed by these unavoidable uncertainties. This, of course, includes all methods of constrained and regularized inversion. The number of coherence times 10^5 at which fractions as large as 10% of the distribution may lie anywhere within a range of sizes over a factor of 2 represents, say, an experiment of duration 100 sec with a mean coherence time of 1 msec. This would be on the high end of normal accuracies and shows what extreme precision is necessary to obtain meaningful inversions. We expose further implications of these results with regard to inversion procedures in Section IV.

IV.   EFFECT OF KNOWLEDGE OF SUPPORT OF THE DISTRIBUTION

From the above considerations we may deduce that an a *priori* knowledge of the support of the distribution will reduce the number of possible distributions that will fit the correlation function within the limits imposed by noise, entirely in accord with the results of the general studies of ill-conditioning due to Bertero et al. [4,5] although these latter studies encompass input functions of arbitrary phase. Positivity of the output will inhibit a number of solutions that may always be generated by the Riemann-Lebesgue theorem as outlined in our previous paper, but there will always be high-frequency oscillatory additions over the support of any probability distribution that will leave the resultant positive, as is indicated in Figure 1.

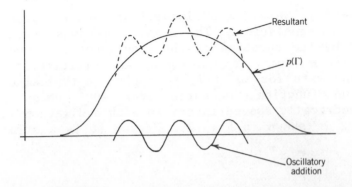

FIGURE 1.   Effect of spurious oscillatory additions
to the linewidth distribution.

The effect is the same as for distributions of arbitrary phase and demands a resolution criterion that
is a function of the a *priori* knowledge of the support. For a given level of noise, if the support is
restricted to a range smaller than that defined by,
say, δ_{10} of Table 1, it will not be possible to set
up oscillatory additions, which require weights outside this support; this restriction improves the
reconstruction within the support given.   Quantifi-
cation of this effect seems difficult within the
framework of the analysis of the previous section but
is effected rigorously in the singular value theory
proposed by Bertero et al. [4,5].

V.   SIMULATIONS AND EXPERIMENT

In an effort to make firm contact with the
theoretical work discussed above, we have carried out
a number of simulated experiments, testing the new
Malvern correlator with the Commodore 8032 microcomputer and extended software. Known input distributions were converted into correlation functions and
noise added according to Eq. (17); these were then
inverted by use of the exponential sampling tech-

nique [3], which is a linear least-squares procedure
with interpolation. The noise level was characterized by the duration of the experiment in numbers of
coherence times N_c, assuming a sufficiently high
photon rate for Eq. (17) to apply. A typical correlation function, measured over 4096 channels at
geometrically spaced intervals with a dilation factor
of $\sqrt{2}$ is shown in Fig. 2. First, it is a straight-

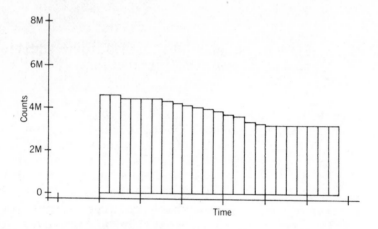

FIGURE 2.   Typical photon correlation function on a
            geometrically spaced progression of delay
            times. Counts are measured in units of
            $M = 10^6$.

forward matter to check the predictions of Table 1 by
creating the difference between theoretical and noisy
single exponential functions and fitting with three
exponentials, two of which are negative. We obtained
satisfactory confirmation in this way of the correct-
ness of the calculations. From many further simulations performed that gave us increasing confidence in
our approach to the essence of this problem, we
select for the moment two for illustration that
clearly point out the features discussed above.   In
both of these examples two exponential functions of

equal weights were used.   In the first example the lower limit was set at a diameter on the scale of the instrument of 200 nm, and the separation was given by a dilation ratio of 1.87 (band limit 5), putting the second component at a diameter of 374 nm.   A noise corresponding to $N_c = 10^5$ was added.   This noise was generated, for reasons of convenience on the micro-computer, from a square distribution of the required RMS width, but it is known [11] that this is a satis-factory procedure.   A fit of 50/50 weights on the given points was then forced and the RMS deviation calculated.   This amounted to 4.2×10^{-3}, a factor of 1.6 larger than the average of Table 1.   The point here is that any individual realization of a noise process does not necessarily have the infinite sample values of mean.   In this case a second fitting of exponentials on the same input points but with a free ratio of weights gave a preferred ratio of 45/55 with the lower RMS error of 3.8×10^{-3}.   When these exponentials were broadened by interpolation with a dilation factor of 1.37 (band limit 10), an even better residual was obtained at 2.6×10^{-3}, which is in satisfactory agreement with the expected fluctua-tions.   We thus see that in a single experiment the criterion of a least-squares fit of a variable input model to the data does not provide as the output best fit the distribution that was put in, no matter how constrained.   A set of possible solutions exist that are indistinguishable, under any given set of con-straints, within the noise on the data.

In the second example the lower diameter was set at 150 nm, the separation governed by a dilation ratio of 3.51 (band limit 2.5) thus putting the upper diameter at 527 nm.   The noise added again corre-sponded to $N_c = 10^5$.   In Figs. 3 and 4 the data are analyzed over two different ranges; least-square fits with the same RMS deviation of 2.08×10^{-3} were obtained for both results.   Again we see the nature of ill-conditioning in practice.   There is no way, without further knowledge, to distinguish which of

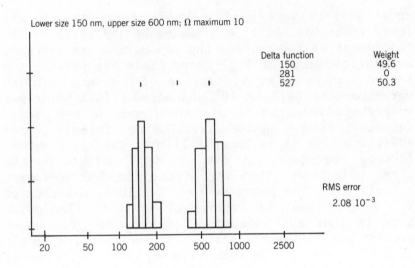

FIGURE 3. Particle size distribution from inversion of simulated data as described in the text.

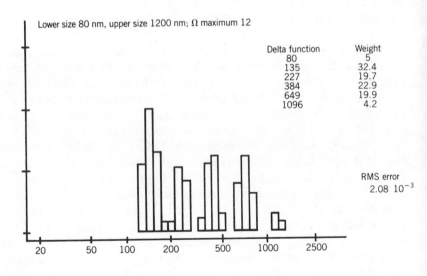

FIGURE 4. As in Fig. 3 with same RMS error.

these two possible distributions is nearer to the input.

What, then, may we conclude so far? It is clear that the fineness of detail or resolution that may be produced in a fitted output distribution is essentially unlimited and that this fineness of detail may be quite unconnected with the actual nature of the object distribution, being solely due to oscillatory additions that are indiscernible in the data. Therefore, we cannot believe any output distribution that has structure on a fine scale than can be so distinguished at some given level of noise, and this latter scale of resolution is also a function of the a *priori* knowledge of the support of the input distribution. Moreover, these conclusions are independent of imposed constraints of a general nature such as positivity or regularization. A linear least-squares fit is as effective as any other to generate output distributions at the allowed level of resolution. In fact, such a linear least-squares fit may be safer since spurious oscillatory additions may be detected as nonphysical negative output weights. We might point out that this feature of ill-conditioning does not seem to affect the accuracy of the mean and variance of the distribution as calculated from the output. The output distribution can be clearly corrupted by spurious oscillatory additions, even running wildly negative, without adversely influencing the mean and variance parameters. This again seems to be an advantageous feature of the linear fitting procedure.

These conclusions lead us then to ask for the quantitative relationship between range and resolution at a given noise level; in other words, given N_c, how does δ, representing the highest resolution obtainable in the object, vary with γ, the ratio of the upper to the lower bounds of support?

Numerical computations of the associated singular value spectra have given such results [4] but these computations were based, for convenience of the theoretical approach used, on fictitious white-noise

objects.   It may be possible to consider the more
limited range of possible "objects" of the present
problem in a future analytic approach, but we have
now used extended facilities of the new Malvern
correlator to run a series of simulations to give
workable answers to this question.   For given values
of γ from 1 to 5, the noise levels, again characterized by N_c, at which two, three, four and five points
could be resolved were found by noting the point of
departure of the output from the known input.   This
is inevitably signaled by spurious oscillatory additions.   The results of this survey are given in
Fig. 5.   Inspection of these results indicates an
approximate empirical connection between the number
of resolvable points and the noise level almost
independent of range.   Thus it seems that to resolve
two points $N_c \cong 10^3 - 10^4$, to resolve three points $N_c
\cong 10^6$, four-points $N_c \cong 10^8$, and so on.   These
values, although disappointing from the point of view
of the light-scattering technique, are in agreement
with previous empirical evidence [3,12] and seem to
have to be accepted.   It would be instructive to be
able to pursue the type of argument given here to
relate, if possible, Fig. 5 to the last column, say,
of Table 1, but we have not yet done this.

Increase of N_c by high-temperature operation (or
otherwise lowering solvent viscosity), high-angle
scattering, and shorter-wavelength operation are the
only apparent measures available for increasing
resolution, apart, of course, from directly increasing N_c by repeated accumulation of correlation functions for integration by use of the self-normalizing
technique [13].

The new Malvern correlator mentioned above has
this latter facility built in, and in conclusion we
give some preliminary experimental results with
$N_c = 5 \times 10^6$ (50 runs of 100 sec), using two sizes of
polystyrene spheres in aqueous suspension, both
separately and in a mixture.   The nominal diameters
were 91 and 220 nm.   The laser line, at 647 nm, was
set at a power level to give a photon rate per coher-

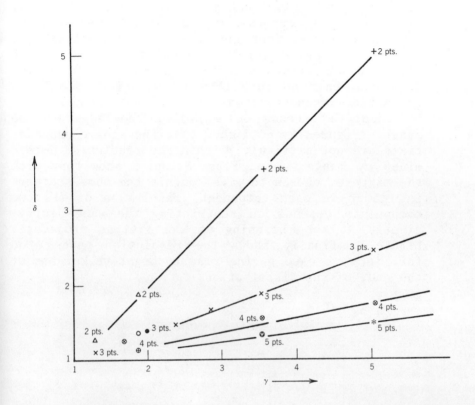

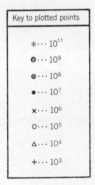

Key to plotted points

$* \cdots 10^{11}$

$\circleddash \cdots 10^9$

$\otimes \cdots 10^8$

$\bullet \cdots 10^7$

$\times \cdots 10^6$

$\circ \cdots 10^5$

$\triangle \cdots 10^4$

$+ \cdots 10^3$

FIGURE 5. Dilation factor δ versus range ratio γ for various experimental durations N_c measured in units of coherence time.

ence time of ~ 100 and per sample time of ~ 0.4. Single clipping was used at a clipping level of zero. Fitting at three, four and five points was attempted, but at the noise level of this experiment, which gives an RMS error on $g^{(1)}(\tau)$ of 3.7×10^{-4}, Fig. 5 indicates that no more than three points should be extracted. Even then, whether the two components can be resolved depends on a *priori* knowledge of the range. Figures 6 and 7 show that the single exponentials are not difficult, within the resolution determined by Table 1, but Figs. 8 and 9 show that with the mixture, chosen to give roughly the same scattering power in each component, resolution of the two components depends on restricting the range sufficiently at the beginning of the fitting procedure. Figure 10 finally shows the resolution pushed too far, in this case giving rise to negative weights at the ends of the distribution.

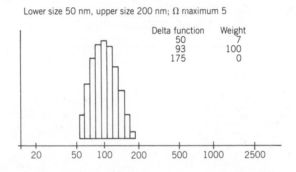

FIGURE 6.   Inversion   of   correlogram   from   nominal 91 nm-diameter polystyrene spheres.

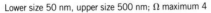

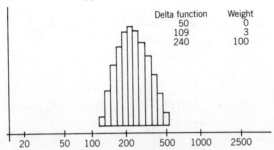

Lower size 50 nm, upper size 500 nm; Ω maximum 4

Delta function	Weight
50	0
109	3
240	100

FIGURE 7. Inversion of correlogram from nominal 220 nm-diameter polystyrene spheres.

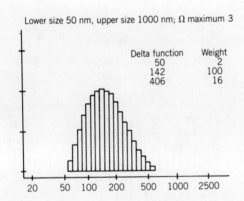

Lower size 50 nm, upper size 1000 nm; Ω maximum 3

Delta function	Weight
50	2
142	100
406	16

FIGURE 8. Inversion of correlogram from mixture of 91 and 220 nm-diameter polystyrene spheres over a range of 50-1000 nm.

125

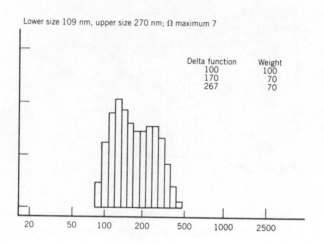

Lower size 109 nm, upper size 270 nm; Ω maximum 7

Delta function	Weight
100	100
170	70
267	70

FIGURE 9.   As in Fig. 8, but with range limited to 109-267 nm.

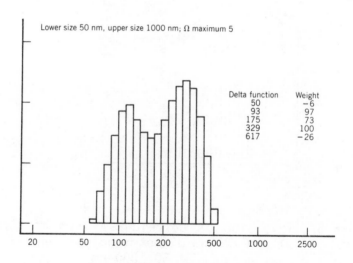

Lower size 50 nm, upper size 1000 nm; Ω maximum 5

Delta function	Weight
50	−6
93	97
175	73
329	100
617	−26

FIGURE 10.   As in Fig. 8, but with attempted higher resolution giving spurious oscillatory additions. Negative values are truncated in the display.

126

ACKNOWLEDGMENTS

We are obliged to Dr. Peter Pusey for supplying the samples used and for his results on these samples with the use of the cumulants method on a linearly spaced correlator, and for valuable discussions. We are also indebted to Dr. Mario Bertero and Dr. Nicole Ostrowski for private communication of results. Finally, we thank Mike Aldritt of Malvern Instruments for assistance with the prototype hardware.

REFERENCES

1. E. R. Pike, in *Scattering Techniques Applied to Supramolecular and Non-Equilibrium Systems*, S. H. Chen, B. Chu, and R. Nossal Eds., Plenum Press, New York, 1981, p. 179.

2. J. G. McWhirter and E. R. Pike, J. Phys. A: Math. Gen. *11*, 1729 (1982).

3. N. Ostrowski, D. Sornette, P. Parker, and E. R. Pike Opt. Acta *28*, 1059 (1981).

4. M. Bertero, P. Boccacci and E. R. Pike, Proc. R. Soc. Lond. A *383*, 15 (1982).

5. M. Bertero, P. Boccacci, and E. R. Pike, (in press, Proc. R. Soc. Lond.).

6. D. Watson and E. R. Pike, (in press).

7. P. N. Pusey, D. E. Koppel, D. W. Schaefer, R. D. Camerini-Otero, and S. H. Koenig, Biochemistry *13*, 952 (1974).

8. D. E. Koppel, J. Chem Phys. *57*, 4814 (1972).

128   E. R. Pike et al.

9.   J. C. Brown, P. N. Pusey and R. Dietz, J. Chem. Phys. *62*, 1136 (1975).

10.   E. Jakeman, E. R. Pike and S. Swain, J. Phys. A *4*, 517 (1971).

11.   P. Pusey, private communication.

12.   S. W. Provencher, Makromol. Chem. *180*, 201 (1974).

13.   C. J Oliver, in reference 1, p. 121.

CHAPTER 5

Polydispersity Analysis of QELS Data by a Smoothed Inverse Laplace Transform

S. BOTT
Stanford University
Stanford, California

CONTENTS

130   S. Bott

I.   INTRODUCTION

When light is scattered from a liquid containing components whose motions can be characterized by more than one decay mode, the autocorrelation (AC) function of the scattered light can be represented by

$$g(t) = \int_0^\infty \exp[-\gamma t] x(\gamma) d\gamma , \qquad (1)$$

where $g(t)$ is the measured AC function and the decay $\exp[-\gamma t]$ is averaged over all decay times $(1/\gamma)$ weighted by an unknown function $x(\gamma)$, giving the amplitude of each decay mode. For a solution containing a macromolecular population of two distinct sizes, for example, the function $x(\gamma)$ would be

$$x(\gamma) = \alpha^2(\gamma_1) f(\gamma_1) m(\gamma_1) \delta(\gamma - \gamma_1)$$
$$+ \alpha^2(\gamma_2) f(\gamma_2) m(\gamma_2) \delta(\gamma - \gamma_2)$$

where $\delta(\gamma)$ are Dirac delta functions, α is the polarizability per particle, $f(\gamma)$ is the form factor for intramolecular interference, and $m(\gamma)$ is the number concentration.

Often information concerning the form of $x(\gamma)$ is of vital interest; therefore several methods have been devised to extract from $g(t)$ information about $x(\gamma)$. The two most widely used techniques in use today are:

1.   Assume a specific form for $x_p(\delta)$ as a function of several parameters p, then use a least-squares fit to find the best values of the parameters. For example, a single exponential decay might be

assumed, and the best amplitude, baseline, and decay constants found; or a Gaussian distribution of unknown mean and variation might be fit for the best values of these two parameters.

2)   Employ the method of cumulants, which gives the moments of the distribution.

The first method is the best method if the assumed form of $x(\gamma)$ is known to be correct but otherwise can give false results. The second method is useful for known monomodal distributions of unknown exact width or form, but not for bimodal, multimodal, or unknown distributions since in practice no more than the first three moments can be found, and these together do not provide enough information about more complex distributions.

A third method is the direct inversion of Eq. (1), which may be recognized as the Laplace transform of $x(\gamma)$ with respect to γ. Until recently, this method, the inversion of the Laplace transform (ILT), was believed to require data of unattainable accuracy, but with current developments, useful information about $x(\gamma)$ can be extracted with a minimum of assumptions about $x(\gamma)$, thus making this method of light scattering data analysis extremely powerful.

Two nearly equivalent parallel developments, one by McWhirter and Pike [1] and Ostrowsky et al. [2], and the second by Phillips [3], Twomey [4], Provencher [5], and Wiff [6], have been mainly responsible for the advancements. The first method uses the truncated expansion of $x(\gamma)$ in the eigenfunctions of the Laplace transform operator to give a filtered version of $x(\gamma)$; the second explicitly restricts the curvature of $x(\gamma)$ to give a smoothed solution. In the following sections we review the ill-conditioned nature of the ILT, briefly compare the ways in which the two methods extract information in view of this ill-conditioning, show some examples of the resolution of the second method with the use of computer generated data, and finally, apply the second method to the analysis of light-scattering data from a system of aggregating and fusing phospholipid vesicles.

II.   NATURE OF THE ILL-CONDITIONING

The Laplace transform is a linear operator useful in solving a variety of problems in mathematics, physics, and engineering. Among its valuable properties are its ability to unravel integral convolutions and its ability to temporarily reduce by one the number of variables in a partial differential equation [7]. In either case the problem transformed into the dummy variable of the Laplace transform may be easier to solve than the original problem. After the solution, the transformed problem must be inverted to recover the solution in the original variable. Despite its common application, the "ill-conditioned" or "ill-posed" nature of the inversion process is often ignored. Mathematically, the ILT is labelled ill-posed because (1) no solution is guaranteed, (2) if a solution exists it may not be unique, and (3) the convergence to a unique solution may not be uniform. These three undesirable mathematical properties are due in part to the extreme smoothing enacted by the Laplace transform. The Laplace transform of a piecewise continuous function, for example, is a differentiable function, and the original function cannot be recovered on inversion [8]. In physical problems in which the smoothness and continuity of the solution are often assured, the inversion may often be obtained analytically and the ill-posedness is not manifest. Problems of this sort occur widely in the field of electrical engineering where the Laplace transform is used to reduce the complexity of an ordinary or partial differential equation describing the behavior of current or voltage in a circuit. However, in cases in which a

numerical inversion is required, either because the
solution is too difficult to find analytically or, as
in the case of many indirect physical measurements
such as QELS, because the Laplace transform is
measured and has no known analytic form, the ill-
posedness of the ILT is a serous obstacle.

The reason for the ill-conditioning of the ILT
can be understood more readily if the problem is
restated in a frequency space. If the original
Laplace transform is

$$g(t) = \int_0^\infty \exp[-\gamma t] x(\gamma) d\gamma , \qquad (1)$$

where $g(t)$ is measured and $x(\gamma)$ is desired, we may,
for example, find a formal solution by expanding $x(\gamma)$
in some set of complete functions spanning the space
of $x(\gamma)$. If a Fourier series is used, then

$$x(\gamma) = \sum_n A_n \cos(w_n \gamma) + B_n \sin(w_n \gamma) , \qquad (2)$$

where the coefficients A_n and B_n are not yet known.
Substituion of this expression into (1) and integration over γ gives the formal solution:

$$g(t) = \sum_n A_n \frac{w_n}{w_n^2 + t^2} + B_n \frac{t}{w_n^2 + t^2} . \qquad (3)$$

It is readily apparent that the coefficients A_n
and B_n approach 0 as $n \to \infty$. Thus the contributions
A_n and B_n of the high frequencies w_n contribute
little to $g(t)$, as may be seen from Eq. (3) but may
contribute much to the solution $x(\gamma)$ as may be seen
from Eq. (2).

Since the Laplace transform of $x(\gamma)$ is a Fredholm integral equation of the first kind with a
symmetrical kernel, it satisfies the Hilbert-Schmidt
theory conditions and thus possesses real eigenvalues

and nondegenerate orthonormalizable eigenfunctions [8]. These, in fact, have been found by McWhirter and Pike [1] and their form is

$$\Psi_\omega^+(\gamma) = \frac{Re[\{\Gamma(1/2 + i\omega)\}^{1/2}\gamma^{-1/2-i\omega}]}{[\pi|\Gamma(1/2 + i\omega)|]^{1/2}} \qquad \omega > 0$$

$$\Psi_\omega^-(\gamma) = \frac{Im[\{\Gamma(1/2 + i\omega)\}^{1/2}\gamma^{-1/2-i\omega}]}{[\pi|\Gamma(1/2 + i\omega)|]^{1/2}} \qquad \omega < 0$$

$$\lambda_\omega = \frac{\omega}{|\omega|} \quad \frac{\pi}{\cosh(\pi\omega)}^{1/2}$$

$$\lambda_\omega \rightarrow \pi^{1/2} \exp[-\pi\omega/2] \quad \text{for } \omega \gg 1 ,$$

where $\Gamma(x)$ is the gamma function and the eigenvalues and eigenfunctions satisfy the eigenvalue equation

$$\int \Psi_\omega(\gamma)\exp[-\gamma t]d\gamma = \lambda_\omega\Psi_\omega(t) . \qquad (4)$$

If $x(\gamma)$ can be expanded in these eigenfunctions, analogous to the Fourier series expansion earlier, it follows that

$$x(\gamma) = \int a_\omega\Psi_\omega(\gamma)d\omega .$$

Substitution of this solution into Eq. (1) again yields a formal solution

$$g(t) = \int a_\omega\lambda_\omega\Psi_\omega(t)d\omega . \qquad (5)$$

Because of the decaying exponential asymptotic behavior of the eigenvalues λ_ω for high ω, it may be seen from Eq. (5) that the amplitudes a_ω of the contribution to $g(t)$ of the frequency ω for high ω is sharply attenuated by λ_ω. If the λ_ω were effectively 0 above some frequency ω_{max}, no frequencies above ω_{max} would contribute to $g(t)$ regardless of what their contributions to $x(\gamma)$ were. Thus the Laplace transform acts as a filter, smoothly attentuating contributions to $g(t)$ of the higher frequency components of $x(\gamma)$.

In anticipation of the numerical studies comprising the remainder of this chapter, we switch to a matrix notation. In this notation, Eq. (1) may be rewritten

$$g = \underset{\sim}{A} \, x \, , \tag{6}$$

where $\underset{\sim}{A}$ is the matrix form of the Laplace transform operator, x is the vector solution, and g is the measured AC vector. In frequency space this would be

$$g = \underset{\sim}{B} \, a \, , \tag{7}$$

where $\underset{\sim}{B}$ and a are the frequency transforms of the Laplace transform operator and the solution vector, respectively. Other equations may be similarly rewritten. These matrix representations can be exact equivalents to their analytical counterparts if the matrices and vectors are allowed to be of infinite dimension. If, as a useful approximation, the upper limits of γ and t are restricted to be large but finite, the resulting eigenvectors of the Laplace transform matrix become discrete and all matrices and vectors finite in dimension. Either representation is adequate for our purposes.

To further demonstrate how the ill-conditioning of the ILT occurs in a practical sense, the following light-scattering experiment is considered. An AC function g is measured at a finite number of delay times. The amount of resolution desired in the solution is incorporated into the problem by choosing the dimension of the frequency transformed solution vector a. The elements of the known matrix $\underset{\sim}{B}$ are computed for the desired resolution, and finally the components of a are allowed to vary to minimize the scalar $(\underset{\sim}{B}a - g)^*(\underset{\sim}{B}a - g)$, representing the sum of the mean-squared deviations from the measured solution. Naturally, **g** represents only a single real-ization of the true AC function and is subject to an error vector **ε**. The matrix $\underset{\sim}{\mathbf{B}}$, because of the filter-ing action of the δ_{ω}, will attenuate the contribu-tions to **g** of the high-frequency components of **a**, and thus an infinite manifold of solutions will satisfy the true **g** to within an error **ε**. A solution with large magnitudes of high-frequency components will give the same **g** as one with much lower magnitudes of the same high-frequency components. Many of the manifold of solutions will thus possess wild, physi-cally unreasonable oscillations. Worse, the greater the resolution demanded (beyond a difficult to determine point), the worse the solution becomes because of the inclusion of more of the highly filtered, high-frequency terms. Although the solu-tion becomes worse, the mean-squared deviation from the measured solution is, of course, reduced. This result emphasizes the danger in obtaining information from light-scattering experiments. If too much information is demanded, false results may easily be obtained and their incorrectness may not be apparent. Similarly, when choosing between two possible candi-dates for the best fit to a given AC function (whether single or double exponentials of specified delay time, cumulants, or smoothed ILT solution vectors), the magnitude of the mean-squared deviation from the measured AC function is most often not a good sole criterion for the choice. This nonintui-tive result is valid because the two candidates may, in many cases *both* fit the data to within the accur-acy of the data although their respective mean-squared deviations from the data differ from each other. For example, it is risky to surmise the existence of an extra mode of decay in an AC function on the basis of the "significantly" smaller mean-squared deviation of a two versus one decaying expo-nential fit to the data since the extra parameters in the two exponential decay fit will always lower the mean-squared deviation. It would be better to ascer-tain whether a fit to one decaying exponential would fit the data to within its error. If this were the case, the accuracy of the data would not be sufficient to verify or deny the existence of an extra decay mode (although it might in fact exist); if not, the assumption of one or more extra decay modes would be warranted. Illustrative examples of divergent solutions obtained by direct inversion of the Laplace transform or by demanding more information than could be delivered in the presence of experimental noise are given by Bellman [7] and Phillips [3].

III. REMEDIES

Two nearly equivalent methods have been devised to extract as much information as possible from experiments using ILTs, in view of the inherent problem of ill-conditioning. The first method involves finding the best truncated solution a or x, thus eliminating the highly filtered, highly oscillatory, high-frequency contributions to the solution. The second method involves "regularizing" the solution by explicitly penalizing overoscillatory solutions. Both methods implicitly assume that the true solution is fairly smooth and as such acknowledge the inherently low resolution obtainable from light-scattering experiments.

In the following discussion, adapted to a large extent from Twomey [4], the second (regularization) method is briefly explained, and the first (truncation) method is viewed from the context of the second method.

To obtain a regularized solution, the original problem $Ax = g$ must be solved subject to one or more additional constraints. To accomplish this formally, we select a quadratic form $Q = x^*Hx$ to quantify the degree of curvature in the solution x. Equation (6) is rewritten to explicitly take account of the error in the measurement

$$\varepsilon = \underset{\sim}{A}x - g ,$$ (8)

where ε is the error vector. Then

$$\varepsilon^2 \geq \varepsilon^* \cdot \varepsilon$$ (9)

is the upper bound to the sum of the squares of the errors and denotes the upper limit to the uncertainty in the experiment; its value is a constraint on the solution x since the larger the value of exp 2, the larger the manifold of solutions x that will yield an error within the bound $\varepsilon^* \cdot \varepsilon \leq \varepsilon^2$. In general, Q will not be minimized by an x satisfying Eq. (9); that x closest to minimizing Q while satisfying Eq. (9) will lie at the upper bound of the error

$$\varepsilon^* \cdot \varepsilon = \varepsilon^2 ;$$ (10)

thus the regularization problem is to find that x minimizing Q subject to the constraint of Eq. (10). This x must satisfy the conditions

$$\frac{\partial}{\partial x} [(\underset{\sim}{A}x - g)^* \cdot (\underset{\sim}{A}x - g) + \alpha x^* \underset{\sim}{H}x] = 0$$ (11)

with α a Lagrangian multiplier that is determined later.

The matrix $\underset{\sim}{H}$ of the quadratic form Q might be chosen, for example, to sum the squares of the second-order differences in x and thus would have the desired high value for oscillatory solutions. The term $\alpha x^* \underset{\sim}{H} x$ may be viewed as a penalty for overoscillatory solutions, with the degree of penalty controlled by α. Instead of second-order differences, Q might be chosen to sum the squares of the nth-order differences in x or the sums of squares of differences of x from an approximate solution of some known shape. Each of these choices would tend to eliminate oscillatory solutions and stabilize or regularize the solution. In general, n such penalty terms may be added, each with different α_i, where the added terms penalize the sums of the squares of the nth-order differences [6]. However, it has been found that for ILTs, penalizing with more than one quadratic form does not improve the solution [9].

In frequency space the problem could be regularized in a similar manner, this time using the quadratic form $R = a^*Ka$ and yielding a condition similar to Eq. (11). For limitation of the oscillations in the solution x, K must be chosen to filter the solution in a. Thus R must penalize solutions that contain appreciable magnitudes in the components of a that correspond to high-frequency eigenvectors of the Laplace transform. In cases where the filtering involves no coupling between eigenvectors, K will have the form of a diagonal matrix with diagonal matrix elements of magnitude $0 \rightarrow 1$ according to the amount of filtering desired at the frequency corresponding to the position along the diagonal. With truncation, a severe form of filtering, K would be all zeros except for ones along the diagonal past the eigenfunction of truncation. In smoother filtering schemes the magnitudes of the diagonal elements would increase slowly toward the lower right.

The solution to Eq. (11) is [4]

$$x = (A^*A + \alpha H)^{-1} A^*g \tag{12}$$

or in frequency space

$$a = (B^*B + \alpha K)^{-1} B^*g \tag{13}$$

The frequency space solution Eq. (13) can be rewritten in the space of x by using the relationship $AS = B$, where S is the known matrix of eigenvectors of the Laplace transform operator:

$$\mathbf{\underset{\sim}{S}} = \begin{bmatrix} \Psi_1(\gamma_1) & \Psi_2(\gamma_1) & \cdot & \cdot & \cdot & \cdot \\ \Psi_1(\gamma_2) & \cdot & & \cdot & & \\ \cdot & & \cdot & & & \\ \cdot & & & & & \\ \cdot & & & & & \end{bmatrix}$$

The solution then has the exact form of Eq. (12) if the identification

$$\mathbf{\underset{\sim}{H}} = \mathbf{\underset{\sim}{S}}^{*-1} \mathbf{\underset{\sim}{KS}}^{-1} \tag{14}$$

is made. This equation shows the way in which the regularization and truncation methods compare. In general, both $\mathbf{\underset{\sim}{H}}$ and $\mathbf{\underset{\sim}{K}}$ will be dominated by diagonal terms; thus the effect of the two methods will be similar.

IV.   SMOOTHED SOLUTION

To find the relationship between the true and filtered solutions, we note that for the true solution

$$\mathbf{\underset{\sim}{g}} = \mathbf{\underset{\sim}{B}}\hat{\mathbf{\underset{\sim}{a}}} \ ,$$

where $\hat{\mathbf{\underset{\sim}{a}}}$ is the true solution; substitution of this expression into Eq. (13) gives

$$(\mathbf{\underset{\sim}{B}}^{*}\mathbf{\underset{\sim}{B}} + \alpha\mathbf{\underset{\sim}{K}}) = \mathbf{\underset{\sim}{B}}^{*}\mathbf{\underset{\sim}{B}}\hat{\mathbf{\underset{\sim}{a}}} \ .$$

Rewriting the left- and right-hand sides separately yields

$$\mathbf{\underset{\sim}{B}}^{*}\mathbf{\underset{\sim}{B}}\hat{\mathbf{\underset{\sim}{a}}} \qquad = \lambda^2\hat{\mathbf{\underset{\sim}{a}}} \tag{15a}$$

$$(\mathbf{\underset{\sim}{B}}^{*}\mathbf{\underset{\sim}{B}} + \alpha\mathbf{\underset{\sim}{K}})\mathbf{\underset{\sim}{a}} = (\lambda^2\mathbf{\underset{\sim}{I}} + \alpha\mathbf{\underset{\sim}{K}})\mathbf{\underset{\sim}{a}} \ , \tag{15b}$$

where λ^2 is the vector of the squared eigenvalues and $\mathbf{\underset{\sim}{I}}$ is the unit matrix. If $\mathbf{\underset{\sim}{K}}$ were the unit matrix, the relationship between the ith component of $\mathbf{a}$ and $\hat{\mathbf{a}}$ would be

$$a_i = \frac{\lambda_i^2}{\lambda_i^2 + \alpha}\, \hat{a}_i \, . \tag{16}$$

From this expression it may be deduced that the amount of filtering depends on the relative magnitude of λ_i^2 versus α. The frequencies for which the eigenvalues λ_i^2 are much smaller than the α are heavily filtered. If $\underset{\sim}{\mathbf{K}}$ were diagonal but not the unit matrix, the relationship would be

$$a_i = \frac{\lambda_i^2}{\lambda_i^2 + \alpha k_i}\, \hat{a}_i \, . \tag{17}$$

and the filtering would depend on the magnitudes of the λ_i^2 versus αk_i, where k_i, is the ith element of the diagonal of $\underset{\sim}{\mathbf{K}}$. For nondiagonal $\underset{\sim}{\mathbf{K}}$, the $\alpha \underset{\sim}{\mathbf{K}}$ term in Eq. (15) will generally mix in some of the neighboring eigenvectors, and thus a given component in the smoothed solution will reflect contributions from the amplitudes of components corresponding to neighboring eigenvectors in the true solution.

The exact amount of smoothing is determined by the choices of $\underset{\sim}{\mathbf{H}}$, $\underset{\sim}{\mathbf{K}}$, and α. With the truncation method, $\underset{\sim}{\mathbf{K}}$ acts like a low-pass filter with abrupt cutoff; α may be set equal to 1; and the choice of n, the first eigenvector to be truncated, is equivalent to the choice of α for the strict regularization problem. The matrix $\underset{\sim}{\mathbf{H}}$ in the regularization method, for most choices of smoothing, is a banded diagonal matrix with center diagonal dominating and thus approximates the smoother cutoff of Eq. (16) or (17). The choice of α in this case is roughly equivalent to the frequency of one-half attenuation in the filter, as may be seen from Eq. (16). Equation (16) indicates that if the components in the true solution $\hat{\mathbf{a}}$

fall off faster than does $\lambda_i^2/(\lambda_i^2+\alpha)$, the filtered solution will closely approximate the true one.

The choices of $\underset{\sim}{H}$ and $\underset{\sim}{K}$ that have been employed primarily to date, that is, second- or third-order difference regularization, and truncation, are not necessarily the best choices for $\underset{\sim}{H}$ and $\underset{\sim}{K}$; and although practical results do not depend too heavily on the details of $\underset{\sim}{H}$ and $\underset{\sim}{K}$, further work in optimization of $\underset{\sim}{H}$ or $\underset{\sim}{K}$ may improve on results obtained to date. Any a *priori* information concerning, for example, the form of x or the statistics of ε can generally be incorporated into $\underset{\sim}{H}$ or $\underset{\sim}{K}$ to give the best solution utilizing all known information [10].

V. CHOICE OF α

Since we have now obtained expressions for the smoothing as accomplished by the regularization and truncation (filtering) methods and shown how they compare, the remainder of the discussion concerns the regularization method; from the previous expressions, extension to the equivalent topics in the other method is straightforward.

Once the choice of $\underset{\sim}{H}$ is made, the important remaining step is the choice of the value of α. Indeed, the choice of α (or, equivalently, n in the truncation method) is probably more important than the exact choice of $\underset{\sim}{H}$ since the magnitude of α will substantially alter the form of the solution. An objective algorithm for choosing α is thus critical.

One manifestation of the ill-conditioning in ILTs is that the matrix $\underset{\sim}{A}$ is nearly singular (otherwise, the problem could be solved by a simple matrix inversion) and possesses eigenvalues spaced over many orders of magnitude. Equation (16) indicates that the value of α will not substantially change the relative magnitude of the smoothed versus true component of $\underset{\sim}{A}$ unless α is near $\hat{a}_i\lambda_i^2$. The magnitude of α will thus range between the largest and smallest λ_i^2, but a change in its exact value will affect the solution x only when it is in the neighborhood of one of the widely spaced λ_i^2. Between the eigenvalues λ_i and λ_{i+1}, the value of α will not change the solution. A linear increase in α beginning at the smallest λ_i would affect the degree of smoothing of x in a sequence of steps as each successive eigenfunction is filtered out with increasing α. For a problem whose true solution is smooth, components of the true $\hat{a}$ will decrease rapidly as i increases, and the smoothing parameter α may acquire a high value without oversmoothing the solution; conversely, for a solution that is rather oscillatory, the value of α can reach only a relatively smaller magnitude before over smoothing and distorting the solution. The preceding discussion is, of course, an oversimplification since the $\underset{\sim}{H}$ will generally involve coupling between eigenfunctions.

Several statistical methods have been devised for proper selection of the smoothing parameter α in various circumstances. One simple method based on the Fisher statistic has been used by Provencher [5] for application when little is known about the form of the solution or the statistics of the errors.

The Fisher test, or F test, is a statistic used to determine whether the variances of two populations are equal. It gives a probability, based on the ratio of sample variances taken from each population and the degrees of freedom of each sample, that the true variances of the two populations are the same. In the ILT problem the variance (mean-squared deviation of solution from measured data points) of the unsmoothed ($\alpha = 0$) problem is compared to the variance of the solution to the smoothed problem with the sample degrees of freedom calculated in each case by Gram analysis [11] or by the method proposed by Mallows [12] . Because of the smoothing penalty term, the variance of solutions obtained with increased amounts of smoothing will be higher than the variance with no smoothing. The F test indicates when the value of α has reached such an extent that the variance in the unsmoothed case (which is due to experimental noise) is so exceeded by the variance in the smoothed case (which is due to experimental noise plus the smoothing penalty term) that the variance due to smoothing alone is becoming appreciable and oversmoothing may begin to occur. In practice, α is first assigned a very small value that is then increased until the Fisher statistic indicates that oversmoothing is beginning to occur. Some difficulties with the method and further detail are discussed elsewhere [5,12].

VI.   COMPUTER SIMULATIONS

To give the reader an idea about the resolution achievable with the use of smoothed ILTs on data from light scattered from systems such as those studied in our laboratory, we have used a computer to generate AC functions that simulate AC functions of light scattered from polydisperse distributions of hollow spheres 100-800 Å in radius suspended in a liquid and undergoing Brownian motion. The data were generated to three- or four-figure accuracy and thus represent close approximations to an upper limit in the resolution obtainable. No noise was added to the simulated AC functions, but the finite accuracy to which they were computed implies a noise contribution. Since the variance of the noise as a function of delay time in an AC function is known in practice and since this knowledge is used to weight the squared residuals when the ILT (or any other) curve fitting is done, the resolution obtained by using appropriately rounded data with the correct (even) weighting of the points should be very nearly identical to that obtained in the presence of experimental noise of known statistics.

The computer-generated AC functions were then fit (without knowledge of the known solution) by an ILT program "CONTIN" kindly provided by Dr. Stephen Provencher. CONTIN uses a sum of squares of second-order difference for $\underset{\sim}{H}$, the Fisher statistic to select α, and, in addition, uses a quadratic programming algorithm for the curve fitting, allowing the addition of linear constraints on the solution **x**. The solution **x** represents the amplitude of light scattered from various decay modes, and the constraining of this solution to be nonnegative corresponds to a further incorporation into the problem of a *priori* knowledge. The use of this algorithm, unfortunately, increases the computer time and power necessary for solution of the problem but seems to improve the solutions obtained. Similar constraints could, of course, be applied to a truncation-type solution.

All the distributions attempted were single or sums of Schulz number-average distributions [14] of various widths and separations. The form of the Schulz distribution is

$$f_z(r) = \frac{1}{z!} \left[\frac{z+1}{\bar{r}} \right]^{z+1} r^z \exp \left[\frac{-(z+1)r}{\bar{r}} \right]$$

where $\bar{r}$ is the number-average size and the width of the distribution $\to 0$ as $z \to \infty$ and $\to \infty$ as $z \to 0$. The parameter z, not to be confused with "z-averaged" quantities in light scattering, characterizes both width and skew of the Schulz distribution.

Figure 1 shows the true and derived solution for monomodal Schulz distributions of various widths, each with number-average radius 100 Å. The hollow spheres are weighted according to the volume of their interior: $[r^3 (r - 40)^3]^2$, where r is the radius. For simplicity, no form factors were used in the simulated data since their inclusion would not affect the determination of resolution. The figures show the z-averaged amplitudes of scattering from different-sized particles.

Since a monomodal distribution is smooth, the fits in Fig. 1 are uniformly close to the known forms. Figures 2-4 show the fits as a function of separation of the two peaks of bimodal distributions

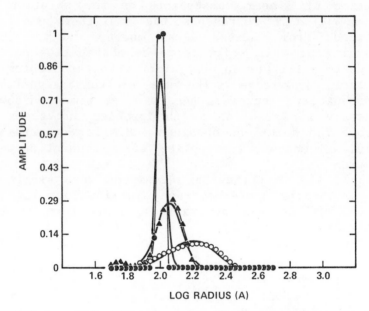

FIGURE 1. ILT fits (●, ▲, o) to computer-generated AC functions simulating light scattered from three monomodal Schulz number-average distributions of hollow spheres. The distributions each have a number-average radius of 100 Å and Schulz z parameters of 200, 25, and 5, respectively, for the narrow, intermediate, and broad distributions (actual distribution are shown by the solid lines). The logarithm of the radius is on the abscissa; the scattered light intensity is on the ordinate. Computer-generated data were rounded to 0.01% of the amplitude of the AC function (four-figure accuracy). The displacement of the peak positions from 100 Å of the broader distribution is due to z averaging.

146

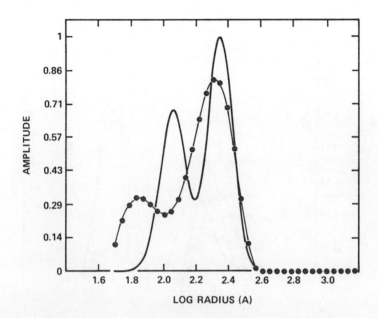

FIGURES 2-4. ILT fits to computer-generated data representing bimodal (sum of two Schulz distributions each with $z = 25$) distributions. Figures 2-4 have number-average radii for the first and second populations of 100 and 200 Å, 100 and 400 Å, and 100 and 800 Å, respectively. All the generated data were rounded to 0.01% of the amplitude of the AC function (four-figure accuracy), except the (▲) in Fig. 3, which shows the ILT fit to data of three-figure accuracy. Solid lines show the actual distributions.

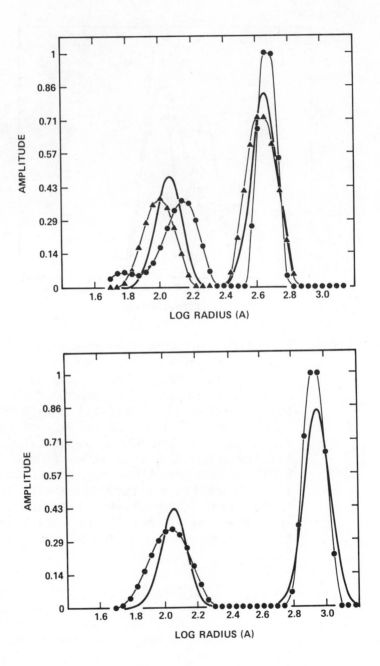

with fixed widths of the separate peaks. When the separation is small (differing in number-average radius by a factor of 2), the presence of systematic error is clear. This error originates in the high sum of squares of second-order differences present in the actual solution, a sum that is penalized in the smoothed solution, leading to the obvious distortion. Viewed from frequency space, the distortion arises from the unavailability of high-frequency eigenfunctions in the AC function, even when these eigenfunctions are present in the distribution. In short, the resolution is low. As the separation of the peaks increases, the solution improves greatly. McWhirter and Pike [1] have noted that if a *priori* knowledge restricting the extent of the solution is possessed, the resolution can be increased. In other words, if the solution were known to lie along a smaller portion of the abscissa and if the fit were constrained to lie in this region, the resolution would be increased. The resolution of the peaks in Fig. 2 could be substantially improved in this way.

The second/first peak number of particles ratios are 0.125, 0.0156, and 0.002 in Figs. 2-4, respectively. These ratios were chosen such that the z-averaged amplitudes of the two peaks would be roughly the same size. These ratios indicate, as most researchers are well aware, that light scattering is sensitive to smaller numbers of larger particles in the presence of larger numbers of smaller ones, but not vice versa.

VII.   APPLICATION TO PHOSPHOLIPID VESICLES

Phospholipids are a group of similar molecules consisting of a zwitterionic head group to which two 14-20-carbon hydrocarbon tails are attached. They comprise a large percentage of the bulk of biological cell membranes. When placed in water and subjected to ultrasonication, the molecules organize into small, hollow, spherical bilayers no smaller than

120 Å in outer radius but ranging up to many micro-meters in size. These "vesicles" can be visualized as thick-walled (40-Å walls) bubbles surrounded by and containing solvent, with the hydrophilic head-groups at the inner and outer solvent interfaces, and with a bilayer interior composed of the hydrocarbon chains. Above a certain temperature T_ϕ (which is dependent mostly on the chain length) the interiors of the bilayers are relatively disordered, but below T_ϕ a phase transition to a lyotropic liquid crystal phase is observed. These vesicles are widely studied as cell membrane analogs.

Surprisingly, the vesicles are stable to fusion (formation of one large spherical vesicle from two or more smaller ones) and aggregation (formation of a connected cluster of two or more small vesicles from individual vesicles, without mixing of interiors) over a wide range of concentrations above T_ϕ, but below T_ϕ the vesicles undergo spontaneous changes. Whether these changes are fusion, aggregation, or both, and the mechanism for these changes are still controversial issues, and resolution of the problem is of great biological interest. We are studying the fusion-aggregation behavior of DMPC (dimyristoyl-phosphatidylcholine) vesicles by light scattering and briefly present here some results demonstrating the utility of the smoothed ILT for analyzing the data.

VIII.   EXPERIMENTAL

During all portions of the preparation of the DMPC vesicles, the temperature of the vesicle solu-tion was maintained above 35°C (the phase-transition temperature for DMPC vesicles is ~ 21°C). Dimyris-toylphosphatidylcholine vesicles were produced by sonication for ~ 1 hr of a 1% w/v solution of DMPC in Hepes buffer (pH 7.6) in a 40-W Ultrasonics bath sonicator. The sonicate was diluted to 0.02% w/v in the same buffer, spun at 10,000g for 1 hr to remove dust and large vesicles, and then filtered into clean

1 cm^2 glass cuvettes through Millipore 0.2 μm "Millex" filters prewashed with 0.5 liters of DD H_2O to remove extractables. The light scattered at 90° (~ 1 coherence area) was collected, and the autocorrelation function was calculated by a Malvern 48-channel single clipped correlator with a clipping level of 1. The light-scattering setup has been described elsewhere [15]. Data sets at three delay time intervals differing by a factor of 2 were pieced together by finding the best amplitude and baseline of the overlapping data points. (This combination of data sets is necessary to obtain accurate information over a sufficient time scale. The mean-squared errors of the overcapping points using the fit amplitude and baseline was typically < 0.1%. The computer-generated data were similarly produced by merging three individually computed AC functions at different time scales.) Radii of vesicle samples were determined assuming the Stokes-Einstein relationship.

IX. RESULTS

To follow the aggregation-fusion reactions of the vesicles, the vesicle size distribution was measured first at a temperature above T_ϕ (T_ϕ ~ 21°C for DMPC). The temperature was then lowered below T_ϕ (19.2°C) for 15 min and the distribution remeasured; the temperature was raised above T_ϕ for 20 min and measured to check for reversible aggregation (fusion would presumably not be reversible), and the temperature was once again lowered and a final measurement was taken.

Figure 5 shows the original population and the population after aggregation-fusion has begun to occur below T_ϕ. The population above T_ϕ shows a surprising bimodality that has been recently verified elsewhere [16]. The existence of this stable bimodal distribution is of great interest since it indicates that two discrete stable states exist when vesicles are formed by the random sonication process; one

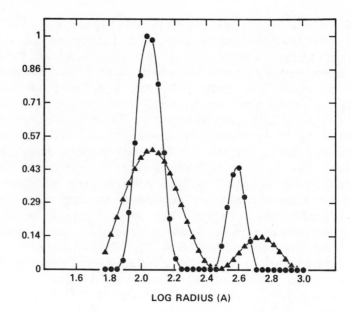

LOG RADIUS (A)

FIGURE 5.   Populations of sonicated DMPC vesicles above (●) and below (▲) T_ϕ from ILT analysis of light-scattering data. The distribution above T_ϕ shows a surprising bimodality indicating the emergence of two stable populations from sonication. The cooled sample shows broadened peaks indicating the onset of fusion-aggregation products shortly (~ 10 min) after lowering the temperature below T_ϕ.

might instead expect one small stable size range at the limit (smallest) size because of the sonication process or a very broad monomodal distribution of stable vesicles of different sizes.

When the temperature is lowered below T_ϕ, the widened distributions indicate that aggregation-fusion is occurring in possibly one, or probably both, of the populations. Since the simplest fusion or aggregation products are not greater than a factor of 2 different in diffusion constants from the unreacted vesicles, we expect from the computer simulation results to be unable to resolve the new fusion or aggregation populations forming, but instead to see a broadening of the peaks. The broadening of each peak also toward smaller sizes is probably an artifact since it is not expected that vesicles in either population would shrink; instead, the broadening toward the larger sizes allows additional smoothing and causes an apparent broadening toward smaller sizes as well. Similarly, a broadening in either peak would cause an increase in the amount of allowed smoothing, which would then tend to make both peaks appear broader; however, the broadening of each peak is so dramatic in this case that aggregation-fusion is probably occurring in both populations. The position of the peak at larger size has shifted toward a still larger size.

Figure 6 shows the return to above T_ϕ and finally the second venture below. Above T_ϕ both peaks narrow to their original width, indicating that reversible aggregation has certainly occurred. The peak at larger size has shifted slightly further toward larger size, and this trend continues in the presence of renewed aggregation when the sample is again lowered below T_ϕ. This shift of the peak at larger size indicates that irreversible aggregation or possibly fusion is occurring to some extent in this peak. Further studies in which the populations are separated and studied independently are in progress.

X.   CONCLUSION

A comparison of two powerful new ILT methods for analysis of light-scattering data has been presented along with reasons for their inherent limitations in resolution and some caveats concerning traditional methods of analyzing light-scattering data. The two methods are similar and provide a way of maximizing the information available from the data without

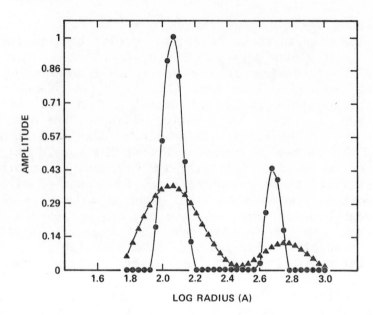

FIGURE 6.   Population of vesicles after raising the
temperature of the same sample of vesicles
(Fig. 5) above T_ϕ (●) and then relowering
below T_ϕ (▲).   The narrowing of the width
of the populations above T_ϕ [as compared
to Fig. 5 (▲)] shows that the broadening
noted in Fig. 5 is primarily due to
reversible aggregation. The slight shift
of the peak of the larger-sized population
in this figure (both above and below T_ϕ)
as compared to Fig. 5 indicates a small
amount of irreversible aggregation-fusion
in this population.

requiring a *priori* knowledge about the form of the
particle size (or decay mode) distribution function.
    Analysis of computer-simulated data and real
light-scattering data has shown that a smoothed ILT
method can yield useful information about particle
size distributions. By combining the usual advantages of light scattering, namely, speed, small

sample size, nonperturbing experiments, and relatively low concentrations with use of these new ILT methods, light scattering may in many cases provide a viable alternative to other particle sizing measurements.

ACKNOWLEDGMENT

The computer program CONTIN was kindly provided by Dr. S. Provencher of the European Molecular Biology Laboratory in Heidelberg. Discussions with Dr. N. Ostrowsky concerning data analysis are gratefully acknowledged. The work was supported by National Institutes of Health grant No. 5RO1 GM 22517 and National Science Foundation grant No. CHE 79-01070.

REFERENCES

1. J. G. McWhirter and E. R. Pike, J. Phys. A: Math. Nucl. Gen. *11*(9), 1729 (1978).

2. N. Ostrowsky, D. Sornette, P. Parker and E. R. Pike, Acta *28*(8), 1059 (1981).

3. D. L. Phillips, J. Assoc. Comput. Mach. 9(1), 84 (1962).

4. S. Twomey, J. Assoc. Comput. Mach. *10*(1), 97 (1963); J. Franklin Inst. *279*, 95 (1965).

5. S. Provencher, Makromol. Chem. *180*, 201 (1979); S. Provencher, J. Hendrix and L. DeMaeyer, J. Chem. Phys. *69*, 4273 (1978).

6. D. R. Wiff, J. Polym. Sci. Polym. Symp. *43*, 219 (1973).

7. R. Bellman, R. E. Kalaba, and J. A. Lockett, *Numerical Inversion of the Laplace Transform*, American Elsevier, New York, 1966; G. F. Miller, in *Numerical Solutions of Integral Equation*, L. M. Delves and J. Walsh, Eds., Clarendon, Oxford, 1974.

8. R. Courant and D. Hilbert, *Methods of Mathematical Physics*, Vol. 1, Interscience, New York, 1953.

9. W. M. Gentleman, J. Inst. Math. Appl. *12*, 329 (1973).

10. O. N. Strand and E. R. Westwater, J. Assoc. Comput. Mach. *15*(1), 100 (1968); S. C. Sahasrabudhe and A. D. Kulkarni, J. Assoc. Comput. Mach. *24*, 624 (1977); G. H. Golub, M. Heath and G. Wahba, Stanford Computer Science Rept., doc Stan-CS-77-622 (1977).

11. D. G. McCaughey and H. C. Andrews, IEEE Transact. Acoust. Speech Signal Process. *25*(1), 63 (1977).

12. C. L. Mallows, Technometrics *15*, 661 (1963).

13. S. Provencher, European Molecular Bio. Lab. Tech Report, EMBL-DA02, 1980.

14. S. R. Aragon and R. Pecora, J. Chem. Phys. *64*, 2395 (1976).

15. A. Flamberg and R. Pecora, Dynamic Light Scattering Studies of Dodecyldimethylammonium Chloride Micelles, in *Scattering Techniques Applied to Supramolecular and Non-Equilibrium Systems*, S. H. Chen, B. Chu, and R. Nossal, Eds., Plenum Press, New York, 1981, pp. 803-808.

16. C. F. Schmidt, D. Lictenberg, and T. E. Thompson, Biochemistry *20*, 4792 (1981); S. E. Schullery, C. F. Schmidt, P. Felger, T. W. Tillack and T. E Thompson, Biochemistry *19*, 3919 (1980); D. Sornette, Doctoral Thesis, University of Nice, France, 1981; D. Sornette and N. Ostrowsky, Kinetics of Growth of Phospholipid Vesicles, in *Scattering Techniques Applied to Supramolecular and Non-Equilibrium Systems*, S. H. Chen, B. Chu, and R. Nossal, Eds., Plenum Press, New York, 1981, pp. 351-362.

CHAPTER 6

Determination of Particle Size and Speed Distributions by Use of Photon Correlation Spectroscopy and First-Order Splines

JEFFREY H. GOLL
Texas Instruments, Inc.
Dallas, Texas

GREGORY B. STOCK
Transaction Technology, Inc.
Santa Monica, California

CONTENTS

This work was performed when the authors were with the Thomas C. Jenkins Department of Biophysics at the Johns Hopkins University, Baltimore, Maryland.

159

I.   INTRODUCTION

For more than 10 years photon correlation spectroscopy (PCS) has been used to determine the sizes
and velocities of particles in solution [1-18]. The
technique is simple and powerful when sample preparations are uniform. However, most particle
dispersions encountered in the laboratory exhibit
nonuniformities of size (polydispersity) or shape
that have led to significant complications. Motile
particles (for example, bacteria and spermatozoa) can
deviate substantially from uniform translational
motion. Complications include rotational motion and
independent motion by different parts of the bodies.
Considerable effort has been devoted to expanding the
capabilities of PCS for characterizing such samples
and defining its limitations. This chapter reviews
the work conducted at Johns Hopkins University with
the use of a mathematical technique called the *method
of splines* [19-20] for extraction of estimates of
particle size and velocity distributions. A brief
introduction to splines is given in Section II.
Sections III and IV contain descriptions of the work

on size and speed distributions, and Section V provides a concluding discussion.

II. SPLINES

The term splines is derived from a draftsman's curve-fitting tool of the same name. The draftsman's spline is a thin, flexible piece of wood that is bent so that it matches the shape of a curve to be drawn. The spline is anchored at selected points by attaching lead weights called "ducks". By varying the attachment points of a sufficient number of ducks, the spline can be bent so that it passes through an arbitrary number of points on a curve.

The mathematical spline is an abstraction and generalization of the draftsman's spline. It is defined by the following:

> Given a strictly increasing sequence of real numbers $(R_0 < R_1 < \cdots < R_m)$, the class of functions that are polynomials of degree $\leq k$ on the intervals $(-\infty, R_0]$, $(R_0, R_1]$, $(R_1, R_2]$, $\cdots$, $(R_m, \infty]$ and have derivatives of orders 0 to $k - 1$ that are continuous everywhere are called *splines of degree* k with knots at $(R_0, R_1, \cdots, R_m)$.

The thin elastic draftsman's spline is an excellent physical approximation of the abstract cubic spline [20].

In the work that follows, particle size distributions $N(R)$ and velocity distributions $P(v)$ are approximated by splines. Any spline $N(R)$ of degree k with knots $(R_0, \cdots, R_m)$ has an expansion of the form

$$N(R) = \sum_{i=0}^{k} a_i R^i + \sum_{i=0}^{k} b_i B_k(R; R_i) \, , \qquad (1a)$$

where

$$B_k(R;R_i) = \begin{cases} 0 & \text{for } R \leq R_i \\ (R - R_i)^k & \text{for } R > R_i \end{cases}. \tag{1b}$$

A set of $m - k$ linearly independent basis splines can be generated [21] by using sequentially, from $p = 1$ to $k + 1$, the following recursive relationship for $i = 0$ to $m - p$:

$$B_k(R;R_i, \cdots, R_{i+p}) \tag{2}$$

$$= \frac{[B_k(R;R_i, \cdots, R_{i+p-1}) - B_k(R;R_{i+1}, \cdots, R_{i+p})]}{(R_{i+p} - R_i)}$$

The resultant $m - k$ functions, denoted by $B_k(R;R_i, \cdots, R_{i+k+1})$ for $i = 0$ to $m - k - 1$, form a basis of local support: each element is identically zero outside the interval $[R_i, R_{i+k+1}]$. This property simplifies computations significantly. Only linear, or first-order, splines were utilized in the following work. The basis functions for linear splines are illustrated in Fig. 1. The basis function $B_1(R;R_i, R_{i+1}, R_{i+2})$ is the triangular function that is nonzero between R_i and R_{i+2}. It is easy to see that a linear combination of the basis functions will give a piecewise linear function that is completely determined by the set of values

$$\{N(R_i) \mid i = 1, 2, \cdots, m - 1\}.$$

III.   PARTICLE SIZE DISTRIBUTIONS BY PCS

A.   Theory

In a PCS experiment a photon-counting photomultiplier tube detects the light scattered by a sample, which consists of a large number of particles

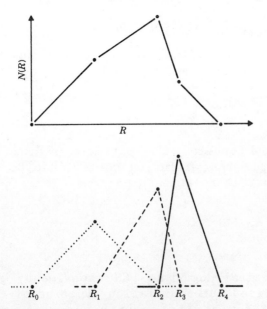

FIGURE 1.   Bottom curve: the three basis functions
for linear splines with knots at $(R_0, R_1,$
$\cdots, R_4)$ that are zero for $R < R_0$ and $R >$
R_4.   Top trace: typical function $N(R)$
expressed as a linear combination of the
basis functions.

in solution. As the particles move about, the relative phases of the various contributions to the light
scattered to the phototube change. Therefore the
diffraction pattern associated with the sample varies
as a function of time and the intensity detected by
the photomultiplier changes with time at a rate
determined by the speed of the motion. The electric
field of the scattered light is similarly related to
sample motion.   For sample particles undergoing
Brownian motion, the autocorrelation function of the
electric field at scattering angle θ is given by
[1,2]

$$|g^{(1)}_{theory}(\tau,\theta)| = \exp[-Dq^2\tau] \ , \qquad (3a)$$

where D is the diffusion coefficient and the scattering wavevector amplitude q is given by

$$q = \frac{4\pi}{\lambda} \sin \frac{\theta}{2} \ . \qquad (3b)$$

For a hard sphere, the particle radius R can be derived from the diffusion coefficient by using the Stokes-Einstein relationship:

$$D = \frac{k_B T}{6\pi\eta R}. \qquad (3c)$$

In the digital homodyne PCS experiment, $g^{(2)}(\tau,\theta)$, the normalized autocorrelation function of the intensity scattered at the angle θ, is measured. If it is assumed that the scattered **E** field amplitude is Gaussian-random distributed, the Siegert relationship [22] may be used to obtain an experimental estimate of the normalized field autocorrelation function $|g^{(1)}(\tau,\theta)|$ from $g^{(2)}(\tau,\theta)$. For a monodisperse sample, this estimate of $|g^{(1)}(\tau,\theta)|$ can be used together with Eqs. (3) to give the Stokes radius of the sample particle.

When the sample exhibits polydispersity, the simple relationships [Eqs. (3)] must be replaced by an integral over a weighting function $W(R,\theta)$ that describes the contribution of the particles of radius R to $|g^{(1)}(\tau,\theta)|$:

$$|g^{(1)}_{theory}(\tau,\theta)|$$

$$= \frac{\displaystyle\int_0^\infty W(R,\theta) \ \exp[-D(R)q^2(\theta)\tau]dR}{\displaystyle\int_0^\infty W(R,\theta)dR} \qquad (4)$$

The denominator is a normalizing factor that ensures that the value of $|g^{(1)}_{theory}(\tau,\theta)|$ is 1 at $\tau = 0$. This integral has the form of the Laplace transform; the weighting function can be determined by inversion of the transform.

The weighting function $W(R,\theta)$ is given by

$$W(R,\theta) = N(R)I(R,\theta) , \tag{5}$$

where $N(R)$ is the number of particles of radius R and $I(R,\theta)$ is the intensity scattered in direction θ by a particle of radius R. Since the particles in solution contribute, in general, different relative weights as the scattering angle varies, the quantity directly available by inversion of the Laplace transform, $W(R,\theta)$, depends on θ. The particle size distribution function $N(R)$ can be obtained from the PCS experiment only if $I(R,\theta)$ is known. The phospholipid vesicle solutions that were studied were modeled as polydisperse populations of spherical shells of 40 Å wall thickness [23-25]. The Rayleigh-Gans form factor for spherical shells was used:

$$P(R,\theta) = (\frac{3}{qR})^6 \{[\sin(qR) - qR\cos(qR)]$$

$$- [\sin(qR_i) - qR_i\cos(qR_i)]\}^2 , \tag{6a}$$

and $I(R,\theta)$ is given by

$$I(R,\theta) = C(\frac{4}{3}\pi R^3)^2 P(R,\theta) . \tag{6b}$$

Here R and R_i are the radii of the outer and inner walls of the vesicle, respectively, and C is a constant involving the polarizability of the particle (assumed uniform in the lipid region of a vesicle).

The distribution function $N(R)$ can be written as a linear combination of basis splines:

$$N(R) = \sum_{i=0}^{m-2} n_i \, B_1(R;R_i,R_{i+1},R_{+2}) \; . \tag{7}$$

Combination of Eqs. (4)-(7) yields

$$|g_{theory}^{(1)}(\tau,\theta)| = \tag{8}$$

$$\frac{\displaystyle\sum_{i=0}^{m-2} n_i \int_{R_i}^{R_{i+2}} R^6 P(R,\theta) B_1(R;R_i,R_{i+1},R_{i+2}) e^{-D(R)q^2(\theta)\tau} dR}{\displaystyle\sum_{i=0}^{m-2} n_i \int_{R_i}^{R_{i+2}} R^6 P(R,\theta) B_1(R;R_i,R_{i+1},R_{i+2}) dR} \; .$$

The integrals in Eq. (8) involve slowly varying functions of R and are readily evaluated by numerical integration, leading to

$$|g_{theory}^{(1)}(\tau,\theta)| = \frac{\displaystyle\sum_{i=0}^{m-2} n_i I_{num}(\tau,\theta)}{\displaystyle\sum_{i=0}^{m-2} n_i I_{den}(\theta)} \; . \tag{9}$$

Only $m - 2$ of the $m - 1$ variables n_i are independent since $|g_{theory}^{(1)}(\tau,\theta)|$ depends only on relative particle numbers. The dependent variables may be eliminated by setting the denominator of Eq. (9) equal to a constant and then solving for one of the coefficients (for example, n_{m-2}). (This corresponds to supposing that the unknown total intensity scattered in the direction θ takes a given value.) The expression for $|g_{theory}^{(1)}(\tau,\theta)|$ is then linear in the parameters $\{n_i | i = 0,1, \cdots, k - 3\}$, and a linear least-squares technique [26] may be used to extract from PCS data the best-fit $N(R)$. With the use of simple linear least squares it is not possible to restrict $N(R)$ to physically meaningful (non-negative) values, as was demonstrated by a trial case that gave a best-fit $N(R)$ with regions having negative numbers of particles. To avoid this difficulty, fitting was based on a nonlinear least squares algorithm [26,27].

In the nonlinear technique $N(R)$ is restricted to non-negative values by replacing Eq. (7) by

$$N(R) = \sum_{i=0}^{m-2} s_i^2 B_1(R; R_i, R_{i+1}, R_{i+2}) \tag{10}$$

so that

$$|g_{\text{theory}}^{(1)}(\tau,\theta)| = \frac{\displaystyle\sum_{i=0}^{m-2} s_i^2 I_{\text{num}}(\tau,\theta)}{\displaystyle\sum_{i=0}^{m-2} s_i^2 I_{\text{den}}(\theta)} \tag{11}$$

and using the s_i as the variables in the fitting procedure. (Thus $s_i^2 = n_i$.) Any first-order spline that is everywhere non-negative can be represented in the form of Eq. (10). The dependent parameter in the set $\{s_i\}$ is removed by simply fixing the value of one parameter, s_f. For simplification of the computations, the instrumental constants (amplitudes) associated with data gathered at the various angles are evaluated by cumulants analysis.

The nonlinear technique also makes possible the *simultaneous* fitting of data taken at several different scattering angles with a single parameterized $N(R)$. (Simultaneous fitting using simple linear least squares could be achieved only if the total intensity scattered at each angle were known.) Simultaneous fitting provides a more stringent test of the validity of the model used to describe the scattering population. It would be preferable, because of experimental difficulties such as dust particles in solution, to gather the data simultaneously as well. This was not possible in our setup.

B. Data Collection

Prescaled intensity autocorrelation functions were obtained by use of a system described elsewhere [16]. Sample preparation [28] was designed to produce the most uniform fraction of small unilamellar phospholipid vesicles possible. For each θ, autocorrelation data were gathered during many equal and consecutive short ($\sim$ 1 sec) periods of time. The final $g^{(2)}(\tau,\theta)$ was an average of the results of selected short experiments. Short experiments were included in the average if their calculated theoretical backgrounds fell within a preselected number ($\sim$ 1) of standard deviations of the mean background. Thus data accumulated when the scattering was unduly influenced by "dust" or other anomalous conditions could be rejected. The effects of possible fluctuations of the laser intensity were minimized by the use of the short-term estimates of the theoretical background. The selected short experiments were used in calculating the mean values and standard deviations made of $g^{(2)}(\tau,\theta)$. The standard deviations made it possible to apply a χ^2 test to the adequacy of the model selected for $N(R)$.

C. Error Analysis

The variance associated with the best-fit $N(R)$ that results from these procedures was estimated as follows. The best-fit values of the s_i were denoted s_{i0} and the minimum value of χ^2 was χ_0^2. The free-fitting parameters $s_i \neq s_f$ were selected one at a time for error evaluation. With the value of s_i fixed at a new value:

$$s_i = s_{i_0} (1 + \delta_i) .$$

$$(12)$$

The other free parameters s_j $(j \neq f, j \neq i)$ were freely varied by the nonlinear least-squares program until the best fit with the restriction of Eq. (12) and the corresponding value of $\chi^2(s_j)$ were found. For each $i \neq f$ this procedure was repeated for a sufficient number of values of δ_i for definition of the set

$$\{s_i | (\chi^2(s_i) - \chi_0^2) < 1\}$$

This set defines the standard deviation associated with s_j [26].

The particle size distributions are presented as normalized mass density functions $M(R)$, easily derived from $N(R)$. The integral of $M(R)$ over all R is 1. This normalization condition couples the parameter fixed in $N(R)$ to the other parameters. Consequently, there is a standard deviation associated with each parameter in the mass distribution.

This procedure gives a direct statistical measure of the uncertainties in the parameters characterizing the distribution, which proved essential in determining the limitations of the resolution of the method. Note that the nonlinear least-squares program [26,27] provides approximate estimates of these uncertainties. These estimates proved to be substantially different from the directly calculated error estimates described above and were not used.

D. Simultaneous Fitting of Data Gathered at Several Angles

A crucial test for any method of sample characterization is that the result be independent of the parameters of the experiment. Initial attempts to characterize vesicle size using cumulants were frustrated by an inability to pass this test. The results of cumulants analysis for a series of mixed-lipid composition vesicles are given in Table 1. The values of $\bar{\Gamma}/q^2$, from which z-averaged radii can be determined, vary with scattering angle for all samples. The dependence of the moments on scattering angle and the deviation from zero of the second moments indicate that the samples were not monodisperse and that some non-Rayleigh scatterers were present. This problem was dealt with in the spline method by adding the assumption that the particles are spherical shells with constant 40 Å wall thickness and using the Rayleigh-Gans form factor Eq. (6). If these assumptions were invalid, the distributions obtained with different scattering angles would not agree. In a few cases such discrepancies did occur. In most cases, however, it was possible to derive a distribution function consistent with data from all scattering angles.

The use of multiple scattering angles is illustrated in Fig. 2, in which the mass distributions obtained with a vesicle sample at varying scattering angles are illustrated. The best-fit distributions obtained by use of a single set of knots and data from low angles only (22° and 29°) and from high angles only (61° and 90°) are plotted together. The two distributions agree fairly closely: they fall within one standard deviation for three of the five knots and within two standard deviations for all five. (For clarity, the standard deviations in the large radius portion of Fig. 2 are deleted; the data are tabulated in Table 2.) Thus the characterization of the distribution does not depend strongly on the scattering angle selected.

When comparing distributions derived from different samples, data from all four or five scattering angles were utilized. Simultaneous fitting to data from all angles minimizes the standard deviations of the distribution parameters and yields a distribution that is consistent in a statistically weighted sense with the largest available data set. For the sample illustrated in Fig. 2, the resulting distribution is shown as part of Fig. 7. The data from five scatter-

TABLE 1. Cumulants Analysis of Phospholipid Vesicles

Sample: 100% Egg Phosphatidylcholine

Angle	22.01°	32.02°	59.98°	90.00°	118.82°
$\bar{\Gamma}/q^2$ [a]	1.90 ± 0.04	2.01 ± 0.02	1.97 ± 0.06	2.02 ± 0.02	2.01 ± 0.01
Q [b]	0.154 ± 0.022	0.145 ± 0.029	0.103 ± 0.015	0.041 ± 0.022	0.022 ± 0.002

Sample: 25% Phosphatidylethanolamine:75% PC

Angle	22.01°	28.81°	61.19°	90.00°	118.81°
$\bar{\Gamma}/q^2$	1.597 ± 0.007	1.633 ± 0.022	1.618 ± 0.006	1.574 ± 0.023	1.541 ± 0.022
Q	0.170 ± 0.010	0.166 ± 0.009	0.127 ± 0.008	0.072 ± 0.015	0.111 ± 0.012

Sample: 50% PE:50% PC

Angle	22.01°	28.81°	61.19°	90.00°	118.81°
$\bar{\Gamma}/q^2$	1.228 ± 0.017	1.331 ± 0.012	1.408 ± 0.011	1.389 ± 0.013	1.369 ± 0.016
Q	0.319 ± 0.012	0.198 ± 0.013	0.110 ± 0.015	0.061 ± 0.010	0.053 ± 0.013

[a] $\bar{\Gamma}$ is the intensity-weighted average of the decay constant $D(R) \, q^2(\theta)$.

[b] Q is the value of $M_2/\bar{\Gamma}^2$, where M_2 is the intensity-weighted average value of $(\Gamma - \bar{\Gamma})^2$.

171

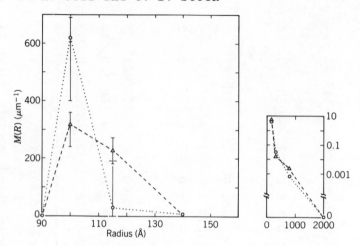

FIGURE 2. Distribution functions for 75% phospha-
tidylcholine:25% phosphatidylethanolamine
vesicles by use of scattering angles 22°
and 29° (·····) and 61° and 90° (-----).
The distribution function obtained by
simultaneously fitting to data from this
sample gathered at five scattering angles
is shown as the dotted curve in Fig. 7.

ing angles gathered by use of the sample shown in
Fig. 2 and the correlation functions derived from the
best-fit distribution are illustrated in Fig. 3. The
standard deviations of the normalized $g^{(2)}(\tau,\theta)$ range
from 0.0019 to 0.0038. The fit is excellent, as
indicated by the value (~ 1.12) of χ^2_ν.

When simultaneously fitting data gathered at
several angles, a problem resulted from our inability
to gather all the data at the same time. Occasionally low-angle data could not be fit together with
the data from higher angles. Sometimes this was due
to "dust" contributions, which varied from one run to
the next. With some samples the problem was systematic, indicating an inability of the model of the
scattering population to account for all the data.

TABLE 2. Mass Distribution Function for Sample in Fig. 2

Set of Angles	Radius (Å)					
	90	100	115	140	300	800	2000
22° and 29°	0	533 ± 125	95 ± 95	5.4 ± 2.2	0.035 ± 0.030	0.0022 ± 0.0006	0
61° and 90°	0	300 ± 60	231 ± 43	6.1 ± 1.1	0.016 ± 0.016	0.0006 ± 0.0006	0
22°, 29°, 61°, 90° and 119°	0	445 ± 30	130 ± 20	7.5 ± 0.3	0.0008 ± 0.0008	0.00185 ± 0.0001	0

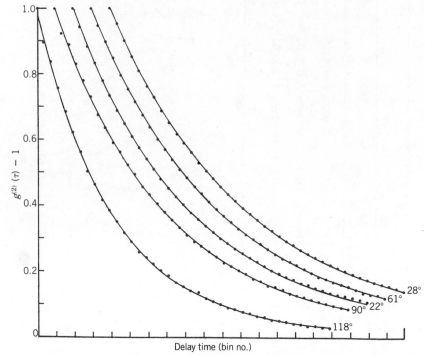

FIGURE 3.   Photon correlation spectroscopy data (·) from five angles for the sample in Fig. 2 and fit (———) to the data derived from the best fit $N(R)$.

Although χ_ν^2 is ~ 1 when data from all five angles or only the higher angles are used, it is unreasonably low (0.19) when only the low-angle data of Fig. 2 are used. This low value of χ_ν^2 probably indicates that in addition to the expected random statistical error, there is a correlated error in the low-angle data. (For example, fluctuations in the number of large particles in the scattering volume from one short experiment to the next would lead to a correlated error.) The effects of any correlated errors could be removed if the data from the individual short experiments were available. In this case they were unavailable. Because of this problem, the simultaneous fit to data taken at all five angles is not as good as that indicated by the calculated value of χ_v^2. This problem and the slight discrepancies noted in Fig. 2 indicate that the model of the scattering population used here is not quite adequate for description of all the data. It is nevertheless significant that the model comes as close as it does.

E. Knot Selection

The fitted $M(R)$ depends to an extent on the set of knots selected. It would be possible to implement an algorithm that, given the number of knots, places these knots at positions that lead to the lowest possible χ^2. This approach was not taken. This would complicate the computations significantly, and the quality of the fit is, in fact, quite insensitive to the specific positions of the knots, provided there are enough knots placed near certain locations. Thus the repeated convergence of a least-squares algorithm to a single set of knot positions would be problematic.

Knots were restricted to the interval between 90 and 2000 Å. These limits were based on results of other measurements [23-25]: they could not be set confidently by using established light-scattering techniques only. The use of these limits considerably reduced the ambiguity in the fitted distributions. An important advantage of the spline-fitting technique is the ease with which such additional information can be incorporated into the fitted function.

Initial fitting utilized 14 knots. The low-radius portion of a typical resulting mass distribution is illustrated in Fig. 4. It is clear that too many degrees of freedom are used in this figure; the 11 fitting parameters are not independent. For example, over a large range of values, the combined effect of contributions at radii 130 and 150 Å are not distinguishable from the effects of a corresponding contribution at 140 Å. The data used in all these figures are rather precise (typical relative errors in the correlation functions are several tenths of 1%) and data from five scattering angles were fit simultaneously. Therefore, it is clear that the level of fine structure that Fig. 4 attempts to define is beyond the precision of the experiment performed. The work of McWhirter and Pike [9] suggests that this limitation is fundamental and would be found in any reasonable extension of our experiment. Four or five independent parameters, as determined from the error analysis, can generally be defined from the experiments. The exact placement of the seven or eight corresponding knots is not important. Provided the knots are near appropriate locations, the data can be fit accurately by the best-fit distributions.

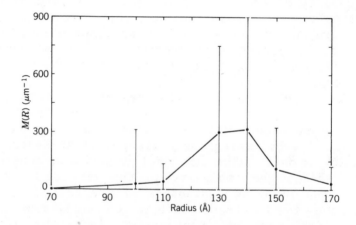

FIGURE 4.   Mass-distribution function (low radius portion only) obtained using 14 knots and data from a 50% PC:50% PE sample. The distribution obtained using 7 knots is shown as the dashed curve of Fig. 7.

The placement of the knots is important in the comparison of samples. This is illustrated in Figs. 5 and 6. Here the fitted distribution functions for

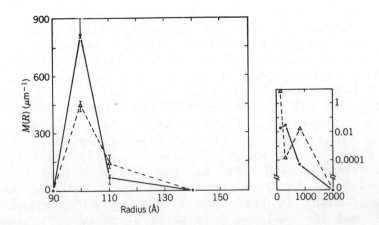

FIGURE 5. Mass distribution functions for (a) PC sample (————) and (b) 75% PC:25% PE sample (------) by use of knot set (90, 100, 110, 140, 300, 800, and 2000 Å).

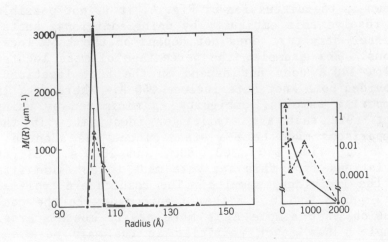

FIGURE 6. Mass-distribution functions for same sample as Fig. 5 by use of knot set with first three knots at 100, 103, and 106 Å.

two different samples are presented with the use of two different sets of knots. Figures 5a and 6a give the distribution functions for a sample of egg phosphatidylcholine (PC) vesicles (sample A), and Figs. 5b and 6b give the distributions for vesicles consisting of 25% egg phosphatidylethanolamine (PE) and 75% PC (sample B). The fits to the data obtained by using either set of knots are essentially identical. For Figs. 5b and 6b, χ^2_ν is given by 1.119 and 1.128, respectively, and the maximum difference between correlation function points derived from the two distributions is less than 0.0004, or a factor of 5 less than the smallest data point standard deviation. Figures 5 and 6 suggest that for either sample there is a main peak and a low-level tail in the distribution functions. Fine-structure details such as the width of the main peak are strong functions of the location of the knots. The data are consistent with the distributions obtained by use of either set of knots or any of a class of similar sets of knots. As shown in the discussion of Fig. 4, it is not possible to resolve this ambiguity by using additional knots. Coarser structure does not depend on the knot locations. For example, the percentage of mass located below 140 Å does not depend on the knot locations, provided both knot sets include 140 Å. Therefore, in comparing samples, ambiguity is minimized by using knot sets that are similar or identical. In the comparisons made here a standard set was adopted — 90, 100, 110, 140, 300, 800, and 2000 Å. Slight deviations from this set were made to optimize fits to the individual samples. The results are typified by Figs. 5 and 6. Either figure leads to the conclusion that sample A is more nearly monodisperse. Sample B has fewer particles in the main peak and more in the intermediate size range (110-300 Å).

F. Comparisons of Size Distributions of Phospholipid Vesicle Solutions

The technique described above was applied to phospholipid vesicle solutions of varying lipid composition. The results are illustrated in Fig. 7, in which the distributions obtained from samples of varying PC and PE content are shown. The distributions differ primarily in the low to middle range of radius. The pure PC vesicles are relatively uniform in size with a large peak near $R = 100$ Å. As the PE content is increased, the distribution exhibits fewer vesicles of this smallest size and more of intermediate sizes. In the 50% PE:50% PC sample, the peak at 100 Å has nearly disappeared in favor of larger particles, with a peak near 130 Å. In addition, the fraction from 150 to 300 Å increases measurably as the PE content increases. The details of the sample preparation and the interpretation of these results can be found elsewhere [8,28].

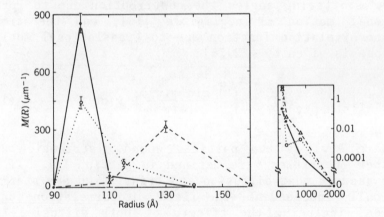

FIGURE 7. Mass-distribution functions for three vesicle samples: 100% PC (———), 75% PC:25% PE (·····), and 50% PC:50% PE (-----).

Comparisons with this level of detail would not be possible by use of the cumulants listed in Table 1. The additional detail of characterization can be attributed to the direct determination of the distribution functions rather than their moments and to the use of external constraints and extensive data for improvement of the accuracy of the approximations to the distribution functions.

IV.   DETERMINATION OF BACTERIAL SPEED DISTRIBUTIONS

A.   Theory

The application of splines to bacterial speed distributions [14-16] is quite similar to the application to particle size distributions described in Section III.   An experimental estimate of the field autocorrelation function is determined from measured values of the intensity autocorrelation function.   At low scattering angles the contribution due to rotational motion is negligible [13], and the field autocorrelation function due to translational motion alone is given by [12,16]

$$|g^{(1)}_{\text{theory}}(\tau)| = \int_0^\infty \frac{\sin(q\tau v)}{q\tau v} P(v)dv \, , \qquad (13)$$

where $P(v)$ is the particle speed distribution and other quantities were defined in Section III.   If it is assumed that diffusion and translation are statistically independent, the field autocorrelation function, including the effects of both diffusion and translation, is given by

$$|g_{theory}^{(1)}(\tau)| \tag{14}$$

$$= \exp[-D_T q^2 \tau] \; [\beta + (1 - \beta) \int_0^\infty \frac{\sin(q\tau v)}{q\tau v} \, P(v) dv] \; ,$$

where β is the fraction of nonmotile bacteria. As in Section III, the speed distribution function $P(v)$ may be approximated by a linear combination of basis functions for first-order splines,

$$P(v) = \sum_{j=0}^{m-2} c_j B_1(v; v_j, v_{j+1}, v_{j+2}) \; , \tag{15}$$

defined in terms of a selected set of knots $\{v_0, v_1, \cdots, v_m\}$. First-order splines were used since, as in the case of particle size distributions, it appeared that no additional information about the samples would be derived by use of higher-order splines. With this expression of the speed distribution, the field autocorrelation function is given by

$$|g_{theory}^{(1)}(\tau_i)|$$

$$= \exp[-D_T q^2 \tau_i] \left(\beta + (1 - \beta) \sum_{j=0}^{m-2} c_j I_{ij} \right) \; , \tag{16}$$

where

$$I_{ij} = \int_{v_j}^{v_{j+2}} B_1(v; v_j, v_{j+1}, v_{j+2}) \frac{\sin(q\tau_i v)}{q\tau_i v} \, dv \; . \tag{17}$$

The diffusion coefficient D_T was measured independently by substantially lowering the pH of the preparation and was not included as an independent fitting parameter.

B.   Synthetic Data Experiments

The capabilities and limitations of the technique were first evaluated by use of synthetic data. Various assumed forms were used for the speed distributions with, typically, upper bounds of 50 µm/sec and no discontinuities or sharp peaks. The autocorrelation functions associated with these distributions were first calculated. These functions were modified by the addition of independent random noise and then fitted by use of various spline basis sets. Comparison of the distributions retrieved by the fitting procedure with the original distributions were then made. The quality of the fits was measured by the value of χ^2_ν. The noise added to each individual point of $g^{(2)}(\tau_i)$ was normally distributed with zero mean and had a standard deviation of 1% of the value of $g^{(2)}(\tau_i)$. For these synthetic data experiments, the value of χ^2_ν was equal to 1 within the fitting error. Error estimates in the fitting parameters and in χ^2_ν were made by adding to the ideal $g^{(2)}(\tau_i)$ generated by each assumed distribution function 10 different sets of random noise. The variations in the fits to these 10 functions defined the error estimates.

1.   **Number of Basis Functions.**   The maximum number of independent parameters that could be extracted from the autocorrelation functions was evaluated by fitting the synthetically generated autocorrelation functions with the use of a progressively decreasing number of spline basis functions until the quality of the fits, as indicated by χ^2_ν, began to deteriorate. For the distributions considered, four basis splines were always sufficient to obtain $\chi^2_\nu = 1$ within experimental error, three were usually sufficient, and two never were. These conclusions were unchanged even with the noise level reduced to 0.1%.

2.   **Knot Selection.**   The influence of the locations of the knots on the details of the retrieved distri-

butions is illustrated in Figs. 8 and 9.   Figure 8
shows two assumed starting speed distributions.   The
autocorrelation function (with noise) generated by
the speed distribution given in Fig. 8b was fitted by
using four different knot sets.   (One set was the
same as was used in generating Fig. 8b.)   The resulting distributions are illustrated in Fig. 9.   In all
four cases, χ_ν^2 was equal to 1 within its error
limits.   As in the case of particle size distributions, the details of the distribution are strongly
influenced by the positions of the knots.   Nevertheless, in this particular case the variations with
knot positions of the first four moments of v are
minor (see Table 3).   In particular all four determinations of the mean speed are within 1% of the
correct value, determined from the original assumed
distributions.   Thus, although the autocorrelation
data do not contain sufficient information for speci-
fication of the fine structure of the speed distribution, they are sensitive to the grosser features of
the distributions, reflected by the moments.   Within
limits the ability to reproduce these grosser features is not affected by the positions of the knots.

3. Distribution Mean and Width.   The first two
moments of the velocity distributions generally could
be recovered by the fitting technique.   This is
illustrated in Fig. 10 and Table 3, using the two
distributions of Fig. 8.   Four different knot sets
were used in fitting autocorrelation data generated
by use of these distributions.   The resulting speed
distributions are illustrated in Figs. 8 and 10.   As
shown in Table 3, for all four sets of knots the
means of the two distributions are recovered within
about 1% and the widths within 3%.   The relative
magnitudes of these moments of the two distributions
are well preserved.   A qualitative comparison of the
two distributions would be affected little by which
of the four knot sets (Fig. 8 or Fig. 10a-c) were
used: distribution b has a higher peak at low velocity than does distribution a; there is a crossover at

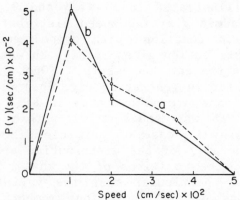

FIGURE 8.   Two different speed distributions were defined by using the knot set 0, 0.001, 0.002, 0.0036, and 0.005 cm/sec. For each distribution function, 10 autocorrelation functions with noise were generated. Fitting to each of these 10 autocorrelation functions, using the same knot set, resulted in the distribution functions shown as curves a and b, which closely approximate the original distributions.

20-30 μm/sec; and distribution a has higher values at the higher velocities.

Simulations of distributions with various shapes [15] show that the technique can recover the second moments of the distributions with reasonable accuracy, except in cases of sharply peaked distributions. In the latter cases the second moments can be recovered successfully only when knots happen to be placed close enough to the peaks. Although higher-order moments are occasionally recovered with good accuracy, as in the case illustrated in Fig. 9, moments of the bacterial speed distributions higher than the second generally cannot be detemrined reliably from the photon correlation spectroscopy data.

TABLE 3. First Four Speed Moments for Figs. 8-10

	χ_ν^2	$\bar{v}$	$\langle(v-\bar{v})^2\rangle^{1/2}$	$\langle(v-\bar{v})^3\rangle^{1/3}$	$\langle(v-\bar{v})^4\rangle^{1/4}$
Fig. 8a		20.3	11.0	8.5	13.5
8b		18.5	10.8	9.5	14.1
Fig. 9a	1.00 ± 0.04	18.5 ± 0.04	10.9 ± 0.19	9.3 ± 0.15	14.2 ± 0.20
9b	1.00 ± 0.05	18.4 ± 0.06	11.0 ± 0.21	9.32 ± 0.16	13.8 ± 0.16
9c	1.00 ± 0.05	18.3 ± 0.09	11.1 ± 0.26	9.2 ± 0.09	13.8 ± 0.18
9d	1.01 ± 0.05	18.7 ± 0.04	10.5 ± 0.14	9.9 ± 0.19	13.8 ± 0.15
Fig. 10a(a)	1.00 ± 0.05	20.2 ± 0.09	11.2 ± 0.22	8.28 ± 0.09	13.7 ± 0.16
(b)	1.00 ± 0.05	18.4 ± 0.06	11.0 ± 0.21	9.32 ± 0.16	13.8 ± 0.16
10b(a)	1.00 ± 0.04	20.2 ± 0.09	11.3 ± 0.28	8.01 ± 0.06	13.8 ± 0.20
(b)	1.00 ± 0.05	18.3 ± 0.09	11.1 ± 0.26	9.2 ± 0.09	13.9 ± 0.18
10c(a)	1.00 ± 0.05	20.4 ± 0.04	10.9 ± 0.16	8.95 ± 0.20	13.6 ± 0.16
(b)	1.01 ± 0.05	18.7 ± 0.04	10.5 ± 0.14	9.9 ± 0.19	13.8 ± 0.15

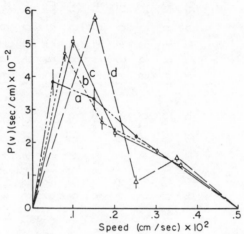

FIGURE 9.   Results of fitting to the 10 autocorrelation functions generated from the distribution of Fig. 8b, using four knot sets:
(a) 0,   0.0005,   0.0015,   0.0025,   0.005
cm/sec;   (b) 0,   0.0008,   0.0017,   0.003,
0.005   cm/sec;   (c) 0,   0.001,   0.002,
0.0036,   0.005   cm/sec;   (d) 0,   0.0015,
0.0025, 0.0035, 0.005 cm/sec.

4. Initial   Guess.   The   nonlinear   least-squares
fitting   routine   requires   an   initial   guess.   When
three basis splines were used, it was found that the
final distribution obtained was not a function of the
initial   guess   used.   When   four   basis   splines   were
used,   the   dependence   of   the   result   on   the   initial
guess was minor.   When more than four basis splines
were used, the final result was significantly dependent on the initial guess, reflecting the fact that
the   set   of   fitting   parameters   was   not   independent.
Generally a uniform speed distribution could be used
as an initial guess.

In the work on particle size distributions the
final   fitted   distributions   were   not   found   to   be
independent   of   the   initial   guesses.   Spurious   minima
were frequently encountered in the search of χ^2 space

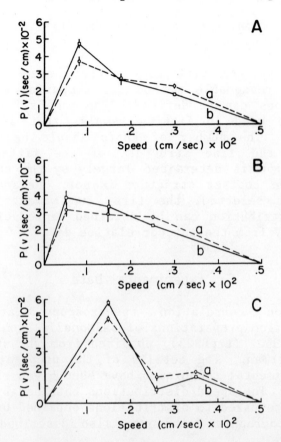

FIGURE 10.   Results of fitting the two distributions of Fig. 8, using three knot sets: (a) 0, 0.0008, 0.0017, 0.003, 0.005 cm/sec; (B) 0, 0.0005, 0.0015, 0.0025, 0.005 cm/sec; (C) 0, 0.0015, 0.0025, 0.0035, 0.005 cm/sec.

conducted by the nonlinear least-squares program. Consequently, much more care was needed in assuring that the best fit had indeed been found. The difference between the two cases may be due to the extended ranges of radii used in the particle size distributions in comparison to the relatively limited range of velocities.

5. Summary.   As with the particle size distributions, a reasonable method for comparing different
preparations can be defined.   The same set of knots
should be used in fitting samples that are to be
compared.   The number of knots should be minimal.
Although the fine structure of the fitted speed
distribution is determined largely by the choice of
knots, the coarser structure is not.   Regardless of
the knots selected, the first two moments of the
speed distribution can be obtained reproducibly and
accurately from the autocorrelation data.

C.   Experimental Data

Photon correlation spectroscopy data were
gathered from preparations of *Salmonella typhimurium*,
strain SB3507 (trpB223), obtained from the stocks of
P. E. Hartman.   The details of the preparation and
the experimental techniques have been given elsewhere
[15, 16].   The speed distributions obtained by use of
PCS were compared to distributions obtained by use of
a cinematographic technique, also described previously.

The speed distributions obtained by the PCS
technique generally consist of three distinct components: a primary peak at 0-50 μm/sec, a small
secondary peak at high speeds (~ 100 μm/sec), and a
nonmotile fraction of variable magnitude.   Seven
parameters were usually required to obtain a fit
characterized by minimum χ^2_ν: four basis splines were
used to represent the main peak, two to describe the
secondary, high-speed peak, and one to describe β.   A
typical result is illustrated in Fig. 11.   Successful
fitting of the autocorrelation data requires a peak
at or above 90 μm/sec: the fits for peaks at 90-
180 μm/sec all have essentially the same value of χ^2_ν.
For high-speed peaks at lower values (Fig. 11a-b),

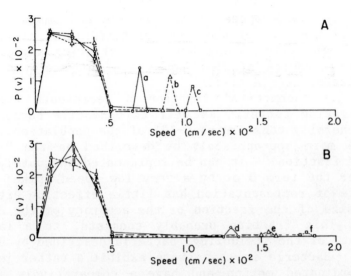

FIGURE 11. Spline fits to the same experimental autocorrelation functions, taken at 30°C and a scattering angle of 7.5°, using several different high speed peaks. The error bars in each case represent the standard deviation of the mean for five sets of data. In all cases the lower knots were 0, 0.001, 0.0025, 0.004, and 0.005 cm/sec. The high-speed knots were

    (a) 0.0065, 0.0070, 0.0075 cm/sec;
    (b) 0.0085, 0.0090, 0.0095 cm/sec;
    (c) 0.0100, 0.0105, 0.0110 cm/sec;
    (d) 0.0125, 0.0130, 0.0135 cm/sec;
    (e) 0.0150, 0.0155, 0.0160 cm/sec;
    (f) 0.0175, 0.0180, 0.0185 cm/sec.

the value of χ^2_ν was significantly higher. The exact position of this peak has little influence on the estimate of either the nonmotile fraction β or the mean speed. The source of this high-speed component is not understood. The cinematographic results strongly suggest that it is not directly related to the translational movement of the bacteria. The secondary peak appeared in all fits and amounted to 4-8% of the area of the speed distribution: its size was essentially independent of scattering angle. A possible source of this peak is light scattered by the bacterial flagella.

The "nonmotile" fraction β represents slowly translating bacteria as well as nonmotile bacteria. It generally constitutes 5-30% of the population. It would more appropriately be described as the "low-speed fraction." It can be represented adequately by either the term β or by a very low speed spike; the choice of representation has little effect on either the size of the fraction or the accuracy of the data fit. This fraction probably consists, to a large extent, of the "twiddling" bacteria described by Berg [29]. Bacteria in this state exhibit a rather jerky uncoordinated motion and have a comparatively slow rate of net translation.

The value of χ_{ν}^2 obtained with experimental data was generally much less than 1. This was probably due to the presence of correlated errors in the successive short experiments ($\sim$ 10 sec) generally used to measure the autocorrelation functions and standard deviations.

The fit to the autocorrelation data obtained by the method of splines is illustrated in Fig. 12. The fit is quite close except for the first few points, corresponding to short delay times, or fast motions. The likely cause of this deviation is inadequate modeling of the unknown causes of fast decorrelation of the intensity. In the current model the fast fluctuations in intensity are accounted for by the "high-speed" peak", which is not confirmed by cinematography. The actual source of the fast fluctuations (possibly flagella movement) is not likely to be adequately described by simple translational and diffusive motion. Therefore, the deviations of the theoretical fit from the experimental data at short delay times are not surprising.

As a final test of the technique, cinematographic and PCS measurements were performed simultaneously on identical aliquots of a single bacterial preparation. The results obtained from the two techniques, illustrated in Fig. 13, were in good agreement except for the high-speed component discussed above.

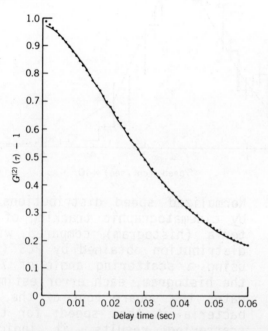

FIGURE 12.   The average of 10 normalized autocorrelation functions, each of 100-sec duration (·) gathered by using a sample of bacteria at 30°C and a scattering angle of 7.5°. The fit to the data is shown as the solid curve.

V. CONCLUSIONS

The extraction of size and speed distribution functions from PCS data with the use of splines has been demonstrated. In both cases the structure of the distributions cannot be revealed with arbitrary detail. Four parameters of the size distributions

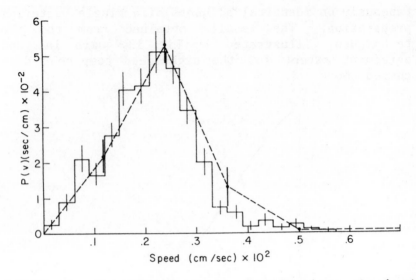

FIGURE 13. Normalized speed distributions obtained by cinematographic tracking of 500 bacteria (histogram) compared with speed distribution obtained by PCS (-----) by using a scattering angle of 7.5°. For the histogram, each error estimate indicates the square root of the number of bacteria at that speed; for the light-scattering results, it indicates the standard deviation of the mean for determinations from five autocorrelation functions of 100-sec duration. For the cinematographic distribution, $\bar{v}$ = 21.4 μm/sec. For the light-scattering distribution, $\bar{v}$ = 22.5 ± 0.9 μm/sec and β = 0.11 ± 0.04. The value of $\bar{v}$ was calculated without including nonmotile bacteria or the high-speed peak of 8.4 ± 0.6%. Knots were evenly spaced at 12, 24, and 36 μm/sec. Salmonella were suspended in NB at 24°C for PCS and 23°C for cinematography.

and seven parameters of the speed distributions were reproducibly extracted by the technique. Statistical analyses based on χ^2 tests were used in assuring that the parameters that were determined were meaningful. This is perhaps the most important feature of the results reported here. The size distributions were determined by use of extensive data gathered at four or five different scattering angles. Simultaneous use of all these data gave size distributions consistent, with relatively minor discrepancies, with all the data. The use of extensive data and the incorporation of information about the samples obtained by other means enhanced the level of detail provided by the technique. The results summarized here also show that PCS can be used as a rapid and accurate assay of bacterial translational movement. Many potential experiments with bacterial samples are rendered practical by this technique.

The technique is quite similar for both size and speed distributions. It allows considerable flexibility in the fitting function used and thus in the types of data that can be fit successfully. External constraints (e.g., upper and lower bounds on sizes or speeds) can be incorporated easily in the model. Specialized distributions such as bimodal or multimodal functions can be used readily in cases when such distributions are anticipated.

ACKNOWLEDGMENTS

The authors are indebted to Professor F. D. Carlson for inspiration and for many valuable suggestions throughout this work, to Dr. R. Haskell for many fruitful discussions, and to Dr. Y. Barenholz, Professor B. J. Litman, and Professor T. E. Thompson of the University of Virginia for providing the vesicle solutions and for much helpful discussion. This work was supported by U.S. Public Health Service Grants GM 05181, AM 12803, AM 16315, Training

Grant 5T01 GM00716, all to F. D. Carlson, and Post-doctoral Fellowship 1-F32-GM06216-01 to G. B. Stock.

REFERENCES

1. H. Z. Cummins and E. R. Pike, Eds., *Photon Correlation and Light Beating Spectroscopy*, Plenum Press, New York, 1974.

2. H. Z. Cummins and E. R. Pike, Eds., *Photon Correlation Spectrometry and Velocimetry*, Plenum Press, New York, 1977.

3. F. D. Carlson, The Application of Intensity Fluctuation Spectroscopy in Molecular Biology, Ann. Rev. Biophys. Bioeng. *4*, 243 (1975).

4. R. Pecora, Quasi-elastic Light Scattering from Macromolecules, Ann. Rev. Biophys. Bioeng. *1*, 259 (1972).

5. D. E. Koppel, Analysis of Macromolecular Poly-dispersity in Intensity Correlation Spectroscopy: The Method of Cumulants, J. Chem. Phys. *57*, 4814 (1972).

6. C. B. Bargeron, The Measurement of a Continuous Distribution of Spherical Particles by Intensity Correlation Spectroscopy: Analysis by Cumulants, J. Chem. Phys. *61*, 2134 (1974).

7. J. H. Goll and G. B. Stock, Determination by Photon Correlation Spectroscopy of Particle Size Distributions in Lipid Vesicle Suspensions, Biophys. J. *19*, 265 (1977).

8. J. H. Goll, Y. Barenholz, B. J. Litman, F. D. Carlson and T. E. Thompson, Photon Correlation Spectroscopic Study of the Size Distribution of Phospholipid Vesicles, Biophys. J. *38*, 7 (1982).

9. J. G. McWhirter and E. R. Pike, On the Numerical Inversion of the Laplace Transform and Similar Fredholm Integral Equations of the First Kind, J. Phys. A: Math. Gen. *11*, 1729 (1978).

10. Es. Gulari, Er. Gulari and B. Chu, Photon Correlation Spectroscopy of Particle Distributions, J. Chem. Phys. *70*, 3965 (1979).

11. N. Ostrowsky, D. Sornette, P. Parker and E. R. Pike, Exponential Sampling Method for Light Scattering Polydispersity Analysis, Opt. Acta *28*, 1059 (1981).

12. R. Nossal, S. H. Chen and C. C. Lai, Use of Laser Scattering for Quantitative Determinations of Bacterial Motility, Opt. Commun. *4*, 35 (1971).

13. G. B. Stock and F. D. Carlson, Photon Autocorrelation Spectra of Wobbling and Translating Bacteria, in *Symposium on Swimming and Flying in Nature*, Plenum Press, New York, 1974, p. 57.

14. G. B. Stock, Application of Splines to the Calculation of Bacterial Swimming Speed Distributions, Biophys. J. *16*, 535 (1976).

15. G. B. Stock, The Measurement of Bacterial Translation by Photon Correlation Spectroscopy, Biophys. J. *18*, 79 (1978).

16. G.B. Stock, *Photon Correlation Spectroscopy. An Assay of Bacterial Motility*, Ph.D. Thesis, Johns Hopkins University, Baltimore, Md., University Microfilms, Ann Arbor, Michigan, 1977.

17. S. H. Chen and F. R. Hallett, Determination of Motile Behavior of Prokaryotic and Eukaryotic Cells by Quasi-Elastic Light Scattering, Quart. Rev. Biophys. *15*, 131 (1982).

196   J. H. Goll and G. B. Stock

18. J. G. McWhirter, A Well-conditioned Cubic
     B-spline Model for Processing Laser Anemometry
     Data, Opt. Acta *11*, 1453 (1981).

19. T. N. E. Greville, Introduction to Spline Func-
     tions, in *Theory and Applications of Spline
     Functions*, T. Greville, Ed., Academic Press, New
     York, 1968, p. 1.

20. J. H. Ahlberg, E. N. Nilson and J. L. Walsh, *The
     Theory of Splines and Their Applications*, Aca-
     demic Press, New York, 1967.

21. C. De Boor, "On Calculating with B-Splines, J.
     Approx. Theory *6*, 50 (1972).

22. L. Mandel, Progr. Opt. *2*, 183 (1963).

23. A. G. Lee, Functional Properties of Biological
     Membranes: A Physical-Chemical Approach, Progr.
     Biophys. Molec. Biol. *29*, 3 (1975).

24. C. Huang, Studies on Phosphatidylcholine Vesi-
     cles: Formation and Physical Characteristics,
     Biochemistry *8*, 344 (1969).

25. C. Huang and T. E. Thompson, Methods Enzymol.
     32, 245 (1974).

26. Donald W. Marquardt, An Algorithm for Least-
     squares Estimation of Nonlinear Parameters, SIAM
     J. Appl. Math. *11*, 431 (1963).

27. P. R. Bevington, *Data Reduction and Error Anal-
     ysis for the Physical Sciences*, McGraw-Hill, New
     York, 1969.

28. Y. D. Barenholz, D. Gibbes, B. J. Litman, J. H.
     Goll, T. E. Thompson and F. D. Carlson, A Simple
     Method for the Preparation of Homogeneous Phos-
     pholipid Vesicles, Biochemistry *16*, 2806 (1977).

29. H. C. Berg, Bacterial Behavior, Nature (Lond.) *254*, 389 (1975).

CHAPTER 7

Particle Size Distributions from Analyses of Quasi-Elastic Light-Scattering Data

E. F. GRABOWSKI and I. D. MORRISON
XEROX Corporation
Webster, New York

CONTENTS

199

I. INTRODUCTION

The mathematical derivations of the fundamental equations for quasi-elastic light scattering (QELS) have been given in complete detail by several authors [1-5]. The aspects of QELS addressed in this chapter are the experimental and mathematical techniques necessary for the study of heterogeneous dispersions. A useful set of definitions on the general types of dispersion of particle properties has been given by Gibbons [6].

> *Homodisperse*: An assemblage of identical particles. (The term "monodisperse" is more frequently used than "homodisperse".)

> *Polydisperse*: An assemblage of particles with unimodal properties.

> *Heterodisperse*: An assemblage of particles with multimodal properties.

Examples of monodisperse systems are well known, with the best-known probably being carefully prepared polystyrene (PS). Polydisperse systems are those systems that are not far from being monodisperse, and many examples can be found. Generally, solution polymerizations and solution crystallizations lead to polydisperse systems. Heterodisperse systems are those that deviate furthest from being monodisperse. Dispersions produced by grinding, such as paints or coating slurries or mixtures of particles of different chemical compositions, are good examples of heterodisperse systems.

The analysis of QELS data for monodisperse systems or slightly polydisperse systems is straightforward and has been described well elsewhere [7]. The analysis of QELS data for systems far from being monodisperse is very much more difficult and possibly cannot be done in general without the assistance of extra information [8]. Nevertheless, the reason for attempting this more difficult and less certain analysis lies in the attractiveness of using QELS as a routine method of determining particle size and particle size distributions for practical applications. Section II in this chapter gives a brief summary of the mathematical formulations of the problem of determining particle size distributions from QELS data. Section III describes some of the methods previously proposed for the solution of the mathematical problem. Section IV describes the method we propose, a non-negatively constrained least-squares solution for which convergence and uniqueness criteria are known. Section V describes the necessary experimental equipment and an example. Section VI shows the experimental data taken on heterodisperse system of known composition and the results of the analyses of the data by the non-negatively constrained least-squares technique.

II. MATHEMATICAL FORMULATION OF THE PROBLEM

In general, the measured photoelectron autocorrelation function $G^{(2)}(\tau)$ as a function of delay time τ is given by

$$G^{(2)}(\tau) = A[1 + \beta|g^{(1)}(\tau)|^2] , \tag{1}$$

where $g^{(1)}(\tau)$ is the normalized first-order autocorrelation function at time τ, A is the baseline constant, and β is an equipment related "instrument constant".

The first-order autocorrelation function is also the Fourier transform of the optical spectrum of the scattered light. The constant β depends on the number of coherence areas viewed by the photomultiplier tube and on the sampling interval [9]. Equation (1) can be rearranged to give

$$g^{(1)}(\tau) = \frac{1}{\sqrt{A\beta}} [G^{(2)}(\tau) - A]^{1/2} . \tag{2}$$

In the mathematical techniques that follow, the assumption has been made that the baseline constant is known. The baseline could be estimated from the average photon count rate [32], or it could be determined as one more of the unknown constants, but in practice it should be measured. Once the baseline has been subtracted from the function and the measured autocorrelation function normalized to be equal to one at $\tau = 0$, the right-hand side of Eq. (2) is an experimentally determinable function.

Equation (2) can be used for any time-varying function. In this chapter we consider only the autocorrelation of light scattered from dispersed particles undergoing Brownian motion for which the autocorrelation function Eq. (2) has been shown to be an integral equation in the particle diffusion coefficients.

For a continuous distribution of decay constants [10],

$$g^{(1)}(\tau) = \int_0^\infty F(\Gamma)\exp[-\Gamma\tau]d\Gamma \, , \qquad (3a)$$

where $F(\Gamma)$ is the normalized linewidth distribution function and Γ is the decay constant.

For a discrete distribution of sizes [11],

$$g^{(1)}(\tau) = \sum_{i=1}^{M} a_i \exp[-\Gamma_i\tau] \, , \qquad (3b)$$

where a_i is the probability of scattering center with decay constant Γ_i.

The only essential difference between Eq. (3a) and (3b) is that the continuous function is smooth in the sense that it has a first derivative everywhere whereas the discrete distribution does not. For narrow distributions, a smooth distribution function is appropriate. For very broad distributions, the first derivative is not so important and the discrete distribution is adequate.

Substitution of Eq. (2) into Eqs. (3a) and (3b) gives the dependence of the distribution of decay constants on experimental data — Fredholm integral equations of the first kind,

$$\int_0^\infty F(\Gamma)\exp[-\Gamma\tau]d\Gamma = \frac{1}{A\beta}\ [G^{(2)}(\tau) - A]^{1/2} \qquad (4a)$$

or

$$\sum_{i=1}^M a_i\ \exp[-\Gamma_i\tau] = \frac{1}{A\beta}\ [G^{(2)}(\tau) - A]^{1/2} \qquad (4b)$$

The experimental problem is to measure the normalized first-order autocorrelation function [Eq. (2)], which is the right-hand side of either Eq. (4a) or (4b). The mathematical problem is to invert the integral or the finite sum to obtain the distribution of decay constants. From the value of the decay constant and the scattering angle, the corresponding diffusion coefficient is given by [3]

$$D = \frac{\Gamma}{k^2}\ , \qquad (5)$$

where $k = (4n\pi)/\lambda)\ \sin(\theta/2)$, n is the index of refraction of the fluid, λ is the wavelength of the incident light, and θ is the scattering angle.

If the particles are noninteracting spheres, the particle diameter can be calculated from the diffusion coefficient by means of the Stokes-Einstein relation and the experimental conditions [12]:

$$d = \frac{k_B T}{3\pi\eta D}\ , \qquad (6)$$

where T is the temperature, k_B is Boltzmann's constant, and η is the viscosity of the fluid.

The number or weight fraction of particles of each size can be calculated from the normalized linewidth distribution function or the probabilities a_i by use of the necessary corrections for Mie scattering [13]. For the analyses discussed later in this chapter, we calculate the distribution functions defined by Eq. (4b) and convert the decay constants to particle sizes, but we do not convert the probabilities to number or weight fractions. An important fact to note is that the conversion of the distribution of decay constants to a distribution of particle sizes is a completely independent problem from the determination of the linewidth distributions from QELS data.

III. METHODS OF INVERTING THE FREDHOLM INTEGRAL EQUATION

The Fredholm integral equation [Eq. (3a)] has been studied extensively [14]. It is an ill-conditioned problem in the mathematical sense that small changes in the experimental data can produce large changes in the solution. For any proposed method of solution, two questions must be always kept in mind: (1) whether the method of inverting the equation quickly converges to a solution and (2) whether the solution is unique. For the analysis of real data, these answers are not as mathematically precise as one might like. First, a method that converges quickly on a large mainframe computer may converge too slowly or require too much storage space to be of practical use on a small computer. Second, a useful solution to the problem may not have to be unique if it does not differ too much from other equally likely solutions. What is too much or not too much is a judgment based on the particular needs of the experimenter. We have found that non-uniqueness may be acceptable for broad or multimodal distributions if each solution contains a sufficient amount of the same information. For example, an approximate description of the heterogeneity of the sample may be more important than the determination of the precise particle size of each component. This is analogous to the use of infrared spectroscopy, where the frequencies of the absorption peaks are usually more important than their absolute intensities. On the other hand, for a very narrow particle size distribution, a unique solution is necessary for a precise measure of the particle size.

Two general mathematical procedures have been described for inverting the integral equation to obtain the distribution of decay constants: the Fourier transform and the method of cumulants. The former has been described in detail elsewhere [8]. The practical limitation of this mathematically precise technique is in the extreme requirements of experimental accuracy. In fact, the general consensus is that the Fourier transform of the data is not useful for much more than monodisperse samples [15,16].

The method of cumulants was first given by Koppel [7], who noted that the logarithm of the autocorrelation function Eq. (3a) is formally equal to the cumulant generating function. Therefore, if the logarithm of the autocorrelation function can be expressed as an infinite series in powers of delay time, the moments of the distribution can be easily calculated from the coefficients of the series [17]. This method of analysis is very powerful if the coefficients of the infinite series in delay time can be determined. In practice, usually no more than the first two cumulants can be determined with any reasonable degree of confidence. For nearly monodisperse samples, this is quite sufficient; for polydisperse samples, it is less sufficient; for heterodisperse samples, it is insufficient. The major problem lies in the fact that the series expansion for the logarithm of the autocorrelation function [Eq. (3a)] does not converge quickly enough for heterogeneous samples [16].

Although these two mathematical techniques are precise, they are seldom practicable. Various simplifications have thus been proposed to enable the inversion of Eqs. (4a) or (4b) practically. Each proposal assumes something about the nature of the distribution. The three most successful assumptions proposed so far are (1) assume a form for the distribution, (2) assume the distribution is the smoothest non-negative form, and (3) assume an histogram distribution. The latter two proposals both assume that the distribution must be non-negative. Clearly, no distribution of diffusion coefficients or particle sizes can have negative values. Introduction of this fact into the mathematical statement of the inversion problem is an important aid in finding a unique solution (for example, the Kuhn-Tucker conditions, reference 18, p. 159). For perfect experimental data, it might not be necessary; for data with experimental error, it is as necessary an assumption as the assumption that the "best" solution is the one with the least-squared error. The fact that the Fourier transform method fails may well be due to the fact that the non-negative constraint is not imposed. A general description of each of these three proposals is now given.

1. *Assume a form for the distribution [19-21].* A simple assumption is that the distribution must be a common distribution function such as a Gaussian, a log-normal, or a Pearson distribution. The great advantage in this kind of assumption is that the number of unknowns is then limited and the function with the best fit to the data can be found easily by the method of nonlinear least squares. For many systems, especially well-characterized systems, this assumption is probably the most practical method of finding an acceptable distribution. If the assumed form of the distribution does not fit the data well, the natural extension is to use a more general function with more adjustable constants. The difficulty

is finding a computational method for minimizing the squared error that converges to a unique solution for nonlinear functions of many variables.

2. *Assume that the distribution is the smoothest non-negative form* [22]. Normally, the best fit of theory to experiment is found by minimizing the sum of the squared deviations between theory and experiment. What is often found is that the best-fit distribution becomes increasingly convoluted as the number of adjustable constants is increased. To minimize this convolution, a penalty function is added to the sum of the squared deviations that reflects the "irregularity" of the distribution. The published method [23] uses the third derivative as a measure of the smoothness of the data so that the function minimized is the sum of the squared deviations plus a constant times the third derivative. The iterative computation decreases the deviations between the calculated curve and measured data, keeping the third derivative of the calculated curve as small as possible. If the distribution becomes irregular, the third derivative becomes significant and the iterative technique searches in a different direction. Since the search for a minimum in error of fit is by an iterative method for nonlinear least squares, the usual pitfalls of nonlinear curve fitting such as large parameter correlations and slow convergence are encountered [24]. This method also includes the important assertion that the real distribution function must be non-negative. The method is capable of describing multimodal distributions only if they are known a *priori*. This last restriction in the use of the method makes it inappropriate for the broad or multimodal problems we consider in this chapter. However, as a systematic method of searching for a unimodal distribution with no restraints on the form of the distribution, the method is useful.

3. *The Histogram method* [11]. This method uses Eq. (4b). If no further assumption about the distribution were made then Eq. (4b) would be a finite sum

of exponentials for which no practical method for finding a solution has been found [15,25,26]. This method reduces Eq. (4b) to a set of linear equations, one for each correlator delay time by assuming a set of decay constants. An arbitrary set of delay constants will not match the actual decay constants in the data, but the assumption is that the set of arbitrary decay constants will be sufficiently similar to the actual set of decay constants that the distribution found will be a sufficient approximation to the real distribution. The essential advantage of this approximation is the ability of a histogram distribution to describe very heterogeneous or multimodal distributions. The same approximation has been used successfully in other problems [27]. The iterative computational method to invert Eq. (4b) is as follows:

1. Assume a reasonable range of decay constants.

2. Assign a fixed number (determined by the quality of the data) of decay constants in the range.

3. Find the constants a_i that minimize the error of the fit by a linear least-squares algorithm.

Two problems must be considered: (1) the range of decay constants is not known a *priori*; and (2) the simple matrix inversion routines to minimize the squared error can give negative portions to the distribution. The first of these two problems is solved by assuming that the range of the decay constants is not known but is a variable to be determined iteratively in a nonlinear least-squares search. This implies that Eq. (4b) has $M + 3$ unknowns; the number of assumed decay constants M, the minimum and maximum of the range of decay constants, and the M values of the a_i. The assumed values of the decay constant are chosen evenly spaced in this range. The second problem is solved by finding the least-squares fit by iterative techniques rather than direct matrix inversion imposing the non-negative constraint during the iterative search. Our experience with this type of iterative search is that extremely large amounts of computer time are used during the time that the fit is slowly improving, that is, when the squared error is decreasing [28]. Because the iterative search never converges, one natural proposal is to terminate the computer program when the difference between the calculated and the measured quantities is within experimental error [11,23]. Another reasonable proposal is to terminate the computer program when the calculated distribution is changing only very slightly with successive iterations [27]. The weakness in either proposal is that without a *priori* information, the limit of experimental error is not known and the experimenter cannot judge whether a particular fit deviates from the data in a systematic (and hence improvable) way or whether the deviation is random and further iteration is useless. In mathematical terms, the weakness of this type of approach lies in the fact that neither convergence of the mathematical procedure nor uniqueness of the final distribution has been demonstrated.

The only example we can find of the use of this method on experimental data of the type we are considering in this chapter, a known bimodal distribution [26], shows the method to be internally consistant; that is, the analysis of data for a mixture of two sizes of particles shows the resulting distribution to be equal to the sum of the analyses for the monodisperse particles. However, the analyses of the monodisperse samples are not consistent with the analyses given earlier in Dyson and Isenberg's paper [26] by the method of cumulants. Nevertheless, the strength of this type of approach is that by assuming independent variables over the range of the experimental data, the form of the distribution is capable of describing any real distribution with only a small loss in the ability to determine individual particle sizes. By imposing the non-negative constraint, the procedure includes an important mathematical tool in searching for physically real solutions.

IV. NON-NEGATIVELY CONSTRAINED LEAST SQUARES

This section describes the method we propose for the solution of the Fredholm integral equation to obtain the distribution of decay constants. Many other interesting problems in science can be formulated in equations of this same form [26]. The determination of the heterogeneities of the surface free energy of solid powders requires a solution to a similar mathematical problem [29]. The non-negatively constrained least-squares solution described here works well for that problem [28]. The same mathematics has been used successfully in the determination of the pore volume distribution in powders [30].

Again we state that the general problem we are considering is determination of the approximate particle size distribution for systems that are very heterodisperse and/or multimodal. For such systems, it is as important to determine the general shape of the distribution as it is to determine any of the particle sizes exactly. For example, a change in the number of very small particles in a dispersion can affect the optical absorbance of a dispersion more than can a slight change in average particle size.

For heterodisperse dispersions, the discrete form of the integral equation given by Eq. (4b) is the most natural expression. The constraint that we impose on the mathematical solution to the problem is the natural one, that no part of the "best" distribution can be negative, that is, the solution is non-negative. This assumption is mathematically significant in considering the questions of convergence and uniqueness [18].

The error of fit r_j between a calculated discrete distribution and the data at each of N experimental delay times r_j is given by

$$r_j = \frac{1}{A\beta} [G^{(2)}(\tau_j) - A]^{1/2} - \sum_{i=1}^{M} a_i \exp[-\Gamma_i \tau_j] \ . \quad (7)$$

The non-negatively constrained least-squares problem for N values of delay time (the number of correlator channels) is

Minimize $\quad \sum_j r_j^2 \ , \ 1 \leq j \leq N \quad\quad\quad (8)$

Subject to $\quad a_i \geq 0$ and $\Gamma_i \geq 0$.

This problem has $2M$ unknowns, the a_i and the Γ_i. The major difficulty in finding a solution to this problem is that the object function Eq. (8) is a transcendental function in $2M$ unknowns. The successful method of solving the problem is to choose a large set of the decay constants within the range determined by the range of the experimental data with the assumption that whatever the real distribution of decay constants is, it can be adequately described by these Γ_i. This assumption of a set of decay constants means that Eqs. (8) have become a set of linear equations since the value of each of the exponential terms is known. The only unknowns are the values of a_i. The statement of the minimization problem now is as follows:

Minimize $\quad \sum_j \left(\frac{1}{A\beta} [G^{(2)}(\tau_j) - A] \right.$

$$\left. - \sum_{i=1}^{M} a_i \exp[-\Gamma_i \tau_j] \right)^2 \quad (9)$$

Subject to $\quad a_i \geq 0, \ 1 \leq j \leq N$.

Equation (9) is similar to the minimization of error for linear equations. However, there is an important difference. The non-negative constraint means that the usual matrix inversion methods cannot be applied. In fact simple application of the matrix inversion routines for solution of the unconstrained problem inevitably leads to physically unreal solutions, namely, negative portions of the distribution curve. The non-negative constraint in practice greatly reduces the number of possible combinations of decay constants that can describe the data. That is, the constrained problem can be shown to have unique solutions even when the unconstrained problem does not [18]. The practical implication of the non-negative constraint is that a large number of decay constants can be considered, but only a few will actually appear in the final distribution. Typically we have used a set of 10-50 assumed decay constants but find that only 2-8 appear in the "best-fit" solution. Although the use of a large number of adjustable constants to fit the data may at first seem unmanageable, the non-negative constraint makes it possible.

Equations such as Eq. (9) have been studied extensively in operations research where solutions to constrained problems are needed. From prior work in evaluating various optimization routines available in the literature [28], we choose the method due to Lawson and Hansen [31], a program called NNLS, to demonstrate a useful procedure for finding the solution to Eq. (9) for actual experimental data taken on samples of known composition.

The NNLS program is guaranteed to converge in a finite number of iterations to a solution with minimum error; therefore, no stopping criterion for the computer calculations is necessary. Tests for the uniqueness of the final calculated distributions have been discussed elsewhere [28]. Our experience is that for problems as large as 104 data points by 50 unknowns, the calculation time on a mainframe computer is less than 1 minute.

In general, the range of decay constants appropriate for the analysis is determined by the data. The only requirement on the range of assumed decay constants is that the range be larger than the range of decay constants in the actual sample. (The data must be taken on the proper time scale.) The measured correlation function [Eq. (1)] is an exponential function of time. The delay time for the first two or three correlator channels determines the fastest detectable decay constant. We determine the fastest decay constant from the time between the correlator channels [23]. The slowest decay constant is determined by the total length of time over which the autocorrelation function is measured. The measured autocorrelation function must decay to within experimental error of the baseline by the last measured channel. We choose the slowest decay constant to be one-twentieth of the initial slope of the logarithm of the square root of the data [see Eq. (2)]. We also include a value of zero in the set of assumed decay constants to approximate the influence of an occasional large dust particle.

Once the range of the assumed decay constants is set, the actual values of the assumed decay constants must be specified. The principle that we have applied in the choice of the decay constants is that the error in uniquely determining each decay constant should be uniform over the entire range of decay constants. We compare the autocorrelation function for a single decay constant to that of the sum of two decay constants just greater and just less than itself to estimate the "uniqueness" of each decay constant. We space the decay constants so that the "uniqueness" of each is the same over the entire range of decay constants. The detectable difference between decay constants is the difference between the autocorrelation functions [Eq. (1)]. We measure the autocorrelation function for a sufficient length of time to reach the baseline so that the detectable difference, or uniqueness, between assumed decay constants is the integral of the difference between the two exponential decays from time equals zero to infinite time.

This separation in assumed decay constants can be derived mathematically as follows. (Other similar approximations are as reasonable.) The measured normalized autocorrelation function for a single exponential decay (above the baseline) is

$$C_1(\tau) = \exp[-2\Gamma\tau] \ . \tag{10}$$

Compare this function to a sum of two exponential decay curves separated by Δ:

$$C_2(\tau) = \{\tfrac{1}{2}\exp[-(\Gamma - \tfrac{\Delta}{2})\tau] + \tfrac{1}{2}\exp[-(\Gamma + \tfrac{\Delta}{2})\tau]\} \ . \tag{11}$$

Note that the difference between the two functions never changes sign, so that the difference can be integrated over time without worrying about a change in sign. The total detectable difference δ^2 between these two functions is the integral of the difference from $\tau = 0$ to $\tau = \infty$:

$$\delta^2 = \int_0^\infty [C_2(\tau) - C_1(\tau)]d\tau = \frac{\Delta^2}{\Gamma(\Gamma^2 - 4\Delta^2)} \ . \tag{12}$$

The object is to keep the detectable difference δ^2 constant for all values of the decay constant. This equation determines the spacing in the assumed decay constants. If Γ_i is the value of the ith decay constant, then (approximately)

$$\frac{d\Gamma_i}{di} \cong \Delta \cong \delta\Gamma_i^{3/2} \ , \tag{13}$$

which is easily solved to give

$$\Gamma_i = \frac{1}{(ai + b)^2} \ , \quad 1 \leq i \leq M \ . \tag{14}$$

The constants a and b are determined by the initial and final values of the assumed decay constants. Equation (14) gives a set of assumed equally "unique" decay constants. Since the decay constants are inversely proportional to particle size, this particular method of spacing the Γ_i spaces the assumed particle sizes quadratically.

In summary, the procedure we propose is as follows:

1. Set the largest decay constant to be the inverse of the delay interval per channel.

2. Set the slowest decay constant to be 1/20th of the initial slope of the logarithm of the square root of the data.

3. Set one decay constant to be zero.

4. Pick M, the number of decay constants, usually in the range of 20-50.

5. Determine the values of the rest of the decay constants by Eq. (14).

6. Calculate the value of each of the exponential terms and the value of each of the normalized experimental data points in the object function, Eq. (9).

7. Call the NNLS program to find the distribution that minimizes Eq. (9) subject to the non-negative constraint.

8. Calculate the particle diameters from the decay constants by Eqs. (5) and (6) and plot the distribution as a histogram.

V.   EXPERIMENTAL PROCEDURE

We assume that the reader is not overly familiar with the operation of digital correlators and spend some time explaining how an autocorrelation function is measured and how one experimentally extracts the baseline. Recall from Section II that the measured homodyne autocorrelation function has the form

$$G^{(2)}(\tau) = A[1 + \beta|g^{(1)}(\tau)|^2] \ . \tag{1}$$

The contant A (the baseline) can be estimated from the average photon count rate but must, in practice, be measured. The instrument constant β depends on the particular experimental setup used and the number of coherence areas viewed by the photo-multiplier and the sampling interval [9].

A digital autocorrelator approximates the auto-correlation function by

$$G^{(2)}(j\ \Delta\tau) = \lim_{N\to\infty} \frac{1}{N} \sum_{i=1}^{N} C_i C_{i-j} \ , \tag{15}$$

for $0 < j < m - 1$

where C_i is the number of photons counted in the ith sampling interval of length $\Delta\tau$. For a 4 bit corre-lator, C_j can vary from 0 to 15. For a 1 bit corre-lator, C_j can be 0 or 1. The total number N of sampling intervals in a given experiment can be obtained from the run time divided by the sampling interval. In practice, the correlator keeps track of N. Most correlators provide N and the contents of the m summation registers separately. Some newer versions with internal microprocessors actually divided the contents of each channel by N if desired. We deal directly with the integer contents of the summation registers. Division by the factor of N would only change the scale and would require the use of floating-point numbers that take up more computer space. When the absolute number of particles of a given size is desired, the factor N must be carried along, but when only the relative distribution of particles is desired, N is not needed.

The number of correlator channels gives one an estimate of m; however, as we see below, some channels are used to determine the baseline. The correlator consists of counting circuitry that counts pulses from the photomultiplier (PMT) for a period of time $m\, \Delta\tau$ and a shift register that holds the values of C_i for the m previous sampling time intervals. One cycle of the correlator consists of accumulating the count C_i, placing the result in the first element of the shift register while simultaneously shifting all the elements of the shift register over one position (with the last element falling off the end and gone forever) and then multiplying the value C_i by each element of the shift register to form the m products $C_i C_{i-j}$. These products are then added into accumulation registers and the cycle repeats.

Real-time operation requires that all these multiplications and additions be done in parallel by hardware, which accounts for the high cost and limited number of channels found in such instruments. If the maximum delay time $m\, \Delta\tau$ is made sufficiently large, the measured autocorrelation function will decay to the baseline.

Ideally, data should be taken only for delay times $m\, \Delta\tau$ over which the measured autocorrelation function is above the system noise level. Some additional time should be allowed for the autocorrelation function to decay to the baseline, and then the baseline should be measured. This is accomplished with delay channels that are included with some correlators and are optional on others. In practice, the delay channels are just extra elements inserted into the shift register before the last few channels (usually eight) with no associated multiplication and summation hardware.

To see how this works, assume that there are m elements of the shift register with associated multiplication and summation hardware, d delay elements, and eight baseline channels with multiplication and summation hardware. As a particular element C is shifted through the first m positions of the shift register, the products $C_i C_{i-j}$ are generated for m values of j. No mathematical operations take place while the given element is shifted through the d delay channels, but it retains its time order with respect to the other elements in the shift register. As C is shifted through the eight baseline channels, the products $C_i C_{i-j}$ for values of j from $m + d + 1$ to $m + d + 8$ are formed, corresponding to time delays of $(m + d)\Delta\tau$ to $(m + d + 7)\Delta\tau$. (The first channel of the autocorrelator is the zero time delay point.)

As a specific example of commercially available hardware, we describe the Malvern K7025 correlator used to acquire the data presented in this chapter. The correlator can be purchased with 64, 128, 192, or 256 basic channels. Our machine has 128 basic channels. Delay channels are added to the correlator by removing a block of eight data channels and replacing them with a 64-element delay card. We have removed the 16 data channels that immediately precede the last eight channels and replaced them with 128 delay channels. The effective correlator is shown in Fig. 1. The autocorrelation function is measured for 104 values of delay time $(0-103\Delta\tau)$. No computations take place for delay times between 103 and $231\Delta\tau$. The eight channels used to represent the baseline are then measured at delay times of $232-239\Delta\tau$.

To accurately measure the baseline, one must select an experimental sample interval $\Delta\tau$ that is sufficiently large to ensure that the correlation function has decayed to the baseline for delay times on the order of $230\Delta\tau$. The sample interval must not be too large, however, as the autocorrelation function will decay to the noise level before the end of the data accumulation channels is reached. Having data at or below the noise level represents a non-optimum use of the correlator and should be avoided as it leads to processing problems when one feeds the measured function into a computer for processing.

Details of the equipment used to perform the measurements presented here are given in Fig. 2. A Malvern RR102 spectrometer consists of a temperature-

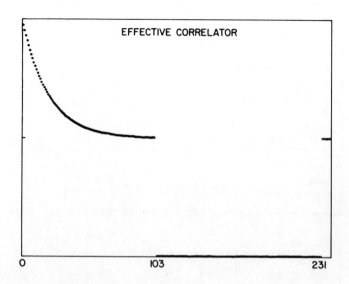

FIGURE 1. Effective correlator: the autocorrelation
function is measured for 104 values of
delay time (0-103Δτ). No information is
retained for 128 delay intervals, and then
eight baseline measurements are made for
delay times of 232-239Δτ.

controlled sample chamber that holds an index matching fluid (distilled water for these measurements)
and a photomultiplier assembly mounted on a movable
arm to allow selection of the scattering angle,
apertures for definition of the volume of sample
viewed to adjust coherence areas, and an amplifier
and a discriminator for generation of electrical
pulses compatible with the input of the K7025 correlator. Light from a 15 mW Spectra-Physics helium-
neon laser (model 124B) is focused into the sample by
a lens (not shown). A Malvern RR56 temperature
controller holds the temperature of the water bath to
± 0.05°C and assures a constant viscosity for the
fluid holding the particles of interest in suspension. High voltage for the PMT is supplied by a

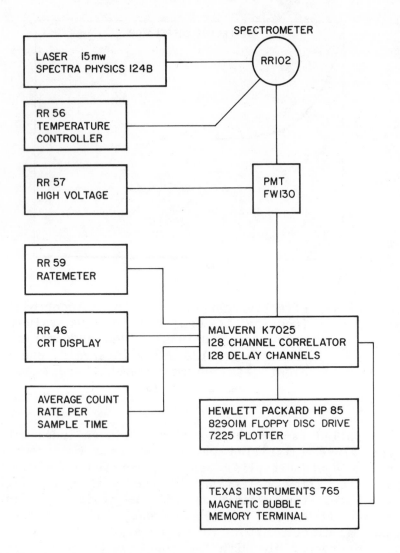

FIGURE 2.   Block diagram of equipment.

Malvern RR57 high-voltage supply. The actual PMT
tube is an FW130 that was specially selected by
Malvern for low after pulsing and dead time.

The autocorrelation function of the scattered light is measured by a Malvern K7025 correlator with 128 channels and 128 delay channels. The PMT count rate is monitored by a Malvern RR59 ratemeter, and the accumulated correlation function is displayed on a Malvern RR46 CRT display. We have also developed an instrument that allows monitoring of the average number of counts in a sample time. This number must be kept less than 15 or the photon count register overflows; in practice, it is maintained at 1 or less.

Acquired autocorrelation data are transferred to a local Hewlett-Packard HP-85 desk-top computer with dual floppy disks and a 7225 plotter for storage and some limited processing. The measured baseline can be extracted by the HP-85, and plots of the log of the measured autocorrelation function minus the measured baseline can be displayed. Extraction of the particle size distribution requires the use of a mainframe computer, and we use a Texas Instruments 765 magnetic bubble memory terminal for the data transfer. The information from the K7025 is transferred into the 765 terminal through the serial RS232 port at a 2400 baud rate, providing rapid storage of data and allowing another acquisition run to be started quickly. The stored data are later transferred to the XEROX RTCC computer system by use of the 300 baud modem in the memory terminal. Processing of the data and generation of graphical output of the results are performed on a SIGMA 9 computer.

An example of the sort of data that can be obtained with this system are now presented. The sample consists of a dispersion of two different polystyrene latices described in Section VI. The accumulation registers in the Malvern correlator are 24 bits in length, so numbers as large as 16 million can be accumulated before the registers overflow. The summation registers in the K7025 correlator are actually 28 bits long, but only the most significant 24 bits are accessible to the user. The actual number of counts accumulated is 16 times larger than

222   E. F. Grabowski and I. D. Morrison

the numbers presented in this chapter. We choose to
not multiply the summation registers by 16 as it
saves a considerable amount of multiplication and
does not alter the mathematical analysis or the shape
of the log plots. The factor of 16 does, however,
become important when a theoretical baseline is
computed or when system noise is estimated [32]. The
data acquired for this example were obtained in an
11-hr run that gave a measured baseline of 8,658,628
and an initial height for the autocorrelation function of 2.7 million counts above baseline. The
sample interval was adjusted so the autocorrelation
function just started into the system noise at the
end of the data acquisition interval. A linear plot
of the measured autocorrelation function minus the
measured baseline is presented in Fig. 3. The actual
data points are represented by the circles, and the
line represents a fit to the data with the NNLS
algorithm. The linear plot does not provide much
information as it is difficult to see what is happen-
ing in the baseline area. A logarithmic plot of the
same autocorrelation function is presented in Fig. 4.
    A plot of the logarithm of the measured auto-
correlation function minus the measured baseline as a
function of channel number provides quite a bit of
information.   As pointed out in Section III, the
logarithm of the square root of the measured autocorrelation function minus the measured baseline is
formally equal to the cumulant generating function.
The initial slope is the weighted average of the
decay constants, and the initial curvature is related
to the spread of decay constants. A monodisperse
system would have a linear decay. Curvature in the
function indicates polydispersity. The log plot also
provides an indication of the noise in the data.
Small systematic variations in the autocorrelation
function such as the periodic fluctuation caused by
convection currents in a sample that absorbs too much
light are readily observed.
    The decay of the autocorrelation function for
this system has been followed for three orders of

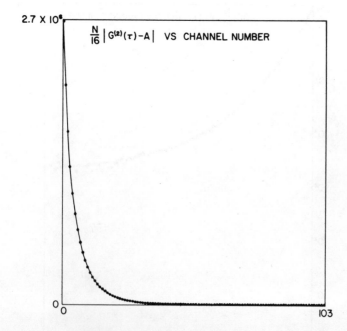

FIGURE 3.   Sample data: output   of   correlator,
$N/16[G^{(2)}(\tau) - A]$ versus channel number.
Details   are   given   in   Table 1, sample G.

magnitude.   For a monodisperse system, we would be
measuring the decay over seven time constants before
reaching the noise level.

The   data   for   the   rest   of   this   chapter   are
presented   in   a   physically   more   compact   form   that
needs   careful   explanation.   We   plot   the   logarithm of
the   contents   of   the   correlator   summation   channels
minus   the   measured   baseline   as   a   function   of   the
channel number.   These   summation   channel   numbers   are
equivalent to

$$\frac{N}{16} G^{(2)}(j\ \Delta\tau) = \frac{NA}{16} [1 + \beta\ |g^{(1)}(j\ \Delta\tau)|^2]\ ,\quad (16)$$

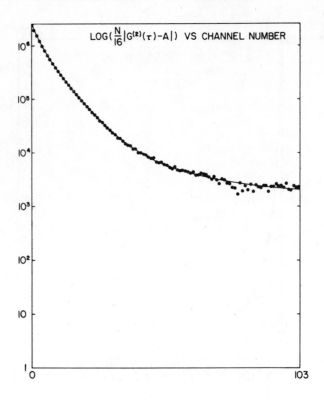

FIGURE 4.   Sample data: same   as   Fig. 3,   except
             semilog plot.

where $G^{(2)}(j\ \Delta\tau)$ is defined by Eq. (15), N is the
number of sample intervals in the given experiment,
and A is the baseline [32].   The semilogarithm graphs
were generated with an autoscale program that adjusts
the scale so that the data fill the available graph-
ing area.   The ordinate will always be logarithm to
the base 10 of the quantity identified in the
figures, but the origin and the maximum value vary
from graph to graph.   Values of the largest and
smallest ordinate points as well as the measured
baseline, the number of sample intervals, and the
total number of photons counted are provided for each

measurement in Table 1. The delay time τ is the channel number multiplied by the sample interval.

VI.   EXPERIMENTAL RESULTS

We now review the preparation and characterization of a heterodisperse sample with known properties. Polystyrene latex spheres were obtained from Dow Diagnostics. The spheres were dispersed in distilled water that had been filtered through a 0.1 μm absolute filter to remove dust. Monodisperse solutions of two different sizes were first prepared. The polystyrene concentration was adjusted so that the solutions scattered about the same amount of light at 90°. These monodisperse samples were characterized and then mixed to form a heterodisperse sample. The 0.085 μm diameter spheres (lot no. SP2N) were dispersed at a concentration of 0.4 mg/g. The 0.261 μm-diameter spheres were dispersed at a concentration of 0.04 mg/g.

All measurements were performed with a 90° scattering angle with the sample temperature held at 30.8°. The measured autocorrelation function for the two monodisperse solutions and their 1:1 mixture are presented in Fig. 5. The sample time was set at 50 μsec/channel with run times of about 7 hr. The particle size distributions used to generate the fit are also presented. The histogram representation was chosen to emphasize the fact that we are selecting a set of discrete particle sizes and asking how many of each particle are present. The vertical lines for each box are drawn at points halfway between the selected particle sizes. The width of the bins varies because the program automatically adjusts the spread of the 50 particle sizes and sometimes reduces the number of particles (and hence increases the spacing) to obtain a good fit.

The measured autocorrelation function for the 0.085 μm sample begins as a linear decay and then reaches the system noise level. This would not be

TABLE 1. Details for Data in Illustrations

	MAX.[a]	MIN.[b]	TOTAL[c]	AVE.[d]	$\Delta\tau$[e]	NUMBER[f]
A	4,279,001	4735	329,426,622	11,700,864	50	584,697,324
B	450,621	398	90,425,837	1,130,874	50	457,038,966
C	119,339	1	52,970,740	353,597	50	500,000,000
D	4,279,001	4735	329,426,622	11,700,864	50	584,697,324
E	1,940,701	1462	174,302,074	6,249,614	100	306,420,593
F	3,430,644	430	236,680,157	12,427,006	150	283,756,648
G	2,699,711	1760	233,800,616	8,658,628	100	397,274,996
H	133,786	1	1,733,815	436,323	100	20,000,000
I	41,885	1	3,560,414	132,918	100	6,000,000

[a]Maximum count.
[b]Minimum count.
[c]Total number of photons counted.
[d]Average of last eight channels (measured baseline).
[e]Sample interval in microseconds.
[f]Number of sample intervals. (The run time is NUMBER × $\Delta\tau$.)

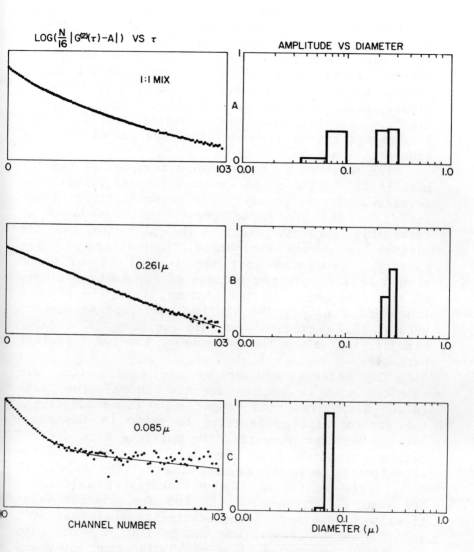

FIGURE 5. Mixing study: the correlator output as a function of channel number and the results of the non-negatively constrained best fit [Eq. (9)] to the data for (a) a one-to-one mix of 0.085 μm PS and 0.261 μm PS, (b) the 0.261 μm PS alone, and (c) the 0.085 μm PS alone. Details are given in Table 1.

considered an optimum data set for this sample because most of the function is at the noise level, but we include it on this time scale for comparison with the other two samples. The 0.261 μm sample represents what we consider an optimum adjustment of sample time. The autocorrelation function is an almost perfect linear decay (indicating a monodisperse sample), and the measured function just reaches the system noise level at the end of the data acquisition period.

The measured autocorrelation function of the 1:1 mixture of the two monodisperse solutions illustrates several important points. The autocorrelation functions for the two monodisperse solutions begin as reasonably straight lines. The plot for the 1:1 mixture is curved everywhere, indicating polydispersity. Note also that the initial slope, which should be the weighted average of the initial slopes of the components of the mixture, does lie between the initial slopes for the two monodisperse components. The particle size distribution is clearly bimodal, indicating the presence of the two different particles.

The effects of varying the sample time are examined next to demonstrate how the relative noise in the data grows with longer delay times and causes the autocorrelation function to decay to the noise level. We also show that the particle size distribution extracted from the data is stable over a significant range of sample times. (The answer is not overly sensitive to the correlator settings.) The same 1:1 mixture of the two monodisperse solutions was used here. The scattering angle was 90°, and the sample temperature was held at 30.8°C. Data were taken for sample times of 50, 100, and 150 μsec/ channel with run times of 8, 8.5, and 11.5 hr, respectively. Results are presented in Fig. 6. As the sample time is increased, the curvature of the autocorrelation function increases and the data points for large delay times seem to deviate more from the computer fit.

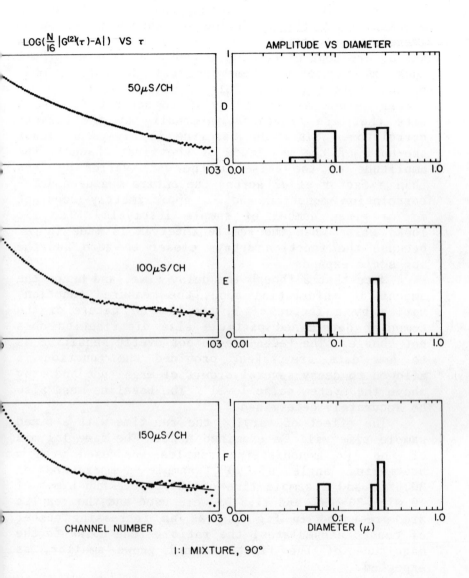

$\text{LOG}(\frac{N}{16}|G^{(2)}(\tau)-A|)$ VS τ

AMPLITUDE VS DIAMETER

50 μS/CH

100 μS/CH

150 μS/CH

CHANNEL NUMBER

DIAMETER (μ)

1:1 MIXTURE, 90°

FIGURE 6. Variation of sample time: effect of variations in sample time on the correlator output and computed least-squares distribution for sample A in Fig. 5. Details are given in Table 1.

229

After studying many samples, we find experimentally that the deviation of the measured autocorrelation function from the computer fit varies as the square root of the total number of counts for all these measurements. Remember that the total number of counts in a given correlator channel is the 24 bit number we use multiplied by 16 to account for the 4 bits that are truncated internally by the Malvern correlator. Since the baseline is always at least one-half of the magnitude of the first channel, the amplitude of the noise in the data varies by less than a factor of $\sqrt{2}$ across the entire measured autocorrelation function and is approximately constant for a given number of sample intervals. The fit looks worse when the sample interval is made longer because the function decays closer to zero and the log scale expands.

Note that although the delay time, and hence the amount of information about the measured function, varies by a factor of 3, the bimodal nature of the computer determined particle size distribution does not change. The technique is not overly sensitive as to how data are taken, provided the function is allowed to decay several orders of magnitude and stay above the system noise level. The baseline must also be accurately determined.

The effect of varying the run time with a fixed sample time will be examined next. The same 1:1 mix of the two monodisperse samples was used with a scattering angle of 90°, sample temperature of 30.8°C, and a sample time of 100 μsec. Run times of 10 min, 33 min, and 11.5 hr were used and the results are presented in Fig. 7. As an increasing number of counts accumulates, the ratio of the noise to the magnitude of the first channel grows smaller, as expected.

Some general characteristics of the new technique are illustrated here. There is a lot of noise in the 10 min data, but the algorithm still extracts the bimodal nature of the sample. A nonlinear program would blow up trying to run a curve through data

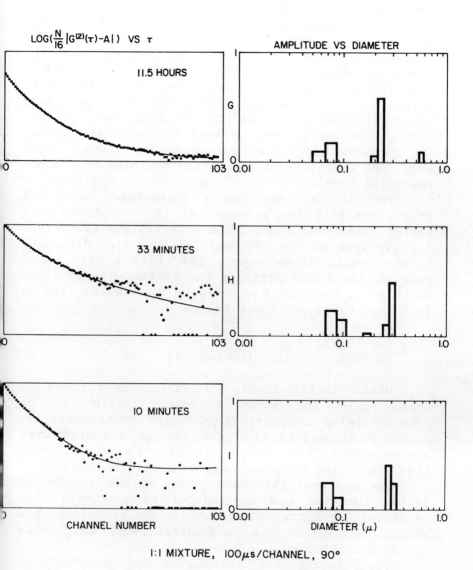

LOG($\frac{N}{16}$|G$^{(2)}$(τ)−A|) VS τ

AMPLITUDE VS DIAMETER

11.5 HOURS

G

0   103 0.01   0.1   1.0

33 MINUTES

H

0   103 0.01   0.1   1.0

10 MINUTES

0   103 0.01   0.1   1.0
CHANNEL NUMBER   DIAMETER (μ)

1:1 MIXTURE, 100μs/CHANNEL, 90°

FIGURE 7.   Variation in run time: effect of correlator run time on the correlator output
and computed least-squares distribution
for the same sample as in Fig. 6. Details
are given in Table 1. The 11.5 hr run is
shown in Fig. 4 in greater detail.

like these. The 33 min run gave a best fit with a small peak, between the two known particle sizes. When noise is present, there will sometimes be a distribution that fits the data better than the actual distribution of particles. This is avoided in practice by keeping the data above the system noise level. If only 30 min were available for data acquisition on this particular sample, the time per channel should be cut in half to avoid the noise. The previous experiment on variation in sample time shows the stability of the answer for data kept above the noise level.

The 11.5 hr run has a reasonable amount of noise, extracts the presence of the two known particles, and finds a small number of larger particles. A close look at the previous particle size distributions reveals three others that have a very small peak at the large particle end of the distribution. There are a few dust particles in our samples, and we feel that these peaks are real.

VII.   CONCLUSIONS

Quasi-elastic light scattering can be used to characterize the particle size distribution of heterodisperse suspensions when the autocorrelation function of scattered light can be measured over approximately three decades of decay and the infinite time limit, the baseline, can be determined.

The mathematical description of the autocorrelation function can be reduced to a simple non-negatively constrained minimization technique for which convergence is guaranteed and uniqueness criteria are known.

The experimental techniques and mathematical analyses are demonstrated on samples of known heterogeneity.

ACKNOWLEDGMENTS

We would like to thank Virginia Dotschkal, who wrote the graphics routines used to generate the plots used in this chapter. We would also like to acknowledge many helpful discussions with Dr. Bruce Weiner and Dr. Walther Tscharnuter of the Brookhaven Instruments Corporation regarding proper selection and operation of the equipment used in this work.

REFERENCES

1. H. Z. Cummins and E. R. Pike, Eds., *Photon Correlation and Light Beating Spectroscopy*, Plenum Press, New York, 1974.

2. B. Crosignana, P. DiPorto, and M. Bertolotti, *Statistical Properties of Scattered Light*, Academic Press, New York, 1975.

3. B. J. Berne and R. Pecora, *Dynamic Light Scattering*, Wiley, New York, 1976.

4. H. Z. Cummins and E. R. Pike, Eds., *Photon Correlation Spectroscopy and Velocimetry*, Plenum Press, New York, 1977.

5. G. D. J. Phillies, J. Chem. Phys. 72 (11), 6123-6133 (1980).

6. R. A. Gibbons, Nature 200 (4907), 665-666 (1963).

7. D. E. Koppel, J. Chem. Phys. 57 (11), 4814-4820 (1972).

8. J. G. McWhirter and E. R. Pike, J. Phys. A: Math. Gen. 11 (9), 1729-1745 (1978).

9. E. Jackeman and E. R. Pike, J. Phys. A *1*, 128-138 (1968); *2*, 115-125 (1969); *2*, 411-412 (1969).

10. H. Z. Cummins, in reference 1, p. 303.

11. Es. Gulari, Er. Gulari, Y. Tsunashima, and B. Chu, J. Chem. Phys. *70* (8), 3965-3972 (1979).

12. H. Z. Cummins, in reference 1, pp. 295-297.

13. G. Grehan and G. Gouesbet, Appl. Opt. *18* (20), 3489-3493 (1979).

14. F. B. Hildebrand, *Methods of Applied Mathematics*, Prentice-Hall, Englewood Cliffs, N.J., 1952, p. 411.

15. M. R. Smith, S. Cohn-Sfetcu, and H. A. Buckmaster, Technometrics *18* (4), 467-482 (1976).

16. C. Y. Cha and K. W. Min, J. Polym. Sci. Polym. Phys. Ed. *19*, 1471-1473 (1981).

17. M. Abramowitz and I. A. Stegun, *Handbook of Mathematical Functions*, Dover Publications, New York, 1965, pp. 927-928.

18. C. L. Lawson and R. J. Hanson, *Solving Least Squares Problems*, Prentice-Hall, Englewood Cliffs, N.J., 1974.

19. D. S. Thompson, J. Phys. Chem. *75* (6), 789-791 (1971).

20. R. L. McCally and C. B. Bargeron, J. Chem. Phys. *67* (7), 3151-3156 (1977).

21. B. Chu, Es. Gulari, and Er. Gulari, Phys. Scr. *19*, 476-485 (1979).

22. S. W. Provencher, J. Hendrix, L. De Mayer and N. Paulussen, J. Chem. Phys. *69* (9), 4273-4276 (1978).

23. S. W. Provencher, Makromol. Chem. *180*, 201-209 (1979).

24. M. J. Gaertner, P. H. Reggio, Cr. Crosby, III, R. L. Schmidt, and J. A. Mayo, J. Colloid Interface Sci. *63* (2), 259-269 (1978).

25. H. Strehlow, Adv. Molec. Relax. Interact. Proc. *12*, 29-46 (1978).

26. R. D. Dyson and I. Isenberg, Biochemistry *10* (17), 3233-3241 (1971).

27. S. Ross and I. D. Morrison, Surf. Sci. *52* (1), 103-119 (1975).

28. R. S. Sacher and I. D. Morrison, J. Colloid Interface Sci. *70* (1), 153-166 (1979).

29. S. Ross and J. P. Olivier, *On Physical Adsorption*, Wiley-Interscience, New York, 1964, pp. 124-125.

30. J. P. Olivier, Micrometrics, Inc., Norcross, Ga., private communication.

31. Reference 18, p. 269.

32. In Eq. (1) the theoretical baseline A is the square of the (average number of photons per sample interval) or the square of the (total number of photons counted/N). The Malvern correlator (1) provides eight summation channels that are averaged to provide a measured baseline A', (2) truncates the last 4 bits of all summation channels, and (3) does not divide the contents of the summation channels by N as

required in Eq. (15). Therefore, the theoretical baseline must be adjusted so that it can be compared to the average of the last eight correlator channels by a factor of $N/16$. Hence $A' = (A \times N)/16$. The total number of photons counted, the measured baseline A', and N are included in the figure captions and Table 1.

CHAPTER 8

Applications of the Histogram Method for Determination of Size Distributions of Suspended Spherical Particles

BRIAN BEDWELL, ERDOGAN GULARI, and DAVE MELIK
University of Michigan, Ann Arbor

CONTENTS

I. INTRODUCTION

The histogram method for analaysis of quasi-elastic light scattering (QELS) data was introduced as a technique for determining size distributions of polymers and suspensions of particles small in comparison to the wavelength of the incident light [1-3]. The major attractions of the histogram method

237

over other methods were its simplicity and the fact that it required no a *priori* assumptions about the shape of the distribution function.

In this chapter we report our results for suspensions of spherical particles where the average particle size *is not* small in comparison to the wavelength of the incident light.

For a dilute suspension of monodisperse spherical particles, the homodyne correlation function, which is proportional to the square of the scattered intensity, is given by

$$C(q,t) - 1 = |I(q,t)|^2 = \beta|\alpha^2 N \exp[-Dq^2 t]|^2 , \qquad (1)$$

where q is the scattering vector, t is time, $C(q,t)$ is the homodyne correlation function, $I(q,t)$ is the scattered intensity, β is an instrument constant, α is the polarizability, N is the number of particles in the scattering volume, and D is the diffusion coefficient of the spherical particles [4]. For a dilute suspension, we can use the Stokes-Einstein relationship to relate D to the particle radius

$$D = \frac{k_B T}{6\pi\eta r} , \qquad (2)$$

where k_B is the Boltzmann constant, T is the absolute temperature, η is the solvent viscosity, and r is the radius of the scattering particle. Since α is the molar polarizability, we can express it in terms of polarizability per unit volume and Eq. (1) can be rewritten as

$$C(q,t) - 1 = \beta|\alpha'^2 (\tfrac{4}{3}\pi r^3)^2 N \exp\left[\frac{-k_B T q^2 t}{6\pi\eta r}\right]|^2 \qquad (3)$$

For a polydisperse system, we need to sum the contributions of each size, and we have

$$C(q,t) - 1 = \beta |\alpha'|^2 (\tfrac{4}{3}\pi)^2 \sum_i r_i^6 N_i \exp\left[\frac{-k_B T q^2 t}{6\pi\eta r_i}\right]|^2 \quad (4)$$

Equation (4) can be written as an integral over the distribution function $f(r)$:

$$C(q,t) - 1 = \beta' |\int_0^\infty f(r) r^6 \exp\left[\frac{-k_B T q^2 t}{6\pi\eta r}\right] dr|^2 \; . \quad (5)$$

If the particles are large, we need to include the Mie factor $P(q,r)$ to obtain

$$C(q,t) - 1 =$$

$$\beta' |\int_0^\infty f(r) r^6 \exp\left[\frac{-k_B T q^2 t}{6\pi\eta r}\right] P(q,r)\, dr|^2 \; , \quad (6)$$

where β' is a new constant combining all the size-independent proportionality constants. From Eq. (6) we see that it is possible to obtain the distribution function $f(r)$ from the autocorrelation function. Unfortunately, inversion of Eq. (6) to obtain $f(r)$ is not easy. It is very sensitive to noise in the data and other problems that plague integral equations of this kind [5,6].

A better method is to fit an $f(r)$ to the data and obtain the parameters defining the distribution function. The details of this approach are given in the literature [1-3,5]. In the histogram method the distribution function $f(r)$ is approximated by a histogram of finite steps:

$$f(r) = \sum_i a_i(r_i) \; . \quad (7)$$

The integral over r in Eq. (6) is not analytic and as a result consumes significant amounts of computational time. To circumvent this problem, we make the following approximation:

For $r_i < r < r_i + \Delta r$,

$$\exp\left[-A/r\right] \cong \exp\left[-A/\bar{r}_i\right] \tag{8}$$

where $\bar{r}_i = r_i + \Delta r/2$ and Δr is the width of each histogram step. The Mie factor is also approximated by its average value for each step.

$$P_i(q,r) = P(q,\bar{r}_i) \tag{9}$$

With these approximations Eq. (6) can be rewritten as

$$C(q,t) - 1 =$$

$$\beta' \mid \sum_i a_i \int_{r_i}^{r_i+\Delta r} r^6 dr \, \exp\left[\frac{-k_B T q^2 t}{6\pi\eta\bar{r}_i}\right] P(q,\bar{r}_i) \mid^2 . \tag{10}$$

We should note that when $\Delta r/r \leq 0.05$, the above approximations are very good.

The coefficients a_i , which are proportional to the number of particles with sizes between r_i and $r_i + \Delta r$, are found by a nonlinear least-squares minimization procedure:

$$\frac{\partial x^2}{\partial a_i} = \frac{\partial}{\partial a_i} \sum_j \frac{1}{\sigma_j} \left[Y_{j,m}(t) - Y_j(t)\right]^2 = 0 , \tag{11}$$

where $Y_{j,m}$ is the measured and Y_j is the calculated value of the correlation function at time t . The output of the least-squares program is a set of coefficients a_i or relative number concentrations. Two points are worth noting: (1) the conversion from the diffusion coefficients to the particle radius is well defined in the absence of interparticle interactions; and (2) the contribution of a sphere of size r

to the homodyne correlation function is proportional to $r^{12} \times P(q,r_i)^2$. Thus for equal concentrations and scattering factors, the ratio of contributions by particles differing in size by a factor of 2 is 4096. This enormous difference in the scattering power is the ultimate limiting factor for the use of QELS as a means of determining size distributions of spherical particles.

II.   EXPERIMENTAL

All correlation functions were measured with a Malvern 64-channel multibit correlator.

Emulsions were prepared by sonicating deionized water and octacosane $(C_{28}H_{58})$ at a temperature above the melting point of octacosane and then cooling the resulting emulsion to room temperature. Scanning electron micrographs were obtained after gold coating the samples.

Latex sphere suspensions were prepared by use of Dow latex spheres of nominal diameters 850 and 4970 Å. Dilutions were with 0.1% sodium dodecyl sulfate in H_2O. All latex samples were centrifuged, filtered, and sonicated to eliminate aggregates. Latex electron micrographs were obtained by transmission electron microscopy.

For light scattering, all latex and emulsion concentrations were less than 0.001% by volume to minimize the effects of interparticle interactions and multiple scattering.

III.   DATA ANALYSIS

Analysis by the histogram method requires the user to establish the positions of the histogram steps in radius space and then to optimize the step heights for that configuration. Histogram step positions and widths are not adjusted automatically.

The computer program written for this chapter takes the specified grid of radii and computes the average Mie factor for each histogram step if the particle is outside the Rayleigh limit. Mie factors are generated from the intensity functions of rigorous Mie theory [7,8], which depend on combinations of the Ricatti-Bessel functions and angular functions related to the associated Legendre polynomials. The procedure used here was modeled after those described by Denman et al. [9] and Penndorf [10].

For each data set, this process was repeated with different choices of histogram step positions and widths. A best configuration was chosen to be the one that gave the best correlation coefficient while keeping the uncertainty in each parameter (step height) to at least one order of magnitude less than the parameter itself. Increasing the number of histogram steps beyond this optimum generally improved the correlation coefficient but also increased the uncertainties.

Some prior knowledge of where to look for a distribution is helpful in choosing a configuration of histogram steps. To this end, all data sets were first analyzed by the method of cumulants.

IV. RESULTS

Our first test was on a narrow distribution in the Rayleigh limit, to determine that histogram steps in radius space were reliably determined without the added complication of Mie factors. Figure 1 shows the results of histogram analysis of an autocorrelation function generated from light scattered by a water-in-oil (W/O) microemulsion of AOT, 2% NaCl brine, and heptane. This distribution has a mean radius that is 29 Å and a standard deviation of 3 Å. Cumulants analysis of the same data gives 31 Å for the radius and 3 Å for the standard deviation.

Now introducing the Mie factors, we attempted to evaluate the accuracy of the method with monodisperse

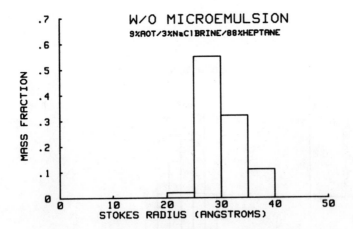

FIGURE 1.   Size distribution of a monodisperse microemulsion system.

latex sphere suspensions. Figures 2a and 2b show the
mass fraction as a function of Stokes radius. These
distributions compare well with the mean and variance
values given by cumulants analysis. The average
sizes measured by light scattering are less than
those claimed by the manufacturer. This is probably
a result of a modification in the distribution when
the samples were centrifuged and filtered to remove
aggregates. Electron micrographs (Figs. 3 and 4)
suggest that the small spheres are more polydisperse
than the large spheres. This is consistent with the
light-scattering results.

Figures 5a-c show the results of histogram
analysis with mixtures of latex spheres. At a scattering angle of 90° the average Mie factor for the
small sphere distribution is about 0.9, whereas that
for the large spheres is about 0.003.

The scattered intensity for any size should be
proportional to

$$P(r,q) \times r^3 \times \text{mass fraction} .\tag{12}$$

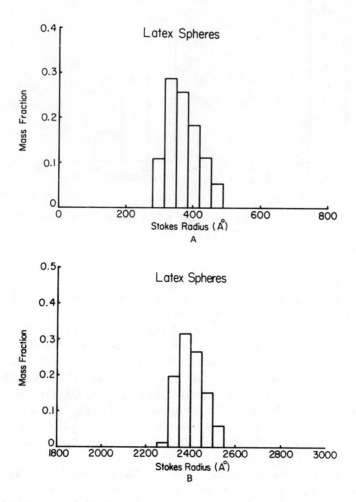

FIGURE 2. (a) The mass average radius for this distribution is 370 Å, and the standard deviation is 47 Å; (b) the mass average radius for this distribution is 2400 Å, and the standard deviation is 59 Å.

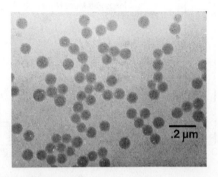

FIGURE 3.   Transmission electron micrograph of small
Dow latex spheres (lot No. 5P2N, nominal
diameter 850 Å).

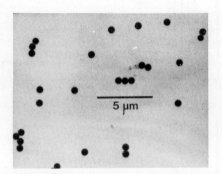

FIGURE 4.   Transmission electron micrograph of large
Dow latex spheres (lot No. 8M4P, nominal
diameter 4970 Å).

For this bimodal system, we have

$$\frac{P(r_1,q)r_1^3}{P(r_2,q)r_2^3} = \frac{(0.003)(2400)^3}{(0.9)(370)^3} = 1.02 \ . \tag{13}$$

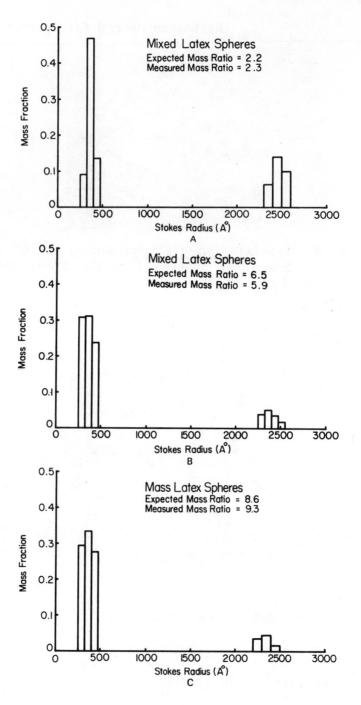

FIGURE 5. Histograms analysis for several different mixtures of two sizes of Dow latex spheres.

Therefore, the ratio of the intensities due to each size is essentially the ratio of the mass fractions.

When the square of the ratio of the intensities increases past the noise limit of the experiment, we would expect the method to begin to fail. The auto-correlation function is influenced by I^2. When I_2/I_1 = 10, we have I_2^2/I_1^2 = 100. The average noise in the data is usually about 0.1%. As the mass ratio in this bimodal system increases, the weaker signal approaches the noise limit.

For the three trials in Fig. 5, the histogram analysis worked well in reproducing the expected distributions. For distributions where the mass ratio exceeded 10, histogram analysis did not succeed in giving the proper mass fractions, although it did reproduce the correct range of sizes.

For a test of histograms on wide unimodal distributions, we chose an emulsion of octacosane in water. The distribution of sizes determined by electron microscopy is given in Fig. 6. Figure 7 shows a typical electron micrograph. The emulsion particles are not all spherical but appear to be close enough to be approximated as such. For this system, histogram analysis indicated the same range of sizes as the scanning electron micrograph (SEM) distribution, but not the same relative amounts of each size. As shown in Fig. 8, histogram analysis suggests the distribution is dominated by the smaller sizes.

This discrepancy may be due to the fact that these emulsions age rapidly by creaming and that the light scattering was performed on an emulsion that was several days older than the sample used for electron microscopy. Alternatively, this could be due to a lack of information about the larger sizes in the autocorrelation function, if it contains only 64 channels spanning a region dominated by short time relaxations. Still another possibility is that in the SEM analysis the magnification ratios used for large particles made the small partiles almost invisible, leaving many of them uncounted.

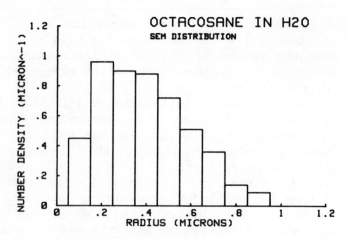

FIGURE 6.   Emulsion   particle   size   distribution
determined by SEM.

The octacosane emulsion was centrifuged at 3000 rpm for several hours and then characterized by light scattering again. Figure 9 shows that, as can be expected, the distribution is considerably narrower and the mean is much lower than before centrifuging.

As a final note, we want to show the difference in the quality of fit between a histogram and a two-parameter cumulants. Figure 10 shows the residuals from these two types of data analysis. These fits are for the system represented in Fig. 5a. The two-parameter cumulants expansion gives only a mean and variance for the distribution and as such does not give an adequate picture of the bimodal distribution.

ACKNOWLEDGMENT

This work was supported by the CPE division of NSF. We would like to thank Dr. George Brooks for the transmission electron micrographs and Mr. Rick Tadsen for his assistance with the scanning electron microscope.

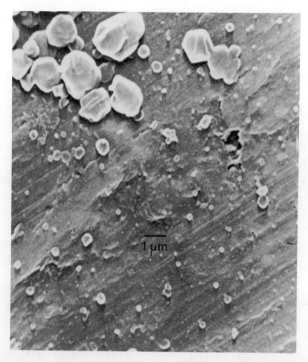

FIGURE 7. Scanning electron micrograph of an octacosane emulsion after drying and gold
coating.

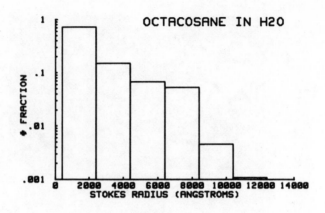

FIGURE 8. Histogram analysis for octacosane emulsion. Note the logarithmic scale.

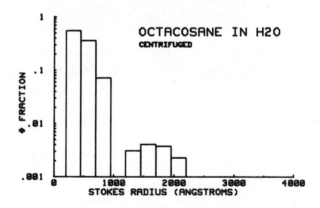

FIGURE 9. Histogram analysis for centrifuged octacosane emulsion.

250

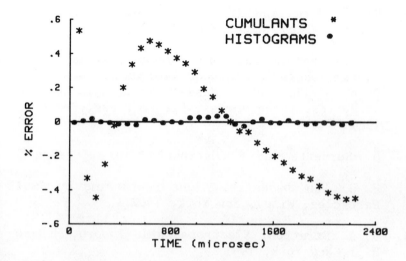

FIGURE 10.   Percent error is defined here as the difference between the measured value at a particular time and the value predicted by the fitted curve for the same time normalized to the sum of the measured value and the baseline.

REFERENCES

1. Es. Gulari, Er. Gulari, Y. Tsunashima and B. Chu, Photon Correlation Spectroscopy of Particle Distributions, J. Chem. Phys. *70* (8), 3965-3972 (1979).

2. B. Chu, Es. Gulari and Er. Gulari, II. Details of Histogram Approach and Comparison of Methods of Data Analysis, Phys. Scr. (Sweden) *19*, 476-485 (1979).

3. Er. Gulari, Es. Gulari, Y. Tsunashima and B. Chu, Polymer Diffusion in a Dilute Theta Solution: I. Polystyrene in Cyclohexane, Polymer *20*, 347-355 (1979).

4. B. J. Berne and R. Pecora, *Dynamic Light Scattering*, Wiley, New York, 1976.

5. E. R. Pike, in *Scattering Techniques Applied to Supramolecular and Nonequilibrium Systems*, S. H. Chen, B. Chu and R. Nossal, Eds., NATO Advanced Study Institutes Series, Plenum Press, New York, 1981.

6. D. Sornette and N. Ostrowsky, in reference 5.

7. H. C. Van de Hulst, *Light Scattering by Small Particles*, Wiley, New York, 1962.

8. J. A. Stratton, *Electromagnetic Theory*, McGraw-Hill, New York, 1941.

9. H. H. Denman, W. Heller and W. J. Pangonis, *Angular Scattering Functions for Spheres*, Wayne State University Press, Detroit, 1966.

10. R. B. Penndorf, New Tables of Mie Scattering Functions for Spherical Particles, Geophysical Research Papers No. 45, Part 6, ASTIA Document No. AD-98772, March 1956.

PART THREE
EXTENDED AND COMPLEMENTARY METHODS

CHAPTER 9

Electrophoretic and Frictional Properties of Particles in Complex Media Measured by Laser Light Scattering and Fluorescence Photobleaching Recovery

B. R. WARE, * **DONNA CYR, SRIDHAR GORTI,**
and FREDERICK LANNI
Syracuse University,
Syracuse, New York

CONTENTS

---

† To whom correspondence should be addressed

I.   INTRODUCTION

The diffusive motion of a particle dissolved or suspended in a solvent may be characterized by the mean-square displacement $\langle X^2 \rangle$ after a period of time t.   In 1905 Einstein [1] demonstrated that this relationship may be expressed in one dimension by the equation

$$\langle x^2 \rangle = 2Dt \ , \tag{1}$$

where D is a constant, called the *diffusion coefficient* or the *diffusion constant*, whose magnitude is determined by the simple relationship

$$D = \frac{k_B T}{f} \ , \tag{2}$$

where k_B is Boltzmann's constant, T is absolute temperature, and f is the friction constant of the particle, which is determined generally by the linear dimension of the particle and the solvent viscosity η.   For a sphere of radius R,

$$f = 6\pi\eta R \ . \tag{3}$$

A second means of characterizing diffusion, and one that is often more accessible experimentally, is to realize that a concentration gradient of particles will relax in time due to the random nature of diffusive motion.   The relevant equation, often called the *diffusion equation* or *Fick's second law* [2], for one dimension is

$$\frac{\partial C}{\partial t} = \frac{\partial}{\partial x} D \frac{\partial C}{\partial x} , \qquad\qquad (4)$$

where C is concentration.

The constant D' is properly called the mutual diffusion constant, and it differs from the tracer diffusion coefficient D because the gradient in concentration of solute particles creates gradients in other thermodynamic quantities that may affect the average motion of a particle at a particular position in the gradient. A general relationship for the mutual diffusion coefficient has been written as [3]

$$D' = \frac{1}{f} \frac{\partial \mu}{\partial \ln C} (1 - \phi) , \qquad\qquad (5)$$

where μ is the chemical potential and ϕ is the volume fraction of the diffusing species. Clearly, in the limit of infinite dilution, the mutual diffusion coefficient becomes equal to the tracer diffusion coefficients; but for the conditions of many experiments, the distinction between the two is significant.

If the dissolved particles bear an electrical charge with respect to the suspending medium, it is well known that the application of an electric field will induce the particles to migrate toward the electrode of opposite polarity with a velocity given by

$$\mathbf{V} = u\mathbf{E} , \qquad\qquad (6)$$

where $\mathbf{E}$ is the electric field and u is the electrophoretic mobility, a constant that is characteristic for a given particle and a given set of solution conditions, whose relationship to more fundamental molecular properties may be written approximately as [4]

$$u = \frac{Ze}{f} \frac{X_1(\kappa R)}{1 + \kappa R} , \tag{7}$$

where Z is the number of charges on the surface of the particle, e is the magnitude of a unit charge, κ is the Debye-Hückel constant [5], and $X_1(\kappa R)$ is Henry's function [6], which ranges from 1.0 for $\kappa R \ll 1$ to 1.5 for $\kappa R \gg 1$. The electrophoretic mobility is thus the hydrodynamic parameter that one measures to characterize the electrical charge on the surface of a suspended particle. A number of classical techniques for both analytical and preparative electrophoresis have been applied over the past several decades for the electrophoretic characterization of biological, natural, and synthetic particles and polyelectrolytes [7-9].

Other chapters in this volume have demonstrated amply the fact that the mutual translational diffusion coefficients of particles in suspension may be determined accurately by use of dynamic light scattering and have illustrated the sophistication of the methodology and data analysis techniques for dealing with such issues as sample polydispersity, molecular rotation and vibration, and interparticle interactions. We seek to demonstrate in this chapter the additional information to be gained with the use of two newer techniques that are related both in concept and in the type of equipment required. The first of these is accomplished simply by the application of an electric field to induce electrophoretic drift in a dynamic light-scattering measurement. This technique, now generally called *electrophoretic light scattering* (ELS), provides new information about the surface charge of the particles and the overall heterogeneity of the sample. The second technique, usually called *fluorescence photobleaching recovery* (FPR), permits determination of the tracer coefficient of specifically labeled particles, even in the presence of many other types of particle or in a complex matrix, and also determines directly the fraction of particles that are not free to move on a given time scale. We believe that the combination of those techniques greatly expands the arsenal of the physical scientist seeking to characterize the physical and chemical properties of charged particles and polyelectrolytes in solution or suspension, and we present preliminary data on several systems of interest that we believe will illustrate this point. Since the ELS and FPR techniques are not discussed elsewhere in this volume, we begin with brief discussions of the principles and methods of each technique, with sufficient citations to permit access to the current literature.

II.   ELECTROPHORETIC LIGHT SCATTERING

The measurement of velocities by means of measuring the Doppler shift of scattered or reflected radiation is as common as highway radar. When laser radiation is the carrier wave, the technique is called laser Doppler velocimetry (LDV). Many applications of LDV and various modifications of the basic methodology have been reported in the voluminous literature on this subject, of which there are several recent reviews [10-12]. The application of the LDV principle to the detection of the electrophoretic motion was first reported by Ware and Flygare in 1971 [13], and the technique and its many applications have been discussed in detail in several reviews [14-18]. We present here only the basic principles and a brief overview of the apparatus and methodology.

The magnitude of the Doppler shift Δv induced in a carrier wave of frequency v_0 and velocity c by a relative motion of velocity V is given by

$$\Delta v = \frac{V}{c} v_0 .$$

(8)

In the case of a light-scattering experiment there are actually two Doppler shifts to consider: one with respect to the source and the other with respect to the detector. The geometry is depicted in Fig. 1. When the difference is taken between the Doppler shifts caused by motion of the scattering particle relative to the source and detector, the net Doppler shift is predicted from the relationship

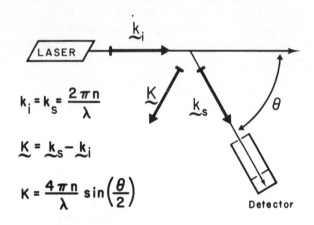

FIGURE 1.   Fundamental geometry of a light-scattering experiment, defining the scattering angle θ and the scattering vector $\mathbf{K}$ in terms of the incident wavevector $\mathbf{k}_i$ and the scattered wavevector $\mathbf{k}_s$. Here n is the average refractive index of the scattering medium and λ is the wavelength of the laser light *in vacuo*.

$$\Delta v = \frac{(\mathbf{k}_s - \mathbf{k}_0) \cdot \mathbf{V}}{2\pi} = \frac{\mathbf{K} \cdot \mathbf{V}}{2\pi} , \qquad (9)$$

where $\mathbf{K}$, the scattering vector, is the vector difference between the scattered and incident wavevectors. The magnitude K is given by

$$K = \frac{4\pi n}{\lambda_0} \sin \frac{\theta}{2} , \qquad\qquad (10)$$

where n is the refractive index, λ_0 is the incident wavelength *in vacuo*, and θ is the scattering angle. When the velocity is due to electrophoretic motion, the Doppler spectrum may be interpreted in terms of the electrophoretic velocities of the species of particles in the scattering medium, so that the ELS spectrum may be interpreted in terms of the electrophoretic histogram of the sample. The magnitudes of the Doppler shifts due to electrophoretic motion are very small (of order 100 Hz), so that the detection of the Doppler shift magnitudes must be accomplished by use of a beating technique in which unshifted laser light, called the *local oscillator*, is mixed with the scattered signal to produce an electronic beat spectrum of the same form and magnitude as the Doppler spectrum, with the carrier frequency shifted to zero by the beating process.

A block diagram of an electrophoretic light-scattering apparatus is shown in Fig. 2. The incident laser light (a low-power, inexpensive laser is generally adequate) is split into two beams, one of which illuminates the sample and the other of which travels around the chamber to be recombined with the scattered light. It is imperative that the local oscillator be collinear with the detected scattered beam and that the wave fronts of the two beams have an optimized match at the surface of the photodetector. The incident beam illuminates the particles to be analyzed in the electrophoretic chamber, which is the major additional specialized equipment necessary for an ELS experiment [17-19]. The electric field in the chamber should be provided by a constant-current power supply, which may be pulsed if desired to obtain higher electric field strengths within the constraints of an acceptable level of Joule heating. The electronic beats from the photodetector are amplified and then analyzed by either an autocorrelator or a spectrum analyzer. The

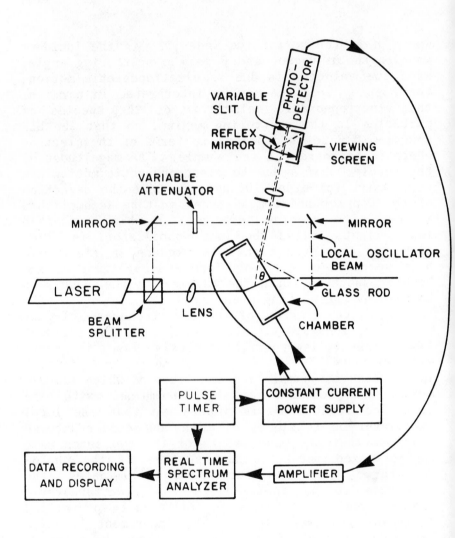

FIGURE 2.   (facing).   Block diagram of an electrophoretic light-scattering (ELS) apparatus.
Incident laser light is split into two
beams, one illuminating the sample undergoing electrophoresis in the chamber and
the other (the local oscillator) bypassing
the chamber to be recombined with the
scattered Doppler-shifted light at the
photodetector. The resulting electronic
beats are amplified and analyzed to obtain
the frequency spectrum. For operation in
a pulsed mode, a pulse timer triggers the
constant-current electrophoresis power
supply and the spectrum analyzer.

264   B. R. Ware et al.

latter is generally preferred because of the direct interpretation of frequency in terms of velocity and hence of electrophoretic mobility. The autocorrelator or spectrum analyzer should be real-time efficient, which for the low frequencies typical for ELS is not difficult to achieve. Successive spectra are collected and signal averaged for improvement of signal/noise ratio and sampling statistics. Electrophoretic spectra are generally completed within a few seconds to a few minutes, depending primarily on the scattering cross section of the sample.

The choice of scattering angle is dependent primarily on the size of the particles to be studied. Each species of characteristic mobility will contribute a characteristic Doppler shift frequency in the spectrum; the width of this peak includes a contribution due to diffusion of magnitude $D'K^2$ as discussed in other chapters in this volume. If one wishes to maximize electrophoretic resolution, it is advantageous to select a scattering angle at which the diffusion width is not greater than the instrumental resolution. For macromolecules on the size scale of proteins, this generally means selecting a low scattering angle, below 10°. For larger particles, this consideration is less critical, and there is an advantage to select a higher scattering angle to increase the magnitude of the Doppler shift. Details of the experimental design and methodology and a survey of applications may be found in recent reviews [15-19].

III.   FLUORESCENCE PHOTOBLEACHING RECOVERY

The technique of photobleaching recovery has evolved over several years, beginning with the experiment by Poo and Cone [20], who bleached half of a single rod outer segment with a short pulse of intense light and monitored the redistribution of the absorbance due to rhodopsin as a function of time. This concept was adapted to the study of motion in cell plasma membranes by labeling specific membrane components with a fluorescent group, photobleaching the fluorescence in a small region of the membrane, and measuring the redistribution of fluorescence in time [21-23]. If the bleaching pulse of light is in the form of a single spot, the recovery of fluorescence by diffusion is a predictable function that depends only on the square of the spot diameter and the diffusion coefficient of the labeled species. If any of the photobleached species are not free to move, the reequilibrium of mobile species leads to incomplete recovery of the fluorescence, and the fractional recovery is a direct measure of the fraction of mobile species. Thus an FPR measurement provides a determination of the fraction of mobile species and the diffusion coefficient of the mobile species.

An alternative to the photobleaching and monitoring of a single spot is the implementation of a periodic pattern [24]. A striped pattern photobleached into the specimen will disappear exponentially in time with a time constant determined by the period of the pattern and the diffusion coefficient of the labeled species. Persistence of some fraction of the pattern indicates an immobility of some of the labeled particles over the time period of the observation. We have modified the periodic pattern technique with the introduction of scanning detection to produce a modulation detection method with a number of significant advantages [25]. The principle of our technique is simply summarized. A periodic photobleaching pattern is produced in the sample by brief, intense illumination through a grating. The grating is then translated at constant speed through the reference beam to produce a moving pattern on the specimen. A modulation of fluorescence emission is produced as the bleached pattern and the illumination pattern fall into and out of phase. The resulting photocurrent contains an alternating-current (AC) component whose frequency is determined by the spacing and velocity of the grating and whose amplitude relative to the direct-current (DC) component is determined by the extent of photobleaching. The decay constant of the AC component of the photo-current is determined by the spacing of the grating and the diffusion coefficient of the labeled species. The major distinct advantages of the technique are: (1) the decay at the fundamental frequency is observed in real time; (2) adventitious photobleach-ing during the monitoring phase of the measurement does not perturb the determined parameters; (3) the measurement is insensitive to DC drift components; and (4) the measurement is insensitive to motion of the specimen.

The modulation detection system is diagrammed schematically in Fig. 3.   In a standard epifluores-cence microscope (Zeiss Universal), the illuminator field stop has been replaced by a coarse diffraction grating (Ronchi ruling) of a period L that is free to slide laterally on Teflon rollers.   The image of the grating is in focus at the specimen, and the grating is oriented so that the lateral motion is perpen-dicular to the parallel slits.   To move the grating at constant speed, a phonograph turntable was modi-fied by the addition of a shallow triangular cam to the perimeter of the disk.   This cam advances and withdraws the grating approximately 1 cm per revolu-tion.   We typically use a 100-line/in Ronchi ruling and set the turntable speed so that the modulation frequency matches the fixed center frequency of the tuned amplifier (83 Hz).   When a 4× microscope objec-tive is in position, the demagnified image of the grating in the plane of the specimen has a period L equal to 0.012 cm.   The output of the tuned amplifier is analyzed by an amplitude envelope detector, which is a simple circuit that consists of a peak voltmeter and zero-crossing detector in parallel.   The output of this device, which is the envelope function at the modulation frequency, is plotted on a strip-chart recorder and digitized for computer analysis.

A plot of the envelope function with time should be a single exponential (for a single labeled species) whose time constant τ is given by

$$\tau = (DK'^2)^{-1} , \tag{11}$$

where D is the tracer diffusion coefficient of the labeled species and $K' = 2\pi/L$, where L is the period of the grating image in the specimen. Incomplete disappearance of a fraction of the envelope function indicates that fraction of labeled material in the sample that is not free to move on the time scale of the measurement.

A more complete description of the theory and circuitry may be found in Lanni and Ware [25].

IV. SAMPLE PREPARATIONS

General considerations for preparation of samples for the various techniques employed have been discussed in the references cited. We describe here only a brief summary of the specific methods used to prepare samples for which data are presented in this chapter.

A. Preparation of Labeled Polystyrene Spheres

Carboxylate-modified polystyrene latex spheres were obtained from Dow Chemical Company, and 10 ml of the 10% w/v suspension of spheres in aqueous surfactant was diluted to 2% w/v, to retard aggregation, by addition of distilled H_2O. The pH of this suspension was adjusted to 7.0 with 1 M NaOH prior to the addition of 3 ml of 0.1 M 1-cyclohexyl-3(2-morpholino-ethyl) carbodiimide, metho-p-toluene sulfonate (Sigma Chemical Company). After approximately 100 min, the excess diimide reagent was removed by centrifugation pelleting and resuspension of the spheres in 40 ml of 1% sodium dodecyl sulfate at pH 7.0. To this suspension was added 400 µl of 0.74 M ethylene diamine;

after 2 hr, excess reagent was again removed by pelleting and resuspension in 40 ml sodium dodecyl sulfate at pH 8.0. This procedure produces spheres, some of whose accessible carboxyl groups have been transformed to amino groups. Fluorescent labeling of the spheres was then accomplished by addition of 20 mg of fluorescein isothiocyanate (Sigma) dissolved in 400 μl of acetone. Reaction was allowed to proceed overnight to ensure labeling of all amino sites. Excess surfactant, dye, and reagents were removed by passing the suspension over anionic and cationic ion exchange resins and by extensive dialysis against the desired medium for each experiment.

B. Preparation of Gel Media

Samples of polystyrene latex spheres suspended in polyacrylamide gels were prepared for the study of particle motion in gel media. The concentration of

FPR Modulation Detector

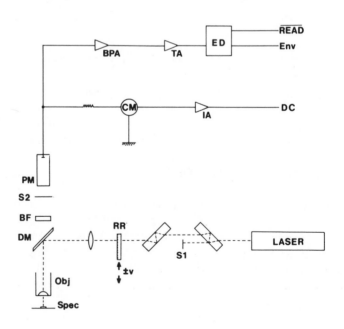

FIGURE 3.   (facing).   Schematic diagram of modulation
detection system.   A standard epifluor-
escence microscope equipped with dichroic
mirror (DM) and barrier filter (BF) is
modified to provide a movable grating
(Ronchi ruling, RR) in the plane of the
illuminator field stop.   The light source
is an argon ion laser with associated beam
control optics for photobleaching (S1). A
photomultiplier (PM) receives fluorescence
emission from the specimen.   The photocurrent is measured by a DC microammeter
(CM) with an isolated recorder output
(IA).   The AC components of the photocurrent in the range 10-300 Hz are amplified by the bandpass amplifier (BPA).   The
resulting signal can be processed by a
spectrum analyzer used as a power envelope
detector.   More simply, the broadband
output can be further amplified by a tuned
stage amplifier (TA) and processed by an
amplitude envelope detector (ED).   The
output of the envelope detector is the
fluorescence recovery signal $E(t)$.   The
signal $\overline{READ}$ is used to trigger a data
recorder when a stable $E(t)$ value is
available during each cycle of the modulation signal.

the gel, the ionic strength of the solvent, and the
size of the spheres were varied to illustrate different points.   As an example of the general method, we
describe the preparation of a 5% gel.   Gel polymerization was initiated by addition of 1.65 ml of 30%
w/v acrylamide, 1.65 ml of 0.8% w/v N,N'-methylene-
bisacrylamide, 2.5 ml Tris buffer (pH 8.0) of varying
concentration (1-50 mM), 100 μl 10% w/v $(NH_4)_2S_2O_8$,
and   10 μl N,N,N',N'-tetramethyl-ethylenediamine.
After the reaction had been initiated, but prior to
gelation, 3.5 ml of the reacting mixture was added to

1 ml of polystyrene sphere suspension in Tris buffer
of the same concentration as employed in the gel
mixture. The resulting solution was loaded into
microslides of 0.1 mm path length that were centrifuged at low speed (10 min, 500 rpm) while polymerization and gelation were completed. This method
produced a homogeneous gel with spheres distributed
uniformly. The ionic strength of the medium was
varied by altering the concentration of Tris buffer
employed.

C.   Preparation of Labeled Poly-L-Lysine

In this procedure 400 mg of poly-L-lysine hydro-
bromide (molecular weight 4×10^4 by viscosity; Sigma
Chemical Co.) was dissolved in 20 ml of 50 mM borate
buffer at pH 8. This solution was then dialyzed
against more borate buffer to which had been added
fluorescein isothiocyanate as a 10% adsorbate on
celite particles (United States Biochemical Company).
The stoichiometry permitted the labeling of 1 of
every 50 of the amino groups on the polymer, but the
extent of labeling was probably significantly less
because of remaining adsorbate, free dye in solution,
and labeling of the dialysis bag. The labeled
polymer was then dialyzed against 5 mM Tris buffer at
pH 7.3 until the dialysate had the same conductivity
as the stock buffer and was then clarified by centrifugation. Ionic strength was varied by adding small
aliquots of a stock solution of 2.5 M KCl.

V.   EXPERIMENTAL RESULTS

Polystyrene latex spheres are a common test
sample for hydrodynamic experiments. We have utilized them in this study to illustrate several aspects
of the combined potential of quasi-elastic light
scattering (QELS), electrophoretic light scattering
(ELS), and fluorescence photobleaching recovery
(FPR). For the FPR experiments, it is of course

necessary to attach a fluorescent label. The procedure is given in Section IV. An immediate question we must ask is whether the labeling procedure alters the surface charge of the particles. For this purpose, we use the ELS technique. Typical data are shown in Fig. 4. The unlabeled sphere had an ELS spectrum with a single major peak of approximately Lorentzian shape and with a width consistent with the theoretical expectation for spheres of diameter about twice as great as the specified diameter, probably indicating some aggregation of the spheres in suspension. The two samples of labeled spheres L1 and L2 were prepared from the same lot by similar methods, but one resulted in a lower electrophoretic mobility, and the other in a higher electrophoretic mobility, than the unlabeled sample. The second procedure, labeled L2, resulted in an ELS linewidth that was about right for diffusion of spheres of this size, indicating little or no aggregation. We have not yet found the change of surface characteristics of these spheres to be reproducible for a given method of labeling. Both signs of alteration of surface charge are possible and reasonable. An excess of unreacted amino groups would tend to decrease the net negative surface charge density, whereas the replacement of a negative charged carboxyl by a fluorescein molecule, which at pH 9 should have a charge of -2, could lead to an overall increase in net negative surface charge density and hence in electrophoretic mobility. In addition to the chemical changes, the labeling procedure may change the surface structure in ways that would alter the electrophoretic mobility, although the net effect would be difficult to predict. Thus in any test experiment for which the surface charge of a labeled sphere may be an important parameter, the measurement of the ELS spectrum provides a direct independent determination through the electrophoretic mobility.

We have performed measurements of the mutual diffusion coefficients and electrophoretic mobilities of these polystyrene spheres in solution as a func-

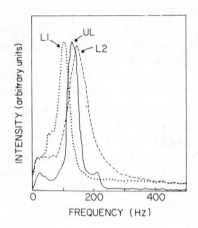

FIGURE 4. Electrophoretic light-scattering spectra of suspensions of 0.038 μm diameter polystyrene latex spheres. The sample UL is an unlabeled sample of carboxylate-modified spheres; L1 and L2 are samples of spheres that have been labeled with fluorescein by the procedure described in Section IV, except with different reaction times. For all three spectra, the experimental conditions were E = 42.2 V/cm, θ = 20°C. The solvent was 10 mM Tris buffer at pH 9. The peak mobility of the UL sample (Δv = 132 Hz), after a correction for electroosmosis [19], was 3.3 × 10^{-4} cm²/sec. The electroosmosis was the result of the adsorption of the spheres to the chamber walls; its magnitude was determined by measuring the flow profile and by calibrating against known mobility standards.

tion of the ionic strength of the suspending medium. As is generally the case, we have found that the electrophoretic mobility rises at low ionic strength, as a result of decreased counterion screening, and the mutual diffusion coefficient rises at low ionic strength, presumably because of interparticle interactions that suppress fluctuations and accelerate their dissipation [26-29]. The behavior of the tracer diffusion coefficient, however, will be expected to be quite different since the interparticle interactions are not the dominating force in a tracer mobility. The primary effect of decreasing the salt concentration should be an effective increase in the hydrodynamic radius of the spheres as the screening length (κ^{-1}) becomes comparable to the radius of the particle [29]. Using FPR to study the tracer diffusion coefficients, we have investigated this phenomenon. A typical data record is shown in Fig. 5. The fluorescence modulation recovery is complete, since all spheres are free to move, and the recovery time can be used to calculate directly the tracer diffusion coefficient. These measurements were performed at constant particle concentration and pH as a function of the ionic strength of the solvent. The results are shown as a plot of D versus κR in Fig. 6. At $\kappa R > 10$, D has the value predicted by the Stokes-Einstein relationship [Eqs. (2) and (3)]; but as κR approaches 1, a decrease of 10% or more in D is evident, indicating the predicted increase in the effective hydrodynamic radius. Future work will be directed toward improving the precision of the data and extending the determination to $\kappa R < 1$.

Linear polyelectrolytes are a system whose properties may change even more drastically as the ionic strength of the suspending medium is varied. Changes in the screening length alter the conformation of the polyelectrolyte molecules as well as the electrokinetic screening, and, except at low polymer concentration, entanglement of neighboring polymers may be altered as well. Electrophoretic light-scattering studies of polyelectrolytes demonstrated these effects previously [30-33]. As an example of the data obtained from a polyelectrolyte solution, we show in Fig. 7 an ELS spectrum of a

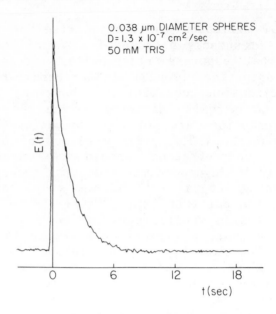

0.038 μm DIAMETER SPHERES
D = 1.3 x 10⁻⁷ cm²/sec
50 mM TRIS

FIGURE 5.   Fluorescence photobleaching recovery data for the diffusion of 0.038 μm diameter polystyrene latex spheres in an aqueous solution of 50 mM Tris buffer at pH 8, T = 26°C. The decay of the modulation envelope of the periodic tracer gradient is recorded as a function of time; the time constant of the exponential decay and the spacing of the grating in the sample (2.45 × 10⁻³ cm) are used to calculate the tracer diffusion coefficient of the spheres. The result is consistent with Eqs. (2) and (3) to within experimental error (2%).

solution of poly-L-lysine and a second spectrum showing the reduction in Doppler shift obtained by addition of 20 mM KCl. The spectra were collected using the same magnitude of electrophoretic current (2.0 mA), so the decreased Doppler shift is the result of the decreased electric field caused by the increase in solution conductivity (58%) and by an

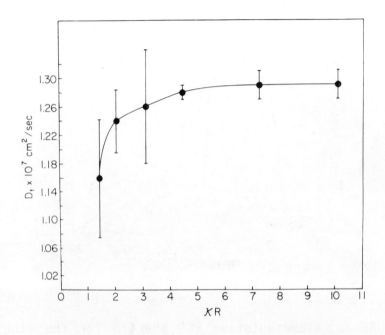

FIGURE 6. Tracer diffusion coefficient of labeled 0.038 μm diameter polystyrene latex spheres as a function of the product of the Debye-Hückel constant and the radius of the particle. The lowest ionic strength ($\kappa R = 1.43$) was 5.2 × 10^{-4} M and the highest ionic strength ($\kappa R = 10.7$) was 2.9 × 10^{-2} M. The high salt limit is the theoretical value for particles of this size; downward deviations as κR approaches one are due to the electroviscous increase of the effective hydrodynamic radius. Error bars in this figure indicate standard deviations.

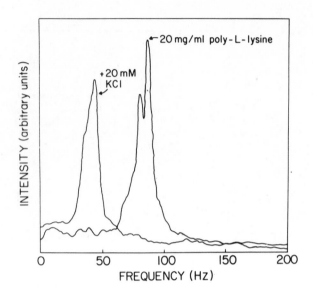

FIGURE 7. Representative ELS spectra for the study of the ionic strength dependence of the electrophoretic mobility of poly-L-lysine. For both spectra, the poly-L-lysine concentration was 20 mg/ml, the scattering angle was 20°, the temperature was 25°C, and the constant electrophoretic current was 2.0 mA. Because of differences in conductivity of the two samples, the electric field was 24.1 V/cm in the absence of added salt and 15.2 V/cm with 20 mM KCl. The respective electrophoretic mobilities in the two cases are 4.6 × 10^{-4} cm^2/V·sec and 3.5 × 10^{-4} cm^2/V·sec. Both samples contained 5 mM Tris buffer at pH 8.

276

actual decrease in the electrophoretic mobility of the polymer molecules (25%). The linewidths in these spectra are essentially the same, although the spectrum of the sample with added salt is closer to the theoretical lineshape for a single species. We have determined the ionic strength of the electrophoretic mobility of poly-L-lysine solutions at 20 mg/ml and 7 mg/ml polymer weight per aqueous solution volume. The magnitudes decrease by about 35% from 0 to 100 mM, so the polyion does not appear to be free draining in this range.

In contrast to the electrophoretic mobilities, the mutual diffusion coefficients of poly-L-lysine show an anomalous behavior with salt and polymer concentration, as reported previously by Schurr et al. [34,35]. At a given polymer concentration, the mutual diffusion coefficients increase as the salt concentration is decreased, down to a critical salt concentration below which D' falls by more than an order of magnitude in a region termed by Schurr and coworkers as the "extraordinary phase." We have verified that our samples of poly-L-lysine studied by ELS were in the extraordinary phase. Quasi-elastic light-scattering linewidths were determined in the absence of an electric field to provide an estimate (~ 10%) of the mutual diffusion coefficients. At 20 mg/ml polymer, we observed D' to increase from 0.9×10^{-8} cm^2/sec to 1.8×10^{-8} cm^2/sec in going from zero to 100 mM added salt. All these values are in the very low range characteristic of the extraordinary phase. At a polyion concentration of 7 mg/ml, the salt dependence of D' shows an abrupt change. Below 20 mM added KCl, $D' \cong 10^{-8}$ cm^2/sec; at 20 mM KCl or above, $D' \cong 9 \times 10^{-8}$ cm^2/sec. These values are still far below D' expected for a particle of this size.

Schurr et al. [34,35] have interpreted these anomalously low values of D' to indicate a dramatic rise in the friction factors of the isolated polyions. However, as we pointed out in Section I, only the tracer diffusion coefficient can be interpreted

directly in terms of the particle friction factor.
Therefore, we have determined the tracer diffusion
coefficient of the identical samples using FPR. A
plot of the tracer diffusion coefficient as a function of added salt is shown in Fig. 8. Astonish-
ingly, the tracer diffusion coefficients are an order
of magnitude greater than the mutual diffusion coefficients and are characteristic of particles of the
size of these polymers and of D' in the ordinary
phase. The slight increase in D with increasing KCl
concentration presumably indicates a reduction in
polyion size because of screened intramolecular
repulsions. There is a hint of a discontinuity at
20 mM that along with our observation of the abrupt
change in D' at this point, may indicate a conformational transition. We conclude that the slow motions
represented by anomalously low values of D' in the
extraordinary phase of poly-L-lysine are not due to
large friction factors of the polyion molecules, but
rather are the result of long-lived fluctuations of
optical refractive index that must represent coupled
motions in the sample.

Another polyelectrolyte effect we have recently
observed is the retardation of the tracer diffusion
of an anionic species in a solution of cationic
polyelectrolyte. As one might expect, this retardation of motion is mitigated by the addition of salt
to the solution (see summary of data in Fig. 9). The
anionic species was fluorescein; the poly-L-lysine
was unlabeled, so the FPR data unambiguously yield
the tracer diffusion coefficient of the fluorescein
in the polyelectrolyte solution. In the absence of
added salt, the tracer diffusion coefficient is
reduced by a factor of 4.6 from the normal solution
value, and even at 100 mM added KCl, D has risen to
only 56% of its free-solution value. Motions of this
type are clearly not related to molecular size
through the viscosity and molecular radius as stated
in Eq. (3), but rather must be dominated by electrostatic attractions to the polyions in the medium.

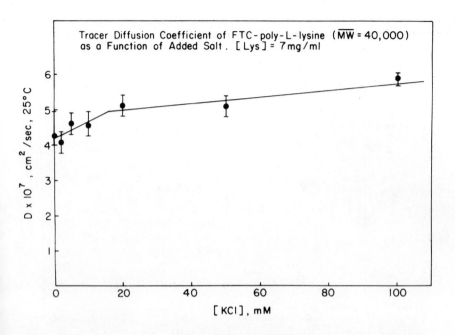

Tracer Diffusion Coefficient of FTC-poly-L-lysine ($\overline{MW}$ = 40,000) as a Function of Added Salt. [Lys] = 7 mg/ml

FIGURE 8. Plot of tracer diffusion coefficient of labeled poly-L-lysine (molecular weight 4×10^4; concentration 7 mg/ml) as a function of the concentration of KCl added to the solution. The possible inflection near 20 mM, combined with a larger change in the mutual diffusion coefficient measured by QELS, may be the result of a molecular conformational transition. The tracer diffusion coefficients are an order of magnitude greater than the mutual diffusion coefficient measured under these conditions. The error bars in the figure represent the total range of repeated measurements.

279

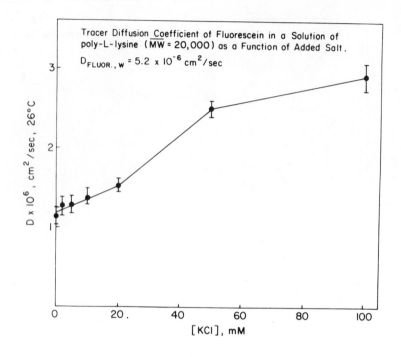

FIGURE 9. Tracer diffusion coefficient of fluorescein in a solution of poly-L-lysine (molecular weight 2×10^4) as a function of added KCl. The solvent included 5 mM Tris buffer at pH 7.4. The fluorescein concentration in all measurements was 1 μM. The error bars represent the total range of repeated measurements. Although added salt reduced the retardation of the negatively charged fluorescein molecules in the polycation solution, in all cases the measured tracer diffusion coefficient was far below the normal aqueous solution value of 5.2×10^{-6} cm²/sec.

An even more complex type of motion, and one that has wide-ranging significance, is the diffusion of particles through gel networks. The size exclusion effects of the network pores, electrostatic interactions between particle and network, and possible effects of the gel network on the effective microscopic viscosity through the pores are physical considerations that have received considerable theoretical interest and speculation but regarding which there is a paucity of experimental data. We are applying the FPR technique to the measurement of the tracer diffusion coefficients of labeled polystyrene spheres of known size and surface charge (from ELS measurements). An illustrative pair of measurements is shown in Fig. 10. The two traces show the decay of the fluorescence modulation envelope for two different sizes of spheres in a 5% polyacrylamide gel. The spheres of 0.038 μm diameter show complete recovery with $D = 1.3 \times 10^{-7}$ cm^2/sec – precisely the value of D in water under the same conditions. The larger spheres (0.254 μm diameter) show a substantial immobile fraction on the time scale of the measurement; the initial slope of recovery is consistent with a D of $\sim 9 \times 10^{-9}$ cm^2/sec, somewhat less than half the diffusion coefficient expected in water.

An even more dramatic effect can be observed by varying the solution conditions used in preparing the gel, as illustrated in Fig. 11. The lower trace shows the recovery for 0.038 μm-diameter spheres in 5% gel prepared by using 1 mM Tris (see Section IV). As in Fig. 10, the tracer diffusion coefficient is seen to be equal to the free-solution value. However, in the upper trace, for a sample prepared under identical conditions except for the use of 50 mM Tris, there is a large immobile fraction and the initial slope is consistent with a tracer diffusion coefficient more than 70 times slower than the free-solution value! It appears that polyacrylamide gel networks are affected by the ionic strength of the polymerization mixture. These drastic changes we

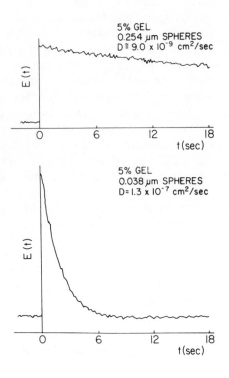

FIGURE 10. Fluorescence photobleaching recovery data for the motion of polystyrene spheres of two sizes in a 5% polyacrylamide gel. The 0.038 μm diameter spheres moved with a tracer diffusion coefficient indistinguishable from that in free solution. The tracer diffusion coefficient of the mobile fraction of the 0.254 μm diameter spheres was a little less than half the free-solution value. Both gels were prepared with 1 mM Tris buffer at pH 8. For these experiments, the period of the bleached pattern in the sample was $L = 2.95 \times 10^{-3}$ cm.

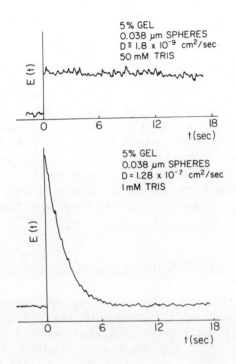

FIGURE 11. Fluorescence photobleaching recovery data for the motion of 0.038 µm diameter spheres in 5% polyacrylamide gels prepared with 1 mM Tris (pH 8), the spheres diffused as if in water; in the gel prepared with 50 mM Tris (pH 8), the diffusion coefficient was more than 70 times lower than the free solution value. For these experiments, the period of the bleached pattern in the sample was $L = 2.95 \times 10^{-3}$ cm.

have observed in the mobility of test particles in
the gel indicate the sensitivity of the microscopic
motions to changes in the network structure. We have
verified this effect for a range of ionic strengths
at two different gel concentrations. In 5% gel the
motion is essentially unretarded when the Tris concentration is below 20 mM. At the transition concentration of 20 mM, we found some regions in the gel
that retarded motion of the spheres and some regions
in which the spheres moved freely. The transition
from mobile to immobile spheres also occurred in 10%
gels, but at a lower Tris concentration (1-2 mM).
From our preliminary evidence these transitions are
very sharp; we have not yet determined their rever-
sibility.

VI.   DISCUSSION AND CONCLUSIONS

The accuracy of QELS methods and the sophistication of accounting for complicating effects such as
polydispersity, interparticle forces, and intramolecular motions have improved and matured for nearly
two decades. The introduction of ELS in 1971
expanded the power of dynamic light-scattering techniques for analyzing mixtures and polydisperse
samples and added the electrophoretic mobility as a
determined parameter so that surface charge density,
as well as hydrodynamic friction, could be investigated. An extensive literature attests to the
success and vigor of this field, but certain major
weaknesses are apparent. Light is scattered by all
refractive index fluctuations, so light scattering is
not inherently able to distinguish among different
species in solution nor between molecular and cooperative modes of motion, nor − in some cases − between
the species of interest and the matrix in which it
moves. Many of these weaknesses are supplemented by
the introduction of FPR methods. The fluorescent
label readily distinguishes one species in a solution
or matrix of any number of other species. Moreover,

the measurement of the tracer mobility leads to direct molecular interpretation, even under conditions of coupled modes of motion.

We have presented preliminary data on several projects that we are currently pursuing that combine FPR and dynamic light-scattering techniques to achieve scientific objectives that could not be attained by these techniques individually. For example, an electroviscous effect on the tracer diffusion coefficient due to interparticle interactions was deduced by Weissman et al. [29] from QELS correlation time and intensity measurements. Schurr [36] has presented a theory of the electrolyte contribution to the friction of charged spheres. These effects could not be measured directly without the introduction of a tracer technique such as FPR, as illustrated in Fig. 6.

The extraordinary phase of poly-L-lysine is a dramatic example of electrostatically constrained systems described theoretically by Weissman and Ware [37] in terms of fluctuation transport theory. In such systems the transport properties of fluctuations in the concentration of the charged polyions may not bear the usual relationships to the transport properties of the individual polyions themselves. In the case of the extraordinary phase of poly-L-lysine, fluctuations in the concentration relax more than an order of magnitude slower than the characteristic time for diffusion of an individual poly-L-lysine molecule over the same distance. Consequently, both the dynamic light scattering and tracer (FPR) measurements are necessary for an elementary understanding of the system.

The study of microscopic motions in gels is important for an understanding of many commercial processes, including photography and various forms of chromatography, and for the characterization of a number of biological systems. A combined FPR-ELS methodology will permit a systemic study of the effects of particle size versus pore size and particle electrical charge density versus gel network charge density and of the behavior of the system near lyotropic and thermotropic transitions. We expect this to be a particularly fruitful area of investigation in the near future.

Finally, we emphasize that dynamic light-scattering and FPR techniques have much in common. The essential equipment and the experimental and theoretical abilities required of the investigators overlap considerably; hence the investment in time and money for the expansion of an experimental program from one of these techniques to the entire range is well worth the capacity to determine an increased scope of observables. There is a perceptible trend in some of these fields to expand the range of approachable systems by increasing the algebraic complexity of data analysis, possibly at the expense of ambiguous, even erroneous interpretation. We believe that it is generally a superior approach to increase the number of independent parameters determined by expanding the experimental arsenal to a range of complementary techniques.

ACKNOWLEDGMENT

The technical assistance of Mrs. Mara Aistars and the advice and suggestions of Ms. Terry Dowd, Mr. James Klein, and Dr. Kee Woo Rhee Yoo are gratefully acknowledged. This work was supported by Grant No. GM 27633 from the National Institutes of Health and Grant No. PCM 80-10924 from the National Science Foundation.

REFERENCES

1. A. Einstein, Ann. Phys. *17*, 549 (1905); *19*, 371 (1906).

2. C. Tanford, *Physical Chemistry of Macromolecules*, Wiley, New York, 1961, pp. 346-363.

3. B. Chu, *Laser Light Scattering*, Academic Press, New York, 1974, p. 215.

4. Reference 2, pp. 412-431.

5. Reference 2, pp. 461-466.

6. Reference 2, pp. 416-418.

7. M. Bier, *Electrophoresis*, Academic Press, New York, 1959, 1967.

8. N. Catsimpoolas, Ed., *Electrophoresis '78*, Elsevier, New York, 1978.

9. O. Gaal, *Electrophoresis in the Separation of Biological Macromolecules*, Wiley, Chichester, England, 1980.

10. H. Z. Cummins and E. R. Pike, Eds., *Photon Correlation Spectroscopy and Velocimetry*, NATO Advanced Study Institute, ser. B, Vol. 23, Plenum Press, New York, 1976.

11. B. R. Ware, in *Chemical and Biochemical Applications of Lasers*, Vol. 11, C. B. Moore, Ed., Academic Press, New York, 1977, pp. 199-239.

12. L. E. Drain, *The Laser Doppler Technique*, Wiley, New York, 1980.

13. B. R. Ware and W. H. Flygare, Chem. Phys. Lett. *12*, 81 (1971).

14. B. R. Ware, Adv. Colloid Interface Sci. *4*, 1 (1974).

15. B. R. Ware, The Study of Biological Surfaces by Laser Electrophoretic Light Scattering, ACS Symp. Ser. *85*, 102 (1978).

288   B. R. Ware et al.

16. E. E. Uzgiris, Ad. Colloid Interface Sci. *14*, 75 (1981).

17. E. E. Uzgiris, Progr. Surf. Sci. *10*, 53 (1981).

18. B. R. Ware and D. D. Haas, Electrophoretic Light Scattering, in *Fast Methods in Physical Biochemistry*, Elsevier, Amsterdam, 1983, in press.

19. B. A. Smith and B. R. Ware, Apparatus and Methods for Laser Doppler Electrophoresis, in *Contemporary Topics in Analytical and Clinical Chemistry*, Vol. 2, D. M. Hercules, G. M. Hieftje, and L. R. Snyder, Eds., Plenum Press, New York, 1978.

20. M.-M. Poo and R. A. Cone, Nature *247*, 438 (1974).

21. R. Peters, J. Peters, K. H. Tews and W. Bahr, Biochim. Biophys. Acta *367*, 282 (1974).

22. K. Jacobsen, E. Wu, and G. Poste, Biochim. Biophys. Acta *433*, 215 (1976).

23. D. Axelrod, D. E. Koppel, J. Schlessinger, E. Elson, and W. W. Webb, Biophys. J. *16*, 1055 (1976).

24. B. A. Smith and H. M. McConnell, Proc. Natl. Acad. Sci. (USA) *75*, 2759 (1978).

25. F. Lanni and B. R. Ware, Rev. Sci. Instrum. *53*, 905 (1982).

26. J. C. Brown, P. N. Pusey, J. W. Goodwin, and R. H. Ottewill, J. Phys. A *8*, 664 (1975).

27. P. N Pusey, J. Phys. A *11*, 119 (1978).

28. B. J. Ackerson, J. Chem. Phys. *69*, 684 (1978).

29. M. B. Weissman, R. C. Pan, and B. R. Ware, J. Chem. Phys. *70*, 2897 (1979).

30. H. Magdelénat, P. Turr, P. Tivant, M. Chemla, R. Menez, and M. Drifford, Biopolymers *18*, 187 (1979).

31. J. P. Meullenet, A. Schmitt, and M. Drifford, J. Phys. Chem. *83*, 1924 (1979).

32. C. Y. Cha, R. L. Folger, and B. R. Ware, J. Polym. Sci., Polym. Phys. Ed. *18*, 1853.

33. M. Drifford, R. Menez, D. Tivant, P. Nectoux, and J. P. Dalbiez, Rev. Phys. Appl. *16*, 19 (1981).

34. W. I. Lee and J. M. Schurr, Biopolymers *13*, 903 (1974).

35. S.-C. Lin, W. I. Lee, and J. M. Schurr, Biopolymers *17*, 1041 (1978).

36. J. M. Schurr, Chem. Phys. *45*, 119 (1980).

37. M. B. Weissman and B. R. Ware, J. Chem. Phys. *68*, 5069 (1978).

CHAPTER 10

Utility of Multidetector Methods in Quasi-Elastic Light-Scattering Spectroscopy

GEORGE D. J. PHILLIES
University of Michigan, Ann Arbor

CONTENTS

I.   INTRODUCTION

In the archetypical light-scattering spectroscopy experiment (Fig. 1a), a sample is illuminated with a laser beam, the light scattered in some
direction is collected with a single detector, and

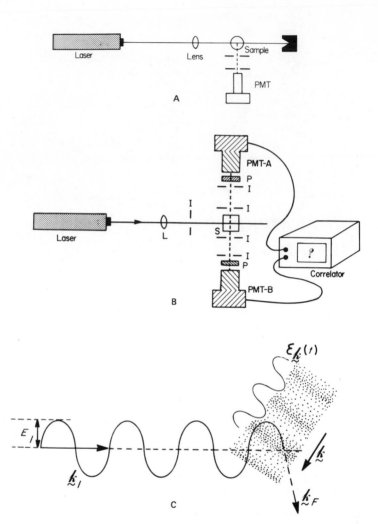

FIGURE 1. (a) Archetypical optical train for QELS apparatus; (b) hypothetical multidetector experiment; (c) notation for describing the scattering event, showing the incident field E_I; the incident, scattered, and scattering wavevectors k_I, k_F, k; and (shading) the index of refraction fluctuation $\varepsilon_k(t)$.

292

the intensity-intensity autocorrelation function of the scattered light is measured. Extensions of this procedure with the use of several illuminating laser beams may readily be envisioned in which the scattering intensity in different directions at different times is compared (Fig. 1b). However, it has long been "well-known" that − except for a few velocimetry experiments − multidetector, multibeam experiments on large samples are without value in that they measure quantities that can already be obtained conventionally. Belief in this "well-known" fact, without recognition of the attendant qualifications, has probably delayed recognition of the experiments discussed here.

It is first convenient to establish our notation (Fig. 1c). The incident light as amplitude E_I and wavevector k_I. The scattered field has amplitude E and wavevector k_F. Scattering occurs because the material in the scattering volume is not uniform. Calling $k = k_F - k_I$ the scattering vector, the scattering in the illustration is due to a nonzero value of the kth spatial Fourier component $\varepsilon(k,t)$ of the index of refraction. Nonuniformities in ε are, in general, due to nonuniformities in the concentration or number density of scattering particles, so that scattering is due to concentration fluctuations of instantaneous amplitude $a_k(t)$. This may be written as

$$E(k,t) = E_I(k_I,t) \; \alpha \; a_k(t) \; , \qquad (1a)$$

$$a_k(t) = \sum_{i=1}^{N} \exp[ik \cdot r_i(t)] \; . \qquad (1b)$$

By convention, E is labeled by the scattering vector k rather than by its own wavevector k_F, and α is a polarizability derivative. Equation (1b) defines $a_k(t)$ in terms of the positions $r_i(t)$ of the N scattering particles. In most cases N itself is time dependent, but often not significantly so.

The scattered field is measured from the intensity I of the scattered light, usually by photon counting. Two detection procedures are in use. In homodyne (self-beat) detection, one simply measures the intensity of the scattered light, so that

$$I_{hom}(\mathbf{k},t) = |E(\mathbf{k},t)|^2 . \tag{2}$$

In heterodyne detection (i.e., with local oscillator), the light scattered is merged with light (field E_L) removed (usually without shift in frequency) from the original incident beam, and the heterodyne intensity is

$$I_{het}(\mathbf{k},t) = |E_L|^2 + 2Re(E_L E^*(\mathbf{k},t)) + |E(\mathbf{k},t)|^2 . \tag{3}$$

The averaged properties of the scattered field can be related to ensemble averages over particle positions by important relationships including

$$<|E(\mathbf{k},t)|^2>$$

$$= |E_I|^2 \alpha^2 < \sum_{i,j=1}^{N} \exp[i\mathbf{k}\cdot\{\mathbf{r}_i(t) - \mathbf{r}_j(t)\}]>$$

$$= |E_I|^2 \alpha^2 NS(\mathbf{k}) , \tag{4}$$

$$<E(\mathbf{k},t)E^*(\mathbf{k},t+\tau)>$$

$$= |E_I|^2 \alpha^2 < \sum_{i,j=1}^{N} \exp[i\mathbf{k}\cdot\{\mathbf{r}_i(t) - \mathbf{r}_j(t+\tau)\}]>$$

$$= |E_I|^2 \alpha^2 NS(\mathbf{k},\tau) , \tag{5}$$

$$\langle E(\mathbf{k},0)E(\mathbf{q},t)E^*(\mathbf{k} + \mathbf{q},\tau)\rangle$$

$$= |E_I|^3\alpha^3 \langle \sum_{i,j,l=1}^{N} \exp[i\mathbf{k}\cdot\mathbf{r}_i(0)$$

$$+ i\mathbf{q}\cdot\mathbf{r}_j(t) - i(\mathbf{k} + \mathbf{q})\cdot\mathbf{r}_l(\tau)]\rangle$$

$$= |E_I|^3\alpha^3 N S^{(3)}(\mathbf{k},\mathbf{q},t,\tau) \,, \tag{6}$$

where $S(\mathbf{k})$, $S(\mathbf{k},t)$, and $S(\mathbf{k},\mathbf{q},t,\tau)$ are the static structure factor, the dynamic structure factor, and the triple dynamic structure factor, respectively [1]. We now enforce the convention that E_L and E_I are real, with phases written explicitly.

All cross-correlation and multibeam experiments are not equal. Most of them don't work. The useful experiments may be characterized by five variables:

1. The number of scattering directions.

2. The number of detectors.

3. The number of laser beams illuminating the sample.

4. The signal analysis method.

5. The sample volume.

Let us consider these variables separately. In any standard experiment a series of pinholes restricts the angles through which light can be scattered and reach the detector. The appropriate unit for describing this range of angles is the coherence area — the diffraction-limited area, at the detector, over which the intensity of light scattered by the sample cannot vary significantly. A single detector can look at one coherence area or a few neighboring coherence areas without great difference in effect. We herein retain the idealization that detectors are limited to single coherence areas.

For many experiments, the starting point is an observation on commercial correlators. The measured spectrum (in time domain) of an intensity $I_A(t)$ is [2]

$$S_{AB}(\tau) = \int_0^T I_A(t)I_B(t + \tau)dt \quad , \tag{8}$$

so that two-detector experiments are not instrumentally inconvenient. It is also possible to define three-time correlation functions (bispectra)

$$S_{ABC}(\tau,\theta) = \int_0^T [I_A(t)I_B(t + \tau)I_C(t + \theta)]dt \quad . \tag{9}$$

Finally, one may dispense with correlation functions, considering instead, for example, the mean time needed for the intensity, initially of some value I_0, to return to the value I_0. The first measurements of a diffusion coefficient D by number fluctuation analysis (made by T. Svedberg [3] and A. Westgren [4] before World War I) used this procedure. In these experiments, an ultramicroscope was used to observe and count, to the ticking of a metronome, the number of particles of a gold sol in a small well-defined volume, with D obtained from the distribution of time intervals between observations of a fixed number n of particles in the volume. (Although Westgren's experiments were not forgotten by some theoreticians [5], the modern reinvention of number fluctuation analysis relied on photon correlation techniques [6].)

By refining a distinction treated by Pusey [7], for example, the size of the sample volume may be used to divide experiments into three classes. The classification works because particle positions are correlated only over microscopic "correlation" volumes V_c, so that scattering from V_c reflects static correlations between the particles in that volume, whereas scattering from particles that are far apart will not involve any static correlations between them. If the scattering volume V_s (the region illuminated by the laser beam and seen by the detector) satisfies $V_s \gg V_c$, and if V_s contains many particles, the scattered field is a Gaussian random variable. Experiments may thus be described as "Gaussian" or "non-Gaussian," depending on the statistics of the scattered field. Some years ago, this author [8] emphasized that one may also consider correlations between fields $E(\mathbf{k},t)$, $E(\mathbf{q},t)$, ... In a system containing many scatterers, with $V_s \gg V_c$, each scattered field is separately a Gaussian random variable. For the condition that $E(\mathbf{k},t)$, $E(\mathbf{k},t + \tau)$ be a Gaussian random *process*, it is also necessary that particle motions have only short-range correlations, so that the *dynamic* correlation volume satisfies $V_c^D \ll V_s$. Furthermore, the fields $E(\mathbf{k},t),E(\mathbf{q},t)^C$, ... need not be *jointly* Gaussian; an experiment that is sensitive to non-Gaussian aspects of the joint distribution of separately Gaussian fields may be called *semi-Gaussian*.

Sections II, III, and IV review results on Gaussian, non-Gaussian, and semi-Gaussian experiments, respectively. Section II begins with the one-beam, single-coherence-area, two-detector method due to Burstyn et al. [9] for photomultiplier tube afterpulsing suppression. Also treated at length in Section II is our [11,12] two-beam, two-detector technique for suppressing multiple scattering spectra. The non-Gaussian effects discussed in Section III may be seen by making the scattering volume small (10 μm^3 in the two-detector, one-beam device described by Griffin and Pusey [25]) or by making the particle correlation volume large, as done in the multibeam fluorescence photobleaching recovery [21] and forced Rayleigh scattering [22] methods. Finally, Section IV examines the proposed semi-Gaussian experiment: the measurement of three-point density correlations by means of heterodyne coincidence spectroscopy.

II. GAUSSIAN EXPERIMENTS

The simplest two-detector quasi-elastic light-scattering (QELS) spectroscopic experiment (Fig. 2) illuminates a sample with a single laser beam and uses a pair of detectors to observe light scattered in a single direction. The optical train is dictated by the requirement that both detectors see the same coherence area. With typical optics, the illuminating beam can be focused down to ~ 100 μ diameter, in which case a coherence area would have an angular extent λ/L ~ 1×10^{-3} radians, that is, 1 mm at a distance of 1 m. To fit two photodetectors into this area, one uses a beam splitter, sending parts of the beam to each detector, and measures the cross-correlation spectrum.

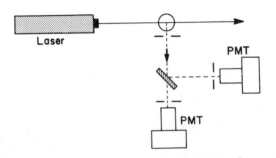

FIGURE 2. Multidetector scheme for suppressing photomultiplier tube afterpulsing effects. After Burstyn et al. [9].

Superficially, this experiment is equivalent to a one-detector experiment. The detectors sample the same coherence area, so they see the same intensity fluctuations, although with some loss of signal. As first recognized by Burstyn et al. [9], this scheme substantially suppresses photomultiplier tube (PMT) afterpulsing effects. In the one-detector experiment the arrival of a photon at $t = 0$ produces — besides the photopulse — disturbances within the phototube that modulate the sensitivity of the PMT for a period of time thereafter. In the two-detector experiment, the arrival of a photon at the first detection at $t = 0$ has no effect on the sensitivity of the other detector at later times; therefore, the sensitivity of the first detector at $t = 0$ and the second detector at $t = \tau$ are the same and afterpulsing effects are thereby eliminated from $C(t)$. This elimination is only approximate. Because a photon reached detector A at $t = 0$, the instantaneous intensity $I(t) = |E(\mathbf{k},t)|^2$ is more likely at $t = 0$ to be above than below its average value; consequently, between $t = 0$ and $t = \tau$, the second detector is more than usually likely to receive a photon. At $t = \tau$, the second detector is then more than usually likely to have been disturbed. Indirect correlations thus lead to afterpulsing effects, even in a two-detector experiment. No exact analysis of these indirect effects seems to have been reported as yet.

The effectiveness of this procedure is illustrated in Fig. 3, based on data obtained by the author. The figure compares a one-detector experiment using an RCA 7265 PMT at room temperature, and the two-detector experiment using a pair of RCA 7265 PMTs. The single RCA tube shows distinct afterpulsing effects [the rising portion of $C(t)$] out to 2 μsec. The pair of RCA 7265 PMTs show a disturbance only in the first five correlator channels — for $t > 500$ nsec there are no signs of afterpulsing effects in trace b. The technique does not allow one to use a given phototube for arbitrarily high frequencies since the travel time of a photopulse down the dynode chain still fluctuates, setting an upper frequency limit for any given tube. A more important restriction on photon correlation studies at very high frequencies is the need for pairs of photons to arrive within a correlation time Γ^{-1} of each other; otherwise there will be no signal. For $\Gamma^{-1} \sim 300$ nsec, $\sim 3 \times 10^6$ counts/sec is needed, which is more than any conventional tube can tolerate.

300   G. D. J. Phillies

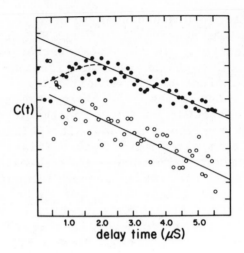

FIGURE 3. $S(k,t)$ from a one-detector experiment and from the two-detector experiment in Fig. 2, showing the reduction in after-pulsing effects (early dip in curve). The correlation time $(2\Gamma k^2)^{-1}$ in this system is much greater than 6 μsec, so $S(k,t)$ in this time domain is expected to have a linear form.

Single-detector measurements at very high frequencies can be performed by high-resolution Fabry-Perot spectrometers; the same is not true for the cross-correlation experiments treated here.

With mirrors or optical fibers, a single detector can simultaneously view light scattered in different directions or from different samples. Crosignani et al. [10] show how such an experiment can determine the scale length of turbulence. A more familiar example of this experiment is the laser heterodyne velocimeter, in which light scattered from a moving sample and light "scattered" by a fixed mirror are compared.

The simplest Gaussian multibeam, multidetector scattering arrangement [11], which was also the first

to work experimentally [12], is indicated in Fig. 4.
Two detectors, set to respond to fluctuations of
wavevector +**k** and -**k**, respectively are placed to view
the same volume. If the incident beams were of
different color, interference filters could be placed
so that each detector responded only to scattering
from one laser beam. For demonstration purposes, two
coaxial antiparallel incident beams and a 90° scattering geometry were employed. Whereas each detector
received photons scattered from each laser beam
through wavevectors **k** and **k'** (-**k** and -**k'** for the
other detector), it is given that |**k**| = |**k'**| = k; so
for single scattering, the scattering vectors all
have the same magnitude (Fig. 5).

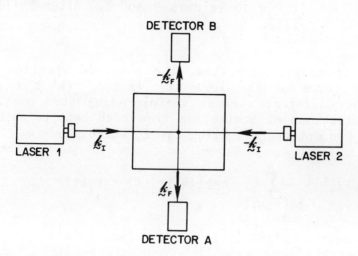

FIGURE 4.   Two homodyne-detector two-beam procedure
for the elimination of multiple scattering. After Phillies [11].

The spectrum obtained in a two-beam, two-
detector experiment may readily be calculated for the
general case in which the scattering vectors from the
first beam to the first detector and the second beam
to the second detector are **k** and **k'**, respectively.

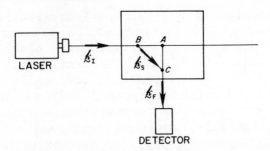

FIGURE 5.   Wavevector geometry for the experiment in
Fig. 4, showing single scattering in the
volume *A* and a representative double-
scattering event of intermediate wavevector *q* in volumes *B* and *C*. After Phillies
[*11*].

With homodyne detection, the observed spectrum has
four terms of the form $\langle I_{hom}(\mathbf{k},t)I_{hom}(\mathbf{k}',t + \tau)\rangle$,
with different terms corresponding to possible
combinations of scattering from each beam to each of
the two detectors. Using Eq. 2, we obtain

$$S_{AB}^{x}(\tau) = \langle \sum_{i,j,l,m=1}^{N} \exp\{i\mathbf{k}\cdot[\mathbf{r}_{i}(t) - \mathbf{r}_{j}(t)\}$$

$$\times \exp\{i\mathbf{k}'\cdot[\mathbf{r}_{l}(t + \tau) - \mathbf{r}_{m}(t + \tau)]\}\rangle. \qquad (10)$$

Equation (10) is simplified by using two rules [*13*]
that are repeatedly invoked in the following:
   i.   In a uniform liquid

$$Q = \langle\exp[i\mathbf{k}_{1}\cdot\mathbf{r}_{1} + i\mathbf{k}_{2}\cdot\mathbf{r}_{2} + \cdots + i\mathbf{k}_{n}\cdot\mathbf{r}_{n}]\rangle \qquad (11)$$

vanishes from translational invariance unless $\mathbf{k}_{1} + \mathbf{k}_{2}$
$+ \cdots + \mathbf{k}_{n} = 0$. For this rule the proof is as follows. If the fluid is uniform, the location of the
origin ($\mathbf{r} = 0$) should be of no importance to internal

properties of the system (those that are independent of the location of the system). However, replacement of $\mathbf{r}_j \rightarrow \mathbf{r}_j + \mathbf{a}$ changes the phase of Q unless $\exp[i\mathbf{k}_1 \cdot \mathbf{a} + i\mathbf{k}_2 \cdot \mathbf{a} + \cdots i\mathbf{k}_n \cdot \mathbf{a}]$ is 1, as is found for arbitrary $\mathbf{a}$ if and only if $\mathbf{k}_1 + \mathbf{k}_2 + \cdots + \mathbf{k}_n = 0$.

ii. In a volume $V \gg V_c$ for n even

$$K_n = \left\langle \sum_{\substack{i,j,l,\ldots m=1}}^{N} \exp[i\mathbf{k}_1 \cdot \mathbf{r}_i \right.$$

n terms

$$\left. + i\mathbf{k}_2 \cdot \mathbf{r}_j + \cdots + i\mathbf{k}_n \cdot \mathbf{r}_m]\right\rangle \qquad (12)$$

is dominated by terms in which the $\mathbf{r}_x$ are found pairwise in different correlation volumes V_c. (When $i = j$, $\mathbf{r}_i$ and $\mathbf{r}_j$ refer to the same particle, so "two $\mathbf{r}_x$ in one V_c" need not imply that two particles are in one correlation volume.) Proof by induction is as follows. For $n = 2$

$$K_2 = \left\langle \sum_{i,j=1}^{N} \exp[i\mathbf{k}_1 \cdot \mathbf{r}_i + \mathbf{k}_2 \cdot \mathbf{r}_j] \right\rangle$$

$$= (N + \left\langle \sum_{i \neq j=1}^{N} \exp[i\mathbf{k}_1 \cdot \mathbf{r}_i + i\mathbf{k}_2 \cdot \mathbf{r}_j] \right\rangle)$$

$$\times \delta(\mathbf{k}_1 + \mathbf{k}_2) , \qquad (13)$$

where the final δ function arises from rule i. The positions of particles in different correlation volumes are independent, so $\langle \exp[i\mathbf{k} \cdot (\mathbf{r}_i - \mathbf{r}_j)] \rangle$ vanishes unless $i = j$ or i and j are distinct particles in the same correlation volume. The final sum in Eq. (13) thus contains Nm terms of order 1, where $m = NV_c/V_s \ll N$ is the number of particles in a typical volume V_c. K_2 thus contains only terms in which pairs $\mathbf{r}_i, \mathbf{r}_j$ are in the same correlation volume, either because $i = j$ or because i and j refer to neighboring particles.

For $n = 4$, K_4 contains three sorts of terms. There are terms that leave at least one position vector r_i isolated in its own coherence volume; these vanish on the average because, for an isolated particle, $\exp[ik \cdot r_i]$ is equally likely to have all possible values. There are $\sim N^2 m^2$ nonzero terms that put distinct pairs r_i, r_j and r_l, r_p ($i = j$ and/or $l = p$ are allowed, but not $i = l$) into each of two correlation volumes. Whereas the pairs i,j and l,p are independent of each other, $\exp[ik \cdot (r_i - r_j)]$ and $\exp[ik \cdot (r_l - r_p)]$ are each separately nonzero because each refers to positions in a single volume V_c. One has nonzero terms that put all four r_x into one V_c. As there are only $\sim N m^3$ terms (all of order 1) of this type, K_4 is dominated by terms that put pairs of position vectors into distinct volumes V_c.

Similarly, in K_n there are $\sim (Nm)^{n/2}$ terms that put vectors pairwise into $n/2$ distinct correlation volumes. Terms with vectors in more than $n/2$ distinct correlation volumes have at least one isolated particle and average to zero; terms with more than two (e.g., three) vectors into the same correlation volume must be of lower order in N than order $n/2$. For placement of a third or higher vector into a volume V_c, a pair of vectors must be removed from some other volume V'_c. Thus, for $V_s/V_c \gg 1$, the dominant terms of K_n (n even) are those in which position vectors r are found pairwise in distinct correlation volumes.

Equation (10) is now simplified by applying rules i and ii. Equation (10) corresponds to K_4 with $k_1 = -k_2 = k$, $k_3 = -k_4 = k'$. For $k \neq \pm k'$, the leading nonzero terms of $S_{AB}(\tau)$ find pairs (r_i, r_j) and r_l, r_m) in distinct correlation volumes. The ensemble average over remote (and hence noninteracting) particles factors into a product of averages, so

$$S_{AB}^{x}(\tau) = E_I^4 \, \alpha^4 \, < \sum_{i,j=1}^{N} \exp\{i\mathbf{k}\cdot[\mathbf{r}_i(t) - \mathbf{r}_j(t)]\}>$$

$$\times \; < \sum_{l,m=1}^{N} \exp\{i\mathbf{k}'\cdot[\mathbf{r}_l(t + \tau) - \mathbf{r}_m(t + \tau)]\}> \; , \quad (14)$$

or, using Eq. (5),

$$S_{AB}^{x}(\tau) = E_I^4 \, \alpha^4 N^2 S(k) \, S(k') \; . \tag{15}$$

However, if $\mathbf{k} = \pm \mathbf{k}'$

$$S_{AB}^{x}(\tau) = |< \sum_{i,j=1}^{N} \exp\{i\mathbf{k}\cdot[\mathbf{r}_i(t) - \mathbf{r}_j(t + \tau)]\}>|^2$$

$$+ \; |< \sum_{i,j=1}^{N} \exp\{i\mathbf{k}\cdot[\mathbf{r}_i(t) - \mathbf{r}_j(t)]\}>|^2 \; , \quad (16)$$

or

$$S_{AB}^{x}(\tau) = |E_I|^4 \alpha^4 N^2 ([S(k,\tau)]^2 + [S(k)]^2). \tag{17}$$

This is identical to the normal homodyne spectrum, except that the experiment uses two different scattering vectors, $\mathbf{k}$ and $\mathbf{k}' = -\mathbf{k}$. The experiment works because the fluctuation $a_k(t)$ causing the scattering is the same from the back and from the front.

The optical train in Fig. 6 is much more elaborate than the conventional one seen in Fig. 1, but both trains give the same spectrum. Why, then, other than as a tour de force, should one bother with the more complicated experiment? The important virtue of the two-detector experiment, as has now been confirmed experimentally, is that multiple scattering does not influence the time dependence of the cross-spectrum $S_{AB}^{x}(\tau)$. If double scattering occurs, for example, both detectors receive single and double-

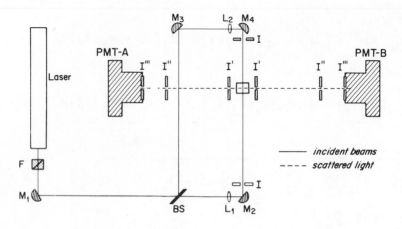

FIGURE 6.   Actual optical train [12] used in the experiment in Fig. 4 (F, filter; M, mirror; BS, beam splitter; L, lens; I, iris; PMT, photomultiplier tube.

scattered photons, so that $\langle I(t)I(t + \tau)\rangle$ obtained by autocorrelation of the signal from either detector includes not only the spectrum due to correlations among single-scattered photons, but also time-dependent terms due to correlations between double-scattered photons. In contrast, the cross-spectrum $S_{AB}^{x}(\mathbf{k},t)$ reflects only the correlations [Eq. (17)] between single-scattered photons. While multiple-scattered photons do reach both detectors, their arrival times are uncorrelated; to order $(V_c/V_s)^2$, they contribute to $S_{AB}^{x}(\mathbf{k},t)$ only a time-independent baseline.

We now give a formal demonstration of this claim. The wavevectors for both experiments are shown in Fig. 5. Single-scattering events occur in the small volume A. In a typical double-scattering event, light is scattered from the incident beam(s) in volume B and scattered again into wavevectors $\pm \mathbf{k}_F$ in volume C. Between B and C the wavevector of the light is $\mathbf{k}_s$. If $\mathbf{q}$ is introduced, where $\mathbf{q}$ is the scattering vector at B [$q = k_s - k_I$], the one-detector double-scattering spectrum is

$$\langle I_2(\mathbf{k},t)I_2(\mathbf{k},t+\tau)\rangle = |E_I|^4\alpha^8$$

$$\times \ |<\sum_{i,j=1}^{N_B} \exp\{i\mathbf{q}\cdot[\mathbf{r}_i(t) - \mathbf{r}_j(t+\tau)]\}>|^2$$

$$\times \ |<\sum_{m,n=1}^{N_C} \exp\{i(\mathbf{k}-\mathbf{q})\cdot[\mathbf{r}_m(t)$$

$$- \mathbf{r}_n(t+\tau)]\}>|^2 \ , \qquad (18)$$

where N_B and N_C are the number of particles in volumes B and C. From rule ii the right-hand-side of this equation is of the order $N^4m^4|E_I|^4\alpha^8$. The double-scattering contribution to the two-detector cross-spectrum is

$$\langle I_A(t)I_B(t+\tau)\rangle = <E_0^4\alpha^8 \ |\sum_{i=1}^{N_B} \exp[i\mathbf{q}\cdot\mathbf{r}_i(t)]|^2$$

$$\times \ |\sum_{j=1}^{N_B} \exp[i(\mathbf{q}+2\mathbf{k}_I)\cdot\mathbf{r}_j(t+\tau)]|^2$$

$$\times \ |\sum_{m=1}^{N_C} \exp[i(\mathbf{k}-\mathbf{q})\cdot\mathbf{r}_m(t)]|^2$$

$$\times \ |\sum_{n=1}^{N_C} \exp[i(\mathbf{k}-\mathbf{q}-2\mathbf{k}_F)\cdot\mathbf{r}_n(t+\tau)]|^2> \ . \quad (19)$$

In Eq. (19) the photons are presumed to have passed from lasers 1 and 2 to detectors A and B, respectively, so that for the first photon the scattering vectors are q and $k - q$, whereas for the other photon the scattering vectors are $\mathbf{k}_s - (-\mathbf{k}_I) = \mathbf{q} + 2\mathbf{k}_I$ and $-\mathbf{k}_F - \mathbf{k}_s = \mathbf{k} - \mathbf{q} - 2\mathbf{k}_F$. It can readily be shown that the leading nonzero terms of Eq. (19) are much smaller than the lead terms of Eq. (18). In the sums, the pairs of characters (i,j) and (m,n), which may be pairwise identical, by definition refer to distinct particles at B and C, which are separated by the vector $\mathbf{k}_s$. If B and C are nonoverlapping, ensemble averages over their contents factor, so that

$$\langle I_A(t) I_B(t + \tau)\rangle$$

$$= |E_I|^4 \alpha^8 \; |< \sum_{i,j=1}^{N_B} \exp\{i\mathbf{q}\cdot[\mathbf{r}_i(t) - \mathbf{r}_j(t + \tau)]$$

$$- 2i\mathbf{k}_I\cdot\mathbf{r}_j(t + \tau)\}>$$

$$\times < \sum_{m,n=1}^{N_C} \exp\{i(\mathbf{k} - \mathbf{q})\cdot[\mathbf{r}_m(t) - \mathbf{r}_n(t + \tau)$$

$$+ 2i\mathbf{k}_F\cdot\mathbf{r}_n(t + \tau)\}>|^2 . \tag{20}$$

By rule i, the $\mathbf{k}_I \cdot \mathbf{r}_j$ and $\mathbf{k}_F \cdot \mathbf{r}_n$ terms force these averages to zero. If regions B and C do overlap, Eq. (19) rearranges to

$$\langle I_A(t) I_B(t + \tau)\rangle$$

$$= |E_0|^4 \alpha^8 \; | \sum_{i,j,m,n=1}^{N} <\exp\{2i\mathbf{k}_F\cdot[\mathbf{r}_j(t)$$

$$- \mathbf{r}_n(t + \tau)]\}$$

$$\times \exp\{i\mathbf{k}\cdot[\mathbf{r}_m(t + \tau) + \mathbf{r}_n(t + \tau) - 2\mathbf{r}_j(t)]\}$$

$$\times \exp\{i\mathbf{q}\cdot[\mathbf{r}_i(t) + \mathbf{r}_j(t)$$

$$- \mathbf{r}_m(t + \tau) - \mathbf{r}_n(t + \tau)]\}>|^2 . \tag{21}$$

Since $j \neq m,n$ and $i \neq m,n$, from rule ii this is nonzero only if i,j,m,n all refer to particles in the same correlation volume; the right-hand-side of this equation is thus of order $N^2 m^6 |E_I|^4 \alpha^8$. By comparison with Eq. (18), the double-scattering contribution to $S_{AB}^x(\mathbf{k},t)$ is less than its contribution to $S_{AA}(\mathbf{k},t)$ by a factor $m^2/N^2 \sim (V_c/V_s)^2$.

As seen in Fig. 7, experiment confirms these predictions [12]. The apparent mutual diffusion coefficient of 0.091 μm polystyrene spheres in 0.4 g/liter sodium dodecyl sulfate-water solution was obtained from $S_{AA}(\mathbf{k},t)$ and $S_{AB}^x(\mathbf{k},t)$ by means of a cumulant-fitting procedure. At low volume fraction ϕ of spheres, the two determinations agree. For $\phi \gtrsim 10^{-3}$, D_m^{AA} obtained from $S_{AA}(\mathbf{k},t)$ increases rapidly with increasing sphere concentration, whereas D_m^{AB} as obtained from $S_{AB}^x(\mathbf{k},t)$ is independent of concentration. These are the expected results, too. For $\phi \lesssim 0.01$, D_m should be nearly independent of concentration; when multiple scattering (as evidenced by the appearance of a one-detector depolarized scattering spectrum) becomes important, D_m^{AA} should increase substantially, as is actually found. Control experiments are consistent with this interpretation of $S_{AB}^x(\tau)$:

1. If either incident beam is blocked, $S_{AB}^x(\tau) = 0$ from the wavevector-matching ($\mathbf{k}' = -\mathbf{k}$) condition.

2. There is no VH depolarized cross-spectrum, even when the one-detector VH spectrum is intense, because all depolarized light is due to multiple scattering, whose intensity fluctuations lack cross-correlations.

This assumes that the incident light is a plane wave. If the incident beams had too short a focal length, their wave fronts would be significantly nonplanar within the scattering volume, a potential difficulty averted by use of long-focal-length ($\sim$ 50 cm) focusing lenses. An identical constraint applies to the

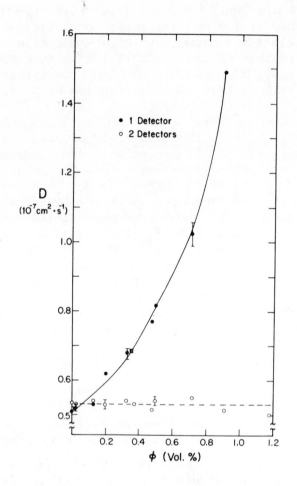

FIGURE 7. Comparison of one-detector measurements (filled circles) of D on a turbid sample with data taken by use of the twin-detector-beam system shown in Fig. 6 (open circles), showing the elimination of multiple scattering effects by the latter method.

collecting optics. In the working experiment, no lenses were placed between the sample cell and the photodetectors.

It was initially expected that it would be necessary to perform spatial filtering of the incident light [11]. This turns out not to be necessary, even though the cross-correlation efficiency [the rate of increase in the amplitude of $S_{AB}^{x}(\mathbf{k},0)$] is about 10% of that predicted theoretically.

This two-beam, two-detector system permits the study of systems that are substantially too turbid to be studied by conventional means. The work by Sorenson et al. [14] has shown that it is also possible to correct for multiple scattering by calculating the double-scattering spectrum from the single-scattering spectrum and the angular scattering cross sections of the particles in solution. Our optical system replaces this mathematical subtraction with a physical one, albeit at an appreciable increase in the optical train. If sufficiently good tables and mirror mounts are available, the physical scheme has perhaps an advantage over the spectral fitting procedure in that the physical method apparently rejects all orders of multiple scattering automatically, whereas the curve-fitting procedure entails further calculations for each order of multiple scattering that is to be rejected. An extension of the above proof to third, fourth, ..., nth-order multiple scattering, implied by the behavior of D_{AB}^{x} when $D_{AA}/D_{0} \geq 3$, needs to be done. Furthermore, to correct the one-detector spectrum mathematically, certain auxiliary data are needed, including $S(k)$ and the beam profiles of the incident and collected light, which are not required for interpretation of the two-detector data. This technique should be of substantial importance in the study of detailed lineshapes of critical systems near T_{C}, concentrated strongly scattering colloid suspensions, and similar solutions.

We briefly note the existence of bispectral analysis: the study of three-time correlation functions [Eq. (9)]. Three-time intensity correlation functions arise most naturally if the intensities involved are not joint Gaussian random variables; for examples, see Sections III and IV. For a Gaussian experiment, bispectral analysis increases the number of independent data points in a correlation measurement, without reducing the bin width (sample time), thereby preserving the signal/noise ratio of each data point. Furthermore, as shown in reference 15, if one uses a single heterodyne detector and single clips the signal at the average intensity, one can effectively measure odd-order correlation functions such as $<|a_k^+(0)a_k(t)a_k(\tau)|>$, with "+" denoting a clipped signal. The odd moment is nonvanishing because of the signal clipping, which for heterodyne detection has the effect of determining the phase of the scattered field and limiting the time average to those periods when the phase of the clipped field is positive. As little theoretical work has been done on the behavior of odd moments of random signals, whether for Gaussian or non-Gaussian variables, the applications of this result are not clear.

III. NON-GAUSSIAN EXPERIMENTS

If V_s/V_c is small, the scattered field E arises from only a few independent sources, so the statistics of E need not be those of a Gaussian random variable; nor need the multitime field distribution have a joint Gaussian form. The study of scattering and other processes in very small regions is the domain of fluctuation spectroscopy, recently reviewed by Weissman [17].

The non-Gaussian experiments mentioned in Section I attain the condition $V_c \sim V_s$ by reducing the scattering volume and making the number of scattering particles small. In these experiments, the time-dependent signal is caused by thermal fluctuations in the sample. Such fluctuations are intrinsically incoherent, having only short-range correlations.

An alternative way of obtaining non-Gaussian behavior is to make a coherent perturbation of the sample, so that either the motions or the positions of the scatterers are correlated over V_s. One example of a coherent perturbation in which $V_c^D \sim V_s$ is provided by light-scattering electrophoresis, in which a uniform electric field is applied to the system [18-20]. Another dynamic perturbation leading to $V_c^D \sim V_s$ is provided by turbulence in the sample [10].

We briefly note a pair of non-Gaussian multi-beam, one-detector methods, fluorescence photo-bleaching recovery [21] (FPR) and forced Rayleigh scattering (FRS) [22]. Fluorescence photobleaching recovery is descended from the non-Gaussian fluores-cence correlation spectroscopy (FCS) technique due to Magde et al. [23], in which fluctuations in the number of fluorescently labeled objects in a small volume of solution are studied by measuring changes in the intensity of fluorescent light emitted from the volume. The number of objects in the volume is inconstant because the objects can diffuse into or out of the volume. The particular translational diffusion coefficient responsible for particle motion depends on the fraction of diffusing particles that are fluorescently labeled. If all the diffusing particles are labeled, fluorescence correlation will measure the mutual diffusion coefficient D_m (in a two-dimensional system, mutual lateral diffusion coefficient); if only a few of the diffusing par-ticles are labeled, the same procedure will determine the tracer diffusion coefficient D_t of the object [24]. Measurement of D_m with FCS in a system in which $D_m \neq D_T$ is difficult. To label a substantial fraction of the molecules in an interacting (that is, concentrated) solution, while simultaneously keeping the total number of labeled molecules sufficiently small so that the fluorescence fluctuations are detectable, is to attempt to satisfy two contra-dictory physical requirements.

Fluorescence photobleaching recovery [21], among its other virtues, provides an escape from this quandary. In this technique the sample volume is illuminated at different times by two laser beams. In an idealized form of the experiment, a first high-powered, pulsed beam acts to eliminate some or all of the fluorophores in the observed volume. A second beam, of much lower power, is used to monitor the recovery of fluorescence as fresh fluorophores diffuse into the (perhaps two-dimensional) volume being observed. This is a coherent non-Gaussian experiment, in which the concentration of labeled molecules is perturbed over all V_s in a uniform way.

A more elaborate multibeam, single-detector experiment is provided by FRS [22] in which the sample is briefly perturbed by a nonparallel pair of incident beams, and the relaxation of the perturbation is monitored by a third beam and a single detector. The nonparallel ("forcing") pair of beams are obtained by dividing the output of a single laser into two parts and focusing both parts on the same volume by means of a mirror train. The forcing beams create in space a three-dimensional interference pattern − a grid − of bright and dark regions. The perturbation of the sample, due to heating, the Soret effect, and other causes, thus varies periodically in space. The third laser beam and detector are so positioned that their scattering vector matches the wavevector of the grid set up by the forcing beams. Forced Rayleigh scattering studies the relaxation of an externally induced periodic inhomogeneity of the sample.

A highly ingenious non-Gaussian two-detector, one-beam experiment has recently been demonstrated by Griffin and Pusey [7,25]. The geometry is that of Fig. 4 with one laser removed. Non-Gaussian statistics are obtained by using a very dilute solution of large scatterers (tobacco mosaic virus), and by using small limits ($V_s \sim 6 \times 10^{-12}$ cm^3) on the incident beam and on the collecting optics. This requires placement of two pairs of "1 µm pinholes" (actually,

30 μm pinholes and 25× magnifiers) so that they view the same volume of solution, which is an extreme test of skill at optical alignment.

The physical basis of the Griffin-Pusey experiment is the fact that the scattered field of a single nonspherical particle need not have conical symmetry. The amplitude of $E(|\mathbf{k}|,t)$ for one particle may depend on the azimuthal angle between the plane fixed by the incident and scattered beams and the plane fixed by the incident beam and a symmetry axis of the scatterer. A particle oriented favorably for scattering toward one detector will, for some particle geometries, be oriented unfavorably for scattering toward the other detector, so that intensity fluctuations at the two detectors will be anticorrelated. The apparent anticorrelation will depend on the difference between the numbers of particles oriented favorably for scattering toward each of the two detectors, which difference tends to increase only as the square root of the number N of particles in the beam, with the total intensity meanwhile increasing linearly with N. The procedure thus is effective only if N is small.

At short times, the predicted spectrum $S_{AB}(\tau)$ for Griffin and Pusey's system rises steeply with increasing τ as anticorrelations in the scattering in different directions are relaxed by the rotational diffusion of the tobacco mosaic virus particles. At longer times $S_{AB}(\tau)$ decays back to zero as fluctuations in N are relaxed by translational diffusion. The measured spectrum shows qualitative but not quantitative agreement with this model; Griffin and Pusey suggest that the published theory for the anticorrelations may not adequately take into account the finite profile of the incident beam.

IV. THE SEMI-GAUSSIAN EXPERIMENT

In closing we turn to an experiment that has never been performed, which because of its complexity and difficulty may prove impractical to perform, and yet — because of its great promise for our understanding of liquids — must be attempted. Even if this procedure is finally found to be ineffective, its consideration has already borne some fruit in that two methods mentioned earlier — multidetector methods for multiple-scattering suppression and bispectral analysis — were considered only as consequences of developmental analysis of this method: heterodyne coincidence spectroscopy [8].

The objective of the experiment is to measure the three-point density correlation function, as obtained [Eq. (6)] from the triple dynamic structure factor $S(\mathbf{k},\mathbf{q},t,\tau)$. The meaning of $S(\mathbf{k},\mathbf{q})$ is suggested by contrasting it with $S(\mathbf{k})$, which is the spatial Fourier transform of the radial distribution function $g(r)$. The term $S(\mathbf{k})$ gives information only on the spherically averaged distribution of scatterers around a given scatterer; $S(\mathbf{k},\mathbf{q})$ gives direct (as opposed to model-dependent) shape information. Thus, for example, the mean-square distance between pairs of scatterers may be obtained from $S(\mathbf{k})$. The term $S(\mathbf{k},\mathbf{q})$ gives average information on which triangles of scatterers are most likely to be found. For example, rodlike groupings of scatterers would enhance $S(\mathbf{k},\mathbf{q})$ for $\mathbf{k}||\mathbf{q}$, whereas disklike configurations would also emphasize $S(\mathbf{k},\mathbf{q})$ for $|\mathbf{k}| \sim |\mathbf{q}| \sim |\mathbf{k} + \mathbf{q}|$.

Three-point correlation functions are of major importance in the theory of fluids since most statistical-mechanical approaches to the properties of liquids, from the Kirkwood-Yvon integral equations [26,27] to the most modern treatments [28], yield hierarchical descriptions of liquids, in which n-body correlations can be calculated only from the (n + 1)-body correlations. A closure scheme of some sort is needed, such as an equation for the triplet correlations in terms of the pair correlations; the closure must also be tested. Egelstaff et al. [29] have related integral averages over the three-point functions to thermodynamic derivatives of $S(\mathbf{k})$, thus allowing testing of closure schemes, but only in a negative sense. Some forms can be rejected as wrong, but which form is right cannot be shown.

An obvious way to try to measure $S(\mathbf{k},\mathbf{q},t,\tau)$ is to study the bispectral form

$$B = <I(\mathbf{k},T)I(\mathbf{q},T + t)I(-\mathbf{k}-\mathbf{q},T + \tau)> \qquad (22)$$

by using three homodyne detectors, but this doesn't work [10]. One has

$$B = |E_I|^6\alpha^6 \quad < \sum_{i,j,l,m,n,p} \exp[i\mathbf{k}\cdot(\mathbf{r}_i - \mathbf{r}_j)$$

$$+ i\mathbf{q}\cdot(\mathbf{r}_l - \mathbf{r}_m) - i(\mathbf{k} + \mathbf{q})\cdot(\mathbf{r}_n - \mathbf{r}_p)]>. \qquad (23)$$

From rules i and ii stated in Section II, B is dominated by terms that put position vectors $\mathbf{r}_x$ pairwise into correlation volumes, doing so in such a way that the total wavevector of particles in each V_c is zero. Equation (23) is thus rewritten

$$B = |E_I|^6\alpha^6 \ [N^3 \ S(\mathbf{k})S(\mathbf{q})S(-\mathbf{k} - \mathbf{q})$$

$$+ N^2 | < \sum_{i,j,l} \exp\{i[\mathbf{k}\cdot\mathbf{r}_i(T) + \mathbf{q}\cdot\mathbf{r}_j(T + t)$$

$$- (\mathbf{k} + \mathbf{q})\cdot\mathbf{r}_l(T + \tau)]\}>|^2] \ , \qquad (24)$$

so B contains a time-independent term of order N^3 plus an interesting time-dependent signal that is weaker than its background by a factor N^{-1}. Actually, only the fluctuations in the background are of significance, but the predetection signal/signal ratio is still like $N^{-1/2}$. Although digital photon correlation in principle affords arbitrary accuracy, to the author's knowledge no one has ever tried this experiment on a bulk sample. [For further analysis on systems with large correlation lengths l, see Swift [30]. In colloidal crystals [31], $l \rightarrow \sim \infty$, so this homodyne experiment might be useful for studying their nonlinear lattice dynamics.]

It is well-known that homodyne detection and heterodyne detection are equivalent for the study of Gaussian scattered fields, in that

$$<I_{hom}(\mathbf{k},t)I_{hom}(\mathbf{k},t + \tau)>$$

$$= a|<I_{het}(\mathbf{k},t)I_{het}(\mathbf{k},t + \tau)>|^2 + b , \qquad (25)$$

where a and b are unimportant experimental constants. Since the three-detector homodyne experiment does nothing useful, a three-detector heterodyne experiment sounds equally pointless. Such a conclusion is not justified because it implicitly assumes that $E(\mathbf{k},t)$, $E(\mathbf{q},t)$ and $E(-\mathbf{k}-\mathbf{q},t)$ are jointly as well as separately Gaussian.

We assumed $V_s \gg V_c$, so each scattered field is separately a Gaussian random variable. The term $<E(\mathbf{k},t)E(\mathbf{q},t)E(-\mathbf{k},\mathbf{q},t)>$ would necessarily vanish, if $E(\mathbf{k},t)$, $E(\mathbf{q},t)$, and $E(-\mathbf{k}-\mathbf{q},t)$ were also jointly Gaussian, but they are not. Whereas each scattered field is separately the sum of many independent random processes, the random processes responsible for scattering from a given V_c through $-\mathbf{k}-\mathbf{q}$ are not independent of the processes responsible for scattering through $\mathbf{k}$ and $\mathbf{q}$. The random processes that cause $E(-\mathbf{k}-\mathbf{q},t)$ are thus collectively not independent of those responsible for $E(\mathbf{k},t)$ and $E(\mathbf{q},t)$, so the central limit theorem does not apply to the joint field distribution function.

A three-detector heterodyne experiment would find the triple heterodyne intensity correlation function

$$C(t,\tau) = <I_{het}(\mathbf{k},T)I_{het}(\mathbf{q},T + t)I_{het}(-\mathbf{k}-\mathbf{q},T + \tau)>$$

$$= E_L^6 + E_L^4 E_I^2 N \alpha^2 [S(\mathbf{k}) + S(\mathbf{q}) + S(\mathbf{k} + \mathbf{q})]$$

$$+ E_L^3 E_I^3 N \alpha^3 S^{(3)}(\mathbf{k},\mathbf{q},t,\tau)$$

$$+ E_L^2 E_I^4 N^2\alpha^4 [S(\mathbf{k})S(\mathbf{q})$$

$$+ S(\mathbf{k})S(\mathbf{k} + \mathbf{q}) + S(\mathbf{q})S(\mathbf{k} + \mathbf{q})]$$

$$+ E_I^6 N^3\alpha^6 S(\mathbf{k})S(\mathbf{q})S(\mathbf{k} + \mathbf{q}) , \qquad (26)$$

with the second equality following from rules i and ii and Eqs. (4)-(6). For each power of E_L, only the nonzero terms of highest order in N are retained. The only time-dependent term is that in $S^{(3)}$, so in principle by measuring $C(t,\tau)$ for small and large values of t and τ, one can perform baseline subtraction to get $S^{(3)}$. This could also be said of the triple homodyne experiment represented in Eqs. (22)-(24). In that experiment, however, the signal s and background b were proportional to N^2 and N^3, respectively. The noise n is dominated by fluctuations in b, so $n \sim N^{5/2}$ and the predetection $s/n \sim N^{-1/2}$, which is highly unfavorable. In the heterodyne experiment, one may show that the signal/noise ratio (s/n) is more favorable. Define $E_L^2 = \rho NE^2$. The factors of $S(\mathbf{k})$, $S(\mathbf{q})$, $\ldots$ are of order 1. The signal is

$$s = \rho^{3/2}E_I^6 N^{5/2} S^{(3)}(\mathbf{k},\mathbf{q},t,\tau) \sim \rho^{3/2}E_I^6 N^{5/2} . \qquad (27)$$

The fundamental noise is of two sorts:

1. Fluctuations effectively in N, modulating $S(\mathbf{k})$, which are down from their original terms by $N^{1/2}$.

2. Laser fluctuations that modulate E_L; these depend on P, the fractional noise/bandwidth of the laser, and $\Delta\nu$, the bandwidth of the experiment.

Rewriting term by term, we find that the noise is

$$n = E_I^6\{N^3\rho^3 P \, \Delta v + 3[\rho^2 P \, \Delta v \, N^3 + 3\rho^2 N^{5/2}]$$

$$+ [\rho^{3/2} P \, \Delta v \, N^{5/2} + \rho^{3/2} N^2]$$

$$+ 3[\rho P \, \Delta v \, N^3 + 3\rho N^{5/2}] + N^{5/2}\} \, . \tag{28}$$

Only noise from the background is important; this has terms such as (1) $\rho^2 N^{5/2}$ and (2) $\rho^2 P \, \Delta v \, N^3$. Conventional heterodyne experiments choose $\rho \gg 1$; to maximize $s \sim \rho^{3/2} N^{5/2}$ relative to the n terms $\rho^2 N^{5/2}$ and $N^{5/2}$, one takes $\rho \sim 1$. If type (1) terms dominate the noise, $s/n \sim N^{5/2}/N^{5/2} = 1$. If type (2) terms dominate the noise, one has $s/n \sim 1/N^{1/2} P \, \Delta v$. For a modern laser, P (RMS value to 1 MHz) is $\sim 10^{-3}$. A 1/2-hr measurement gives $\Delta v \sim 10^{-3}$ Hz, so for type (2) terms, $s/n \sim 1/[N^{1/2}(10^{-3}/10^6) \, 10^{-3}] \sim 10^{12} N^{-1/2}$, which is negligible. The predetection signal/noise ratio due to type (1) terms is $\sim 1/7$.

The postdetection signal/noise ratio $(s/n)_{post}$ must also be considered. For one count of correlation per correlation time Γ^{-1} so that $(s/n)_{post}$ due to Poisson fluctuations in photon counting is of order 1, a single correlated photon per detector per correlation time, or $s \sim \rho^3 E_I^3 \, N^{5/2} \sim 1$, is needed. The background is larger than s by $N^{1/2}$, but b is the product of three intensities. For $(s/n)_{post} \sim 1$, at each detector the ratio of uncorrelated to correlated intensities must be

$$\frac{I_{unc}}{I_{corr}} \sim N^{1/6} \, , \tag{29}$$

a large but not insane ratio.

Which values of N are relevant? Supposing that the scattering volume is a 100 μm cube, for bovine serum albumin [molecular weight (MW) = 66,000 amu] at 200 g/liter, $N \sim 2 \times 10^{12}$. For a larger polymer or a colloidal crystal of large polystyrene latex spheres,

N may be $< 10^6$. For a critical fluid, such as nitrobenzene:n-hexane near its consolute point, the molecules may not be treated as independent scatterers. If the correlation length ξ is ~ 100 Å, $N \sim V_s \xi^{-3} = 10^{12}$. As $S(\mathbf{k},\mathbf{q},t,\tau)$ will have interesting structure only if $k\xi$ is significantly nonzero, $N \sim 10^{12}$ is a useful upper bound. One then has $I_{unc} < 10^2 I_{corr}$; for $I_{corr} > \Gamma^{-1}$ counts/sec, the signal/noise ratio will be $\sim$ unity if the detectors observe $\sim 10^2 \Gamma^{-1}$ counts/sec. Systems in which ξ is large and interacting systems in which N is small generally have relatively slow relaxations. If a single detector can operate at 10^6 counts/sec, the experiment can reach decay times $\Gamma^{-1} \geq 100$ μsec, or frequencies < 1.5 kHz. For the colloid suspensions or larger values of ξ, $N < 10^{12}$, and substantial shorter decay times are accessible. The development of photon-counting detectors with higher limiting photocurrents could improve the frequency limits substantially.

This proposed three-wavevector experiment involves several unique optical problems. Since the angle between the incident beam and the scattering vector is necessarily obtuse, at least two incident beams are required. With the exception of the degenerate case in which the wavevectors are $\mathbf{k},\mathbf{k},-2\mathbf{k}$, three detectors are also needed. The alignment requirements are the same as those solved in our two-beam, two-detector homodyne experiment; the detectors must be positioned to permit observation of a matched set of coherence areas, requiring placement of the detectors with an accuracy of 0.5 mm at 1 m from the scattering volume. This alignment appears to be possible for scattering angles other than 90°, although this has not been shown experimentally.

Because this is a *multibeam* heterodyne instrument, there is the unique further requirement that the relative phases of the heterodyne reference beams must be controlled. This is an obscure idea; the author first notes what is *not* meant:

1. Phase *matching* of the scattered and reference beams over the photodetector, so that the beams are coplanar. This is required but is not more serious than in conventional heterodyne work.

2. Phase *stability* of the reference beams. A twin-beam heterodyne detection system is also a seismometer, recording relative motions of sample and reference mirror.

3. Phase *drift* of the laser field. If the laser frequency wanders rapidly, a difference in the path lengths of the scattering and reference beams will appear as an oscillation in the relative phases of the two beams at the detector.

For a 1 cm difference in the two beam path lengths, this effect would require a significant change in the phase of laser over ~ 30 psec; that is, this phase drift noise vanishes if the laser is stable in the 5 GHz range.

The actual phase matching problem is in the relative path lengths of the reference beams. A homodyne experiment, with no such beams, is sensitive to both sine and cosine wave fluctuations in the density of scatterers. A one-detector heterodyne experiment responds only to sine or cosine wave fluctuations, that is, if there is a fluctuation a_k that gives a maximal signal for its magnitude $|a_k|$ at a heterodyne detector, and if all the scatterers are displaced at a uniform quarter of a scattering wavelength in a direction parallel to the scattering vector, the signal will vanish because $E_L^*E(\mathbf{k}, t)$ will be purely imaginary. Provided the reference beam is constant, the limitation to only sine (or cosine) waves is not important since the detector will respond to waves of the same phase at t and $t + \tau$. However, in a three-detector experiment the relative reference beam phases at the three detectors are significant. This was implicitly assumed in Eqs. (6) and (26), in which a particle at the fixed point $\mathbf{r} = 0$ gave rise to scattered fields of phases $\exp[i\mathbf{k}\cdot0]$

$= 1$, $\exp[i\mathbf{q}\cdot 0] = 1$, and $\exp[-i(\mathbf{k} + \mathbf{q})\cdot 0] = 1$.   To obtain this condition in an experiment, one must modulate the path length of one of the reference beams as by putting a variable pressure gas cell into the system.   If the above-mentioned unique optical problems are soluble, heterodyne coincidence spectroscopy should be a workable experimental technique.

V. CONCLUSIONS

In summary, we have treated a variety of modifications of the familiar QELS spectrometer. By changing the number of incident beams, the sample volume, and the signal analysis, we are in principle able to obtain information about the single-scattering spectrum, particle shapes, and three-point correlations, which would be difficult or impossible to obtain from the usual procedure. We are reminded of a comparison with nuclear magnetic resonance, in which the classical experiment — in which absorption is measured at a fixed RF frequency as the applied field is swept — has been supplanted by Fourier transform, multiple frequency, and multipulse methods that allow one to study systems under otherwise inaccessible conditions. The more elaborate optical techniques discussed in this chapter may never replace the classical QELS spectroscopic experiment. However, the analogy with magnetic resonance and the paucity of groups working on these new methods in light-scattering spectroscopy suggest that there may be significant rewards awaiting those willing to invest the effort to undertake developmental work.

ACKNOWLEDGMENTS

The partial support of this work by the donors of the Petroleum Research Fund, administered by the American Chemical Society, is gratefully acknowledged. This work was in part generously supported by the National Science Foundation under Grant CHE79-20389.

REFERENCES

1. B. Chu, *Laser Light Scattering*, Academic Press, New York, 1974.

2. The replacement of time averaging with ensemble averaging is, with a unique exception, a theoretical calculation, not an operational method. Some references define S as $\lim_{T\to\infty} T^{-1}$ of our S; operationally, experiments run for a finite time T and are seldom normalized by the observation time.

3. T. Svedberg, Kolloid-Zeitschrift 7, 5 (1910).

4. A. Westgren, Arkiv Matematick, Astron. Fys. *11* (8 and 14)(1916); *13* (14) (1918).

5. S. A. Rice and P. Gray, *The Statistical Mechanics of Simple Liquids*, Wiley, New York, 1965.

6. D. W. Schaeffer and B. J. Berne, Phys. Rev. Lett. *28*, 475 (1972).

7. P. N. Pusey, in *Light Scattering in Liquids and Macromolecular Solutions*, V. Degiorgio, M. Corti, and M. Giglio, Eds. Plenum Press, New York, 1980.

8. G. D. J. Phillies, Molec. Phys. *32*, 1695 (1976).

9. H. C. Burstyn, R. F. Chang, and J. V. Sengers, Phys. Rev. Lett. *44*, 410 (1980).

10. B. Crosignani, P. DiPorto, and M. Bertolotti, *Statistical Properties of Scattered Light*, Academic Press, New York, 1975, especially p. 197.

11. G. D. J. Phillies, J. Chem. Phys. 74, 260 (1981).

12. G. D. J. Phillies, Phys. Rev. A 24, 1938 (1981).

13. As found, for example, in reference 10.

14. C. M. Sorenson, R. C. Mockler and W. J. O'Sullivan, Phys. Rev. A 14, 1520 (1976); 17, 2030 (1978).

15. G. D. J. Phillies, J. Chem. Phys. 72, 6123 (1980).

16. G. D. J. Phillies, J. Chem. Phys. 74, 5333 (1981).

17. M. B. Weissman, Ann. Rev. Phys. Chem. 32 (1981).

18. B. R. Ware and W. H. Flygare, Chem. Phys. Lett. 12, 81 (1971).

19. E. E. Uzgiris, Opt. Commun. 6, 55 (1972).

20. G. D. J. Phillies, S.M. thesis, Massachusetts Institute of Technology, Cambridge, Mass., 1971.

21. D. E. Koppel, D. Axelrod, J. Schlesinger, E. L. Elson, and W. W. Webb, Biophys. J. 16, 1315 (1976).

22. D. W. Pohl, S. E. Shwarz, and V. Irniger, Phys. Rev. Lett. 31, 32 (1973).

23. D. Magde, E. Elson, and W. W. Webb, Phys. Rev. Lett. 29, 705 (1972); E. L. Elson and D. Magde, Biopolymers 13, 1 (1974); D. Magde, E. L. Elson and W. W. Webb, Biopolymers 13, 29 (1974).

24. G. D. J. Phillies, Biopolymers 14, 499 (1975).

326 G. D. J. Phillies

25. W. G. Griffin and P. N. Pusey, Phys. Rev. Lett. *43*, 1100 (1979).

26. J. G. Kirkwood, J. Chem. Phys. *3*, 300 (1935).

27. J. Yvon, Actualites Scientifiques et Industrielles, Vol. 203, Herman et Cie, Paris (1935).

28. A. D. J. Haymet, S. A. Rice, and W. G. Madden, J. Chem. Phys. *75*, 4696 (1981).

29. P. A. Egelstaff, D. I. Page, and C. R. T. Heard, J. Phys. C *4*, 1453 (1971).

30. J. Swift, Ann. Phys. *75*, 1 (1973).

31. R. M. Fitch, *Polymer Colloids II*, Plenum Press, New York, 1980; R. Williams and S. Crandall, Phys. Lett. A *48*, 225 (1974).

CHAPTER 11

Simultaneous Measurements of Aerodynamic Diameter and Electrostatic Charge on a Single-Particle Basis

M. K. MAZUMDER, R. E. WARE, and W. G. HOOD
University of Arkansas, Graduate Institute of Technology, Little Rock

CONTENTS

I.   INTRODUCTION

Electrical charge and aerodynamic diameter play important roles in the transport of aerosol particles. If the particle motion is in the Stokes

327

region, the friction coefficient f can be represented for a spherical particle of diameter d_p as

$$f = \frac{3\pi\eta d_p}{C_c} , \tag{1}$$

where C_c is the molecular slip correction factor and η is the viscosity of the medium. For a nonspherical particle, the friction coefficient depends on the orientation of the particle as it moves through the medium. For a particle of any size and shape, the quantity f can be written as

$$f = \frac{3\pi\eta d_a}{C_{ca}} , \tag{2}$$

where d_a is the aerodynamic diameter of the particle and C_{ca} is the slip correction factor referred to the aerodynamic diameter d_a. The electrical mobility of a particle is given by the ratio of its electrical charge q to the friction coefficient f. Thus

$$Z_e = \frac{q}{f} = qZ , \tag{3}$$

where Z is the mechanical mobility of the particle.

Particle charge is usually measured by employing the Faraday cage method originally developed by Masters [1] and later modified by a number of researchers [2]. The Faraday cage method is applicable to measurement of the average charge per unit mass and the presence of bipolar charge cannot be distinguished from unipolar charge distribution. Measurement of diameter and charge of individual particles was first performed by Millikan in his oil droplet experiment. Recently McDonald et al. [3] reported a method for measurement of the charge on individual particles by use of a Millikan cell. They applied this technique to measuring the charge on fly ash particles in the size range of 0.3-1.5 μm in diameter at ambient temperatures in the range of 38-343°C. Although this method produces useful data on the magnitude and polarity of the individual charge, it is inherently slow because it requires the visual observation of individual particles inside the Millikan cell.

The method discussed here using the E-SPART analyzer is capable of measuring a particle's charge and its aerodynamic diameter on a single-particle basis and at a rate of greater than 200 particles/sec. Similar in principle to that of a Millikan cell, the method is based on (1) the measurement of the electrical mobility of charged particles in an electric field of known frequency of oscillation and (2) the measurement of the phase lag of the particle motion with respect to the electric field. The aerodynamic diameter d_a is determined from the measured value of the phase lag. The phase lag depends on the frequency of oscillation and the aerodynamic diameter of the particle and is independent of the magnitude of the charge. Once the aerodynamic diameter is computed from the phase lag, the charge can be determined from the measured value of the electrical mobility.

II.   SIMULTANEOUS MEASUREMENTS OF d_a, q, AND q/m

The basic principle used in the E-SPART analyzer [4] for measuring the aerodynamic diameter, as well as the electrical charge of individual particles, is similar to the principle of operation of the acoustic single-particle aerodynamic relaxation time (A-SPART) analyzer [5]. In the A-SPART analyzer, the aerodynamic diameter d_a of a particle is determined by subjecting the particle to an acoustic field of known frequency and by measuring the oscillatory motion induced on the particle by using a frequency-biased dual-beam laser Doppler velocimeter (LDV) [5]. The particle motion lags behind the acoustic excitation because of the inertia of the particle; consequently, the phase lag ϕ between the particle motion and the acoustic excitation provides a measurement of the aerodynamic diameter of the particle.

In the E-SPART analyzer, which was used for this study, an electrical excitation replaces the acoustic excitation and the particles are electrically charged. Both the velocity amplitude and the phase lag of charged particles inside the analyzer are measured by a frequency-biased dual-beam LDV. The equation of motion of an electrically charged particle suspended in air within a uniform AC electric field can be written as

$$\tau_p \frac{dv_p}{dt} + v_p = Z_e E_d \sin(\omega t) , \qquad (4)$$

where v_p is the particle velocity, E_d is the amplitude of the driving electrical field of angular frequency ω, and τ_p is the aerodynamic relaxation time of the particle, given by

$$\tau_p = \frac{\rho_0 d_a^2 C_{ca}}{18\eta} , \qquad (5)$$

where d_a is the aerodynamic diameter, ρ_0 equals 1 g/cm^3, C_{ca} represents the Cunningham correction factor for molecular slip evaluated as a function of d_a, and η is the viscosity of air.

The electrical mobility Z_e is a function of both the charge q and the particle size [Eq. (3)].

The steady-state solution of Eq. (4) can be written as

$$v_p(t) = \frac{Z_e E_d}{(1 + \omega^2 \tau_p^2)^{1/2}} \sin(\omega t - \phi) , \qquad (6)$$

where the phase lag ϕ is related to the aerodynamic relaxation time τ_p of the particle by

$$\tan \phi = \omega\tau_p \ . \tag{7}$$

The phase lag ϕ is thus independent of both the driving field amplitude E_d and the particle charge q. From Eq. (6), the amplitude of the particle velocity is given by

$$V_p = \frac{Z_e E_d}{(1 + \omega^2\tau_p^2)^{1/2}} \tag{8}$$

which depends on q, E_d, and d_a. The amplitude V_p is determined by measuring the Doppler frequency shift of the LDV signal,

$$f_d = \frac{2V_p}{\lambda} \sin \frac{\theta}{2} \ , \tag{9}$$

where f_d is the Doppler shift, λ is the wavelength of radiation of the laser beam, and θ is the angle of intersection between the two laser beams (Fig. 1).

Since both the phase lag ϕ and the velocity amplitude V_p are simultaneously available for each particle, the E-SPART analyzer can measure the aerodynamic diameter d_a and the electrical charge q of individual aerosol particles in real time. From Eq. (5) and (7),

$$d_a = \left[\frac{18\eta \tan \phi}{\omega\rho_0 C_{ca}}\right]^{1/2} \ , \tag{10}$$

and from Eq. (3), (6), and (7)

$$q = \frac{3\pi\eta d_a\lambda f_d}{2E_d C_{ca} \sin(\theta/2)} (1 + \omega^2\tau_p^2)^{1/2} \ . \tag{11}$$

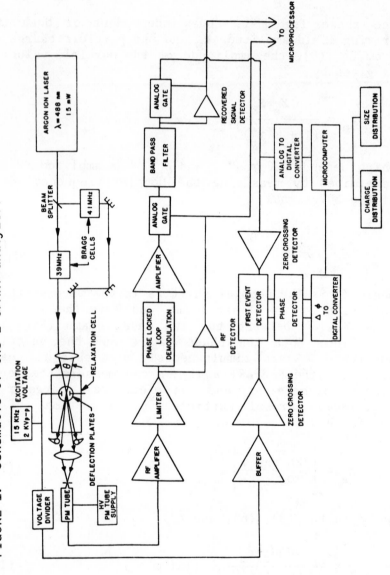

FIGURE 1.   Schematic of the E-SPART analyzer.

332

The relationship between q and f_d is linear in a given LDV geometry. Figure 2 shows the variation of f_d as a function of the number of the elementary charges for two aerodynamic diameters, 0.5 and 1.0 μm. From Eqs. (10) and (11), the charge/mass ratio (q/m_a) can be determined where m_a is the aerodynamic mass of the particle and is expressed as

$$m_a = \frac{\pi}{6} d_a^3 \rho_0 . \tag{12}$$

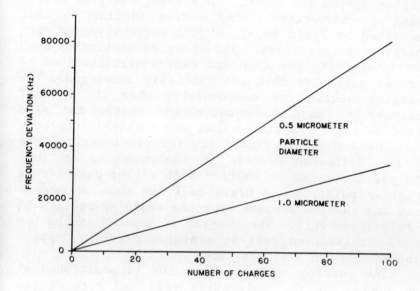

FIGURE 2. Variation of f_d as a function of the number of elementary charges for 0.5 μm and 1.0 μm diameter particles.

For a spherical particle of density ρ_p, the ratio of aerodynamic mass (m_a) to the actual mass (m_p) is given by

$$\frac{m_a}{m_p} = \left[\frac{\rho_p}{\rho_0}\right]^{1/2} \left[\frac{C_{cp}}{C_{ca}}\right]^{3/2} , \tag{13}$$

where C_{cp} is the Cunningham slip factor referred to the geometric diameter d_p.

III. EXPERIMENTAL SETUP

Figure 1 is a schematic diagram of the E-SPART analyzer. The high-voltage drive circuits, the laser Doppler velocimeter, the electronic data processing system, and the microprocessor-based storage and display system are shown. An aerosol charging device using a radioactive ^{241}Am source similar to that described by Fjeld et al. [7] is currently used when charging the particles. The charging section is used for determining the size and charge distributions of aerosol particles that are initially uncharged. The charging section is disconnected when the E-SPART analyzer is used to measure the charge and size distribution of particles that are initially charged.

The aerosol particles acquire electrical charge by the diffusion process in the charging section. The particles can be charged with either positive or negative polarity. A brass delivery tube is used to draw the charged aerosol from the charging section to a relaxation cell. The charger is mounted at the top of the relaxation cell to minimize particle deposition inside the delivery tube.

The sensing volume of the LDV is positioned at the center of the relaxation cell and between two deflection plate electrodes across which a 25 kHz high voltage AC signal (6500 V_{p-p}) is applied. The two electrodes are positioned symmetrically with respect to the LDV sensing volume. A signal generator, operating at 25 kHz, is connected to an RF power amplifier, and a step-up transformer is used to provide the high-voltage drive input to the deflection electrodes. A voltage divider monitors the applied voltage and serves as the reference signal for measuring the relative phase difference between the particle motion and the electrical drive signal.

The charged particles experience the oscillating electric field while passing through the LDV sensing volume. Light scattered by particles transiting the sensing volume is detected by a photomultiplier tube whose output is connected to the signal-processing circuits. These circuits, shown schematically in Fig. 1, recover the phase and amplitude of the induced velocity of each particle.

The phase measuring circuit measures the relative phase difference between the particle motion and the electrical drive signal and assigns a channel number based on the particle size corresponding to one of 30 microcomputer memory locations. Simultaneously an analog-to-digital converter (ADC) is used to measure the velocity amplitude of the particle, and its signal is processed through the microcomputer, which determines the particle charge based on the measurement of V_p and the size information indicated by the channel number. The polarity of the individual charges is measured by detecting the direction of motion of the particle with respect to the applied electric field.

The sensing volume of the LDV is approximately 10^{-5} cm^3. The instrument has 32 channels for size classification, and each channel has the capacity for counting 10^6 - 1 particles. For each of the 32 size channels, there are 128 channels for storing information on particle charge, 64 channels for positive charges, and 64 channels for negative charges. As each particle is sensed by the detector, the aerodynamic diameter, the charge, and the polarity are determined. However, if the particle is uncharged, no size measurement is performed and the particle is recorded as an uncharged particle. A Commodore PET microcomputer and a Hycom digital plotter are used in conjunction with the E-SPART analyzer to obtain the size and the charge distribution plots.

Current capabilities of the E-SPART analyzer include measurements in an aerodynamic size range of 0.3 - 10.0 μm in aerodynamic diameter with a maximum count rate of 200 particles/sec. For a 1.0 μm diameter-particle, the measurable range of elementary charge units is approximately 3-300 electronic charges per particle.

IV. RESULTS AND DISCUSSION

To determine the resolution of the E-SPART analyzer for measuring aerodynamic diameter, monodisperse aerosols containing polystyrene latex spheres (PLS) were used as standards. Figure 3 shows the distribution $dN/d[\log(d_a)]$ of PLS aerosols containing 0.620, 0.822, 1.09, and 2.02 µm particles as measured by the instrument. The distributions show that the resolution obtained was similar to that obtained from the A-SPART analyzer [5]. Figures 4 and 5 show charge distributions measured on PLS particles of 0.822 and 2.84 µm aerodynamic diameter, respectively.

Measurement of the particle charge performed by the E-SPART analyzer is compared with similar charge measurements by use of a Millikan cell. Such comparison is difficult unless aerosol size and charge distributions are both monodisperse because in using the E-SPART analyzer, the charge distribution is measured typically at a rate of 10-30 particles/sec; whereas in using the Millikan cell, several minutes are spent measuring the electrical charge on a single particle. Table 1 shows a comparative measurement between the Millikan apparatus and E-SPART analyzer.

The E-SPART analyzer can measure the aerodynamic diameter and the electrical charge of particles in a wide range by controlling the frequency and amplitude of the excitation voltage. The phase lag ϕ is given by the time interval Δt between the zero crossing of these two signals. The relationship between ϕ and Δt is

$$\phi = \omega \, \Delta t \, , \tag{14}$$

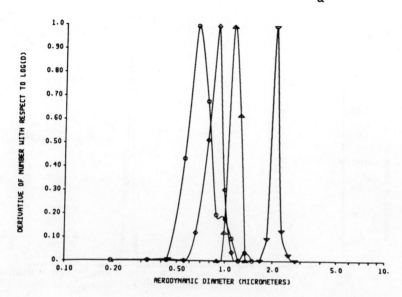

FIGURE 3. Normalized size distributions of mono-disperse aerosols containing 0.620, 0.822, 1.09, and 2.02 μm PLS particles as measured by the E-SPART analyzer.

where ω is the angular frequency of excitation. The signal processing circuit measures Δt by digitally counting the periods of an oscillator, the frequency of which is chosen to be much higher than the drive frequency. The analyzer has its greatest sensitivity in the particle size range where φ changes most rapidly as a function of particle size.

The E-SPART analyzer developed and tested in this study used a drive frequency of 25 kHz. The instrument operating at this frequency can be used to make measurements for particles in the range of 0.3-10.0 μm in aerodynamic diameter, with its greatest sensitivity for particles in the range of 0.8-2.0 μm. However, it is possible to operate the analyzer intermittently or simultaneously at two

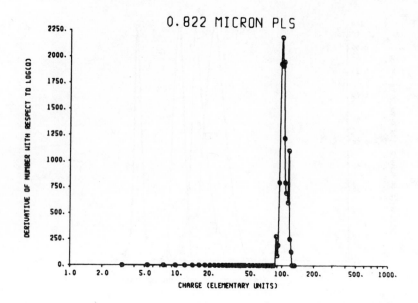

FIGURE 4.   Charge distribution of 0.822 µm diameter PLS particles as measured by the E-SPART analyzer.

frequencies in order to cover a wider range with a higher resolution.

For measuring submicron particles, it is desirable to increase the angle θ between the two incident laser beams resulting in a smaller fringe spacing. On the other hand, for larger particles, the beam angle should be small so as to increase the fringe spacing, which will provide better fringe visibility of the Doppler signal.

V.   APPLICATIONS

Currently, E-SPART analyzers are used in two applications as related to electrostatic filtration of aerosols: (1) electrostatic precipitation of high-resistivity fly ash and (2) measurement of the

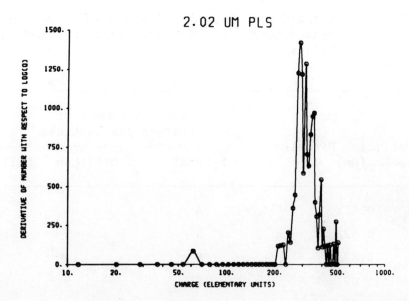

FIGURE 5.   Charge distribution of 2.02 μm diameter
PLS particles as measured by the E-SPART
analyzer.

efficiency of electrified fabric filters as a function of aerodynamic size and particle charge.

Measurement of the charge distribution acquired
by the particles inside an electrostatic precipitator
may give an insight into the effectiveness of charging particles as a function of the following variables: (1) the type of charging device; (2) the
distribution of electric field; (3) the geometry of
the discharge electrode; and (4) the resistivity of
fly ash inside the precipitator. Such measurements
are currently used as a diagnostic tool for determining the presence and the effect of back corona, particle sneakage, and particle reentrainment inside an
electrostatic precipitator.

In an electrified fabric filter the particle
collection efficiency varies with the electrical
charge of the particle as well as with the polarity.

340   M. K. Mazumder et al.

TABLE 1.   Comparison Between Average Charge (q)
Measurements by E-SPART Analyzer and a
Millikan Cell

| Particle Diameter (μm) | Average Number of Charges per particle | |
	E-SPART	Millikan Cell
0.822	4	7
0.822	5	9
0.822	8	11
1.09	7	9
1.09	8	13

Currently the E-SPART analyzer is used to test the
aerodynamic size and electrical charge distribution
of polydisperse test aerosols that are used to test
the efficiency of electrified fabric filters.

REFERENCES

1. J. I. Masters, An Aerosol Analyzer, Rev. Sci.
Instrum. *24*, 586-588 (1953).

2. G. Langer, J. Pierrard, and G. Yamak, Further
Development of an Electrostatic Classifier for
Submicron Airborne Particles, Internatl. J. Air
Water Poll. *8*, 167-176 (1964).

3. J. R. McDonald, M.H. Anderson, and R. B. Mosley.
Charge Measurements of Particles Exiting Electrostatic Precipitators, EPA 600/7-80-077 (1980).

4. R. G. Renninger, M. K. Mazumder, and M. K. Testerman. Particle Sizing by Electrical Single Particle Aerodynamic Relaxation Time Analyzer, Rev. Sci. Instrum. 52(2), 242-246 (1981).

5. M. K. Mazumder, R. E. Ware, J. D. Wilson, R. G. Renninger, F. C. Hiller, P. C. McLeod, R. W. Raible, and M. K. Testerman. SPART Analyzer: Its Application to Aerodynamic Size Distribution Measurement, J. Aerosol Sci. 10, 561-569 (1979).

6. M. K. Mazumder, Laser Doppler Velocity Measurement Without Directional Ambiguity by Using Frequency Shifted Incident Beams. Appl. Phys. Lett. 16, 462 (1970).

7. R. A. Fjeld, R. Graunitt and A. R. McFarland. Aerosol Charging by Bipolar Ions of Unequal Current Densities: Experiments and Low Electric Fields. J. Colloid Interface Sci. 83 (1), 82-89 (1981).

CHAPTER 12

Measurement of Diffusion Diameter Distribution on a Single-Particle Basis by Use of Dual-Beam Laser Doppler Velocimetry

D. K. HUTCHINS and M. K. MAZUMDER
University of Arkansas, Graduate Institute of Technology, Little Rock

CONTENTS

I. INTRODUCTION

Quasi-elastic light scattering (QELS) spectroscopy [1,2] has been successfully used to determine
the diffusion diameter distribution of particles in
suspension. Generally two experimental arrangements
of laser beam geometry are used: self-beating and
heterodyning. Both configurations employ a single
laser beam illuminating the suspension. In the
self-beating mode, the fluctuating intensity of the
scattered light incident on the photodetector is
correlated. In the heterodyning approach, the scattered light field is mixed coherently on the photodetector surface with a light field from a local
oscillator, usually derived from the same laser. In
most cases the scattered light incident on the
photodetector surface originates from a number of
particles illuminated by the laser beam.

If the suspension is monodisperse, the determination of the diffusion diameter is relatively
straightforward. The experimental methods have been
successfully demonstrated by several authors for
particles suspended in a fluid medium [3], including
aerosols [4]. However, if the particle suspension is
heterodisperse, one of the several relatively complex
inversion techniques is used to recover the size
distribution from the measured correlation function.
The complexity arises because the light intensity
correlation function becomes a sum of exponentials
related to each size interval in the distribution,
rather than a single exponential function as obtained
from the translational Brownian motion of monodisperse particles. An inversion technique, such as
Laplace inversion, requires an extensive computational facility, even for a unimodal distribution,
and becomes increasingly complex for multimodal
distributions of particles. Further complication
arises if the scattered signal is generated by particles of unknown size and shape.

If the particle size distribution is wide and
contains a significant number of particles larger

than 1.0 μm in diameter, the presence of an even small number of large particles within the sensing volumes has an overwhelming interference on the measurements of smaller submicron particles.

The present chapter describes a slightly different approach in which the diffusion coefficient D is determined on an individual particle basis. A frequency-biased dual-beam laser Doppler velocimeter (LDV) is used to obtain the correlation function of the translational Brownian motion of individual particles in a suspension. A sample of the suspension is drawn past the LDV sensing volume in a laminar flow field. The LDV sensing volume and the number density of particles in the suspension are both controlled to reduce the probability of the presence of more than one particle within the sensing volume at a given time. A breadboard apparatus has been employed in feasibility studies to measure the diffusion constant D of individual particles of polystyrene latex spheres (PLS) suspended in air as well as in water. The experimental data and the possible application of this method for determining the diffusion diameter distribution in a polydisperse suspension are briefly presented.

II.   EXPERIMENTAL APPROACH

A.   Measurement of Diffusion Coefficient D On a Single-Particle Basis: Single- versus Dual-Beam Optical Geometry

Figure 1 shows a typical beam geometry used in both self-beating (homodyning) and heterodyning systems. In a heterodyning system an additional reference beam is generated. One method of generating a reference beam is to place a small opaque object (stationary) inside the sensing volume to provide a constant intensity of light that is several times more intense than that of the signal. In another method, the reference beam is provided by the

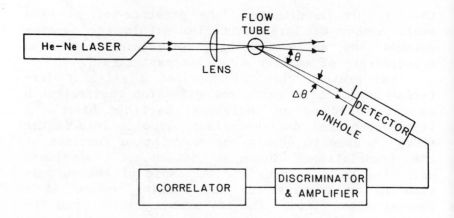

FIGURE 1.   Typical beam geometry used in self-beating and heterodyning spectroscopy.

scattered light from the cell wall or by using beam splitters that mix a portion of the laser beam so that the wavefronts of the signal and reference beams are parallel over the detector surface. Usually the number density of particles in a suspension is controlled so that a large number of particles is present inside the sensing volume, resulting in a Gaussian-Lorentzian light field on the detector surface, and yet the number density is small enough that the multiple scattering effect can be neglected. In the beam geometry shown in Fig. 1, the scattering vector K is defined by

$$K = K_s - K_0 ,\qquad(1)$$

where the magnitude of K can be expressed as

$$|K| = \frac{4\pi n}{\lambda} \sin \frac{\theta}{2} ,\qquad(2)$$

where n is the refractive index of the suspending medium. In this beam geometry the angular spread in light scattered into the detector $\Delta\theta$ must be sufficiently large to provide sufficient signal strength. However, $\Delta\theta$ must be minimized to avoid uncertainties in determining the value of **K**.

If the particle density in the suspension is now reduced to a point where the sensing volume contains only one scattering particle, the intensity of the signal beam will be greatly reduced. The decline in the intensity of the scattered light with the particle number means that both the laser power and the solid angle subtended by the aperture of the detector optics must be increased.

A convenient method of increasing the solid angle without increasing the ambiguity in determining the scattering vector **K** is to use a dual-beam system [5,6] as shown in Fig. 2. In this beam geometry the scattering vector **K** is defined by

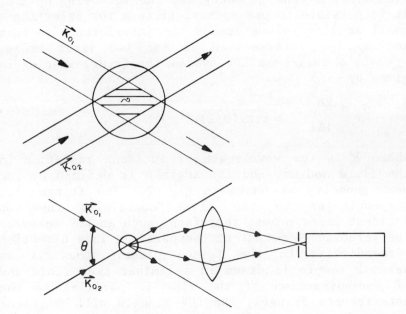

FIGURE 2.   Dual-beam   LDV   geometry   for   diffusion measurement.

$$K = K_{01} - K_{02} \; , \tag{3}$$

and

$$|K| = \frac{4\pi n}{\lambda} \sin \frac{\theta}{2} \tag{4}$$

where θ is the angle of intersection between the two incident laser beams and is independent of the direction of the scattered beam and the magnitude of the solid angle subtended by the receiver. For detection on a single particle basis, the sensing volume is reduced to a minimum in order to reduce the Rayleigh limiting diameter — the lower particle detection limit set by the Rayleigh scattering of light from the fluid molecules present in the sensing volume. Since the sensing volume is defined by both the transmission beam geometry and the receiving optics, it is possible to use spatial filters for selecting a small sensing volume inside the interference pattern set by the intersection of the two laser beams.

In a dual-beam LDV system the fringe spacing is given by

$$d_i = \frac{2\pi}{|K|} = \frac{\lambda}{2 \sin (\theta/2)} \; , \tag{5}$$

where λ is the wavelength of incident radiation in the fluid medium, and the angle θ is defined by the beam geometry as shown in Fig. 2. The fringes are perpendicular to the plane containing the two incident laser beams; therefore, such an LDV measures the velocity component of the particle in a direction perpendicular to the fringe plane. Thus if an aerosol sample is drawn in a laminar flow field and if the direction of the flow is parallel to the interference fringes, the LDV signals will be insensitive to the particle velocity in the direction of sample flow. However, the translational Brownian motion of the particle will have a component U_B in the direction perpendicular to the fringes and will generate a Doppler signal of frequency f_D at the output of the photodetector, given by

$$f_D = \frac{2U_B}{\lambda} \sin \frac{\theta}{2} \ . \tag{6}$$

Since the Brownian displacements are small, particularly for those particles with diameters greater than 0.5 µm, the particles may not traverse a significant number of fringes as they transit through the sensing volume. For example, a 0.5 µm-diameter particle in air will diffuse approximately 10 µm/sec. If the residence time of the particle is approximately 10 msec inside the sensing volume, the displacement is only 1.0 µm. For $\theta = 80°$ (Fig. 2), the fringe spacing is approximately 0.5 µm, and thus the number of fringe crossings (Doppler cycles) will be about two. To detect such small displacement, it is possible to employ a frequency-biased [7] LDV where the frequencies of the two incident beams indicated by the propagation vectors K_{01} and K_{02} are slightly different. Such frequency-biasing techniques are frequently used in laser Doppler velocimetry to avoid directional ambiguity where velocity reversals are possible. For example, if the frequency difference between the two beams is 5 kHz, a particle residing for 10 msec inside the sensing volume will generate 50 Doppler cycles. However, the average number of photons scattered and the number of correlation channels per Doppler cycle must be considered in selecting the bias frequency. As is shown in the next section, such biasing would generate an exponentially decaying cosine function at the output of the correlator as each particle transits through the moving fringes.

B.   Measurement of Translational Diffusion Coefficient

The translational diffusion of a small particle suspended in a fluid is a random process that can be characterized in one dimension by the conditional probability $[P(x,t;0,0)dx]$ that a particle initially located at $x = 0$ at $t = 0$ will be found within a distance dx of position x at some later time t. If the distance x contains many steps in the random walk of the particle, $P(x,t;0,0)$ is given by [8]

$$P(x,t;0,0) = (4\pi Dt)^{-1/2} \exp \frac{-x^2}{4Dt} \qquad (7)$$

where D is the diffusion constant of the particle. For a spherical particle in air, the diffusion constant D is given by [9,10]

$$D = \frac{k_B T C_c}{3\pi \eta d} , \qquad (8)$$

where k_B is Boltzmann's constant, T is the absolute temperature, C_c is the Cunningham slip correction factor referred to the diameter d of the particle, and η is the viscosity of air.

A particle transiting through moving interference fringes within the LDV sensing volume of a frequency-biased dual-beam LDV will scatter light with a time-varying intensity $I(t)$ given by

$$I(t) = I_0 \exp\{i[\omega t + \phi(t)]\} + I_b , \qquad (9)$$

where $\phi(t)$ is a random phase factor due to the diffusive motion of the particle and ω is the angular bias frequency $(\omega_{01} - \omega_{02})$ between the two laser beams. The intensity I_b represents the term centered at zero frequency.

The autocorrelation function $G(\tau) = \langle I^*(t)I(t + \tau)\rangle$ of this scattered intensity is given by

$$G(\tau)$$

$$= I_0^2 \exp(i\omega\tau)\langle\exp\{i[\phi(t + \tau) - \phi(t)]\}\rangle + I_b^2 . \qquad (10)$$

If the motion of the particle in the direction perpendicular to the interference fringes is represented by its instantaneous position $x(t)$, the phase factor $\phi(t)$ is given by

$$\phi(t) = \frac{2\pi}{d_i} \cdot x(t) , \qquad (11)$$

where d_i is the fringe spacing given by Eq. (5). Thus

$$\phi(t) = |K|x(t) = Kx(t) . \qquad (12)$$

Substitution of the value of $\phi(t)$ in Eq. (10) yields

$G(\tau)$

$$= I_0^2 \exp(i\omega\tau)\langle\exp\{iK[x(t + \tau) - x(t)]\}\rangle + I_b^2 . \qquad (13)$$

Since the Brownian motion is a stationary process with respect to the time frame,

$$\langle\exp\{iK[x(t + \tau) - x(t)]\}\rangle = \langle\exp[iKx(\tau)]\rangle. \qquad (14)$$

Thus, from Eq. (7),

$\langle\exp[iKx(\tau)]\rangle$

$$= \int_{-\infty}^{+\infty} (4\pi D\tau)^{-1/2} \exp(iKx) \exp\frac{-x^2}{4D\tau} dx . \qquad (15)$$

The autocorrelation function $G(\tau)$ can be written from Eq. (13) in the form

$$G(\tau) = I_0^2 \exp(i\omega\tau) \exp(-DK^2\tau) + I_b^2 , \qquad (16)$$

or the real part used for the measurement

$$G(\tau) = I_0^2(\cos \omega\tau) \exp(-DK^2\tau) , \qquad (17)$$

where ω is the bias frequency — the frequency difference between the two incident laser beams.

The above analysis can be summarized as follows. The Brownian motion of the particle perpendicular to the direction of laminar flow causes phase changes in the periodic signal scattered by the particle. The scattered light is detected by a photomultiplier tube (PMT), and the output of the PMT is fed to a digital autocorrelator, resulting in an exponentially decaying cosine function. The diffusion constant, and hence the diffusion diameter, of the particle can be obtained from the autocorrelation function of the scattered light intensity.

III.   EXPERIMENTS AND RESULTS

A schematic of the experimental setup is shown in Fig. 3.   A test aerosol is drawn from a 4-liter flask through a glass tube 1.0 cm in diameter and 50 cm in length.   Aerosol flows at a rate of approximately 10 cm^3/min, corresponding to a maximum velocity of 0.5 cm/sec at the center of the tube, where the sensing volume of the LDV is located.   The flow Reynolds number is maintained well below turbulent flow.   The average transit time of the particles through the sensing volume of the LDV is estimated to be 30 msec.

The LDV uses a helium-neon (He-Ne) laser (λ = 632.8 nm) with 15 mW of output power.   The sensing volume is formed by the intersection of two beams derived from the laser and by the geometry of the receiving optics.   The angle θ of beam intersection is 80°.   The geometry of the receiving optics is adjusted to locate the sensing volume at the center of the interference fringe pattern so that the Doppler signal generated by a stationary particle can be approximated by a waveform of constant amplitude irrespective of its location within the sensing volume.   The sampling time for autocorrelation is kept smaller than the average transit time of the particles through the sensing volume.

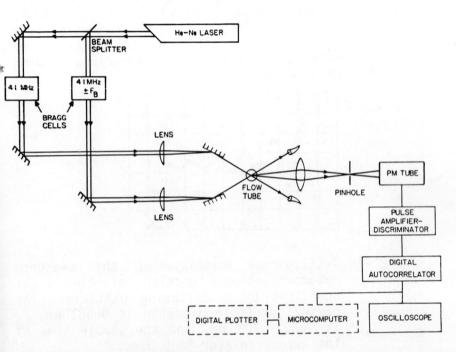

FIGURE 3.   Experimental setup.

Light scattered from a particle as it passes through the sensing volume of the LDV is collected by a lens and focused onto a pinhole in front of the PMT (RCA 7265). The output of the PMT is fed to a pulse amplifier-discriminator where transistor-transistor logic (TTL) compatible pulses are generated in response to the photodetection bursts in the PMT. Pulse sequences from the pulse amplifier-discriminator are analyzed by a 64-channel Langley-Ford digital autocorrelator, and the resulting autocorrelation function is displayed on an oscilloscope. Similar experimental arrangements were used to sample PLS suspensions in water.

The feasibility of measuring the diffusion constants of individual particles was tested by use of monodisperse suspensions of PLS particles in the size range 0.234-1.09 μm in diameter. Figures 4, 5, and 6 show the autocorrelation function derived from three different sizes of PLS particles in aerosols. These aerosols were generated by nebulizing suspensions of PLS in alcohol.

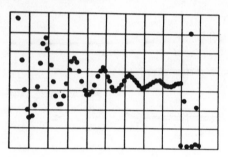

FIGURE 4.   Oscilloscope display of the measured autocorrelation function of the light scattered by a diffusing particle. The nominal particle diameter is 0.481 μm, f_B is 1.82×10^4 Hz, and the sample time of the autocorrelator is 5 μsec.

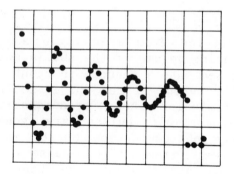

FIGURE 5.   Oscilloscope display of the measured autocorrelation function of the light scattered by a diffusing particle. The nominal particle diameter is 0.822 μm, f_B is 1.43×10^4 Hz, and the sample time of the autocorrelator is 5 μsec.

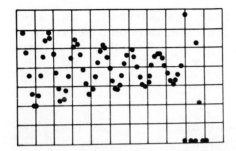

FIGURE 6. Oscilloscope display of the measured autocorrelation function of the light scattered by a diffusing particle. The nominal particle diameter is 1.09 μm, f_B is 1.82 × 10⁴ Hz, and the sample time of the autocorrelator is 5 μsec.

The data are analyzed [12] by computing the logarithmic decrement δ in the amplitude of the measured autocorrelation function. In accordance with Eq. (12), δ is given by

$$\delta = \frac{DK^2}{f_B} ,$$

where f_B = ω/2π is the frequency difference between the two incident laser beams of the LDV. Measured values of DK^2 are compared to DK^2 as calculated from Eqs. (2) and (8) by use of the known diameter of the test particles. Results are shown in Tables 1 and 2 for aerosols and hydrosols, respectively.

TABLE 1. Diffusion Measurement on an Individual Particle Basis on Aerosols Containing Monodisperse Polystyrene Latex Spheres (PLS) at T = 293 K, θ = 80°

Nominal Particle Diameter (μm)	Number of Particles Counted	$DK^2(\sec^{-1})$ Calculated	$DK^2(\sec^{-1})$ Measured
1.09	16	4.1×10^3	$3.4 \times 10^3 \pm 15\%$
0.822	5	5.6×10^3	$4.7 \times 10^3 \pm 17\%$
0.620	21	7.7×10^3	$6.7 \times 10^3 \pm 15\%$
0.481	25	1.1×10^4	$0.8 \times 10^4 \pm 23\%$
0.364	10	1.5×10^4	$1.3 \times 10^4 \pm 20\%$

TABLE 2. Diffusion Measurement on an Individual Particle Basis of Hydrosols Containing Monodisperse Polystyrene Latex Spheres (PLS) at T = 293 K, θ = 80°

Nominal Particle Diameter (μm)	Number of Particles Counted	$DK^2(\sec^{-1})$ Calculated	$DK^2(\sec^{-1})$ Measured
1.09	9	113	122 $\pm$ 14%
0.620	20	199	273 $\pm$ 21%
0.364	14	340	333 $\pm$ 22%
0.234	8	528	624 $\pm$ 14%

IV. DISCUSSION

Tables 1 and 2 show that the mean values of DK^2 can be determined for individual particles. Measurement of the quantity DK^2 from correlograms as observed on the oscilloscope display was somewhat crude. The agreement between the observed and the calculated values of DK^2 is considered satisfactory in view of the possible experimental errors in measuring the angle θ and the value of δ from the oscilloscope displays. Furthermore, no attempt was made to stabilize the temperature of the suspension since the present experimental studies were aimed only to establish the feasibility of the method. To determine the diffusion diameter distribution of polydisperse suspensions, it will be necessary to employ a data processing and storage system for computation and display of the distribution. Currently a microcomputer-based data processing and storage device is being added to this experimental setup. Once this phase of the work is complete, it will be possible to obtain the size distribution of particles in suspension in a given range.

One possible application of this device is to measure the diffusion diameter of aerosol particles in the size range of 0.05-0.5 μm in diameter. The transport of aerosol particles in this size range is primarily governed by the diffusion process. Particles greater than 0.5 μm entering the sensing volume can be rejected by the detection circuitry. The lower limit will probably be set by the detection limit of the experimental setup.

Some of the advantages of a measurement technique that can be used to measure the translational coefficient D rapidly and accurately on a single particle basis are (1) real-time analysis of diffusion diameter distribution, (2) high inherent size resolution, (3) application to both stable and unstable aerosols and hydrosols, allowing determination of such processes as growth and decay of particles, (4) possible size classification of aerosol particles according to their diffusion diameter in a size range (for example, 0.05-0.5 μm in diameter) without any interference from particles with larger diameter.

However, the measurement of diffusion diameter distribution on a single-particle basis poses several experimental constraints. The sampling time available for correlation spectroscopy on a single-particle basis is very limited, to allow rapid measurement of diffusion constants for a large number of particles. For a stable suspension, the application of conventional QELS techniques does not have such a stringent limitation in that the correlation spectrum from the sample can be obtained during a relatively long time period. Because of the relatively low intensity of the scattered light available from a single particle, the lower limit of detection for determining the diffusion diameter will be relatively higher than is possible when a large number of particles are contributing to the scattered light intensity in a self-beating or a heterodyning system.

In summary, the experimental data establish the feasibility of the use of a dual-beam, frequency-biased LDV for measuring diffusion diameter distributions on a single-particle basis.

ACKNOWLEDGMENT

The authors wish to thank J. D. Wilson, R. E. Ware, A. L. Adams, and R. L. Bond for helpful suggestions and technical assistance in the course of this work. The authors are also thankful to B. Dahneke, P. Roberson, J. Cochran, and J. Hunter for editorial assistance. The work was supported in part by an EPSCOR grant from the National Science Foundation.

REFERENCES

1. H. Z. Cummins and E. R. Pike, Eds., *Photon Correlation and Light Beating Spectroscopy*, Plenum Press, New York, 1974.

2. B. J. Berne and R. Pecora, *Dynamic Light Scattering with Applications to Chemistry, Biology and Physics*, Wiley, New York, 1976.

3. S. B. Dubin, J. H. Lunacek and G. B. Benedek, Proc. Natl. Acad. Sci. (USA) 57, 1164 (1967).

4. W. G. Hinds and P. C. Reist, J. Aerosol Sci. 3, 501, 515 (1972).

5. M.K. Mazumder and D. L. Wankum, Appl. Opt. 9, 633 (1970).

6. F. Durst, A. Melling and J. H. Whitelaw, *Principles and Practice of Laser Doppler Anemometry*, Academic Press, New York, 1976.

7. M. K. Mazumder, Appl. Phys. Lett. *16*, 462 (1970).

8. A. Einstein, *Investigations on the Theory of Brownian Movement*, R. Furth, Ed., Dover, New York, 1956.

9. H. L. Green and W. R. Lane, *Particulate Clouds: Dusts, Smokes and Mists*, 2nd ed., SPON Ltd., London, 1964.

10. C. N. Davies, Proc. Phys. Soc. 57, Pt. 4, No. 322 (1945).

11. T. T. Mercer, *Aerosol Technology in Hazard Evaluation*, Academic Press, New York, 1973.

12. R. J. Stephenson, *Mechanics and Properties of Matter*, Wiley, New York, 1952.

CHAPTER 13

The Bigger They Are, The Harder They Fall: Accurate Aerosol Size Measurements by Doppler Shift Spectroscopy

ILAN CHABAY
National Bureau of Standards
Washington, D.C.

CONTENTS

I. INTRODUCTION

Many techniques have been developed to measure aerosol particle size and size distribution. The properties on which the techniques are based include mass, linear momentum or inertia, diffusion coefficient, thermal mass, electrostatic charge/mass ratio, optical and electron imaging, optical transmission,

361

and light scattering [1]. These techniques each depend on a specific definition of size, such as aerodynamic size or total mass. The particle Doppler shift spectrometer (PDSS) is unique in that two different measures of size, aerodynamic and optical, are correlated in a way that provides accuracy in the size measurement of homogeneous, spherical aerosol particles of known index of refraction and density between 4 and 20 μm in diameter. For particles with an arbitrary size distribution, the PDSS measures the size accurately in real time with a maximum error of 0.05 μm. Measurement of the falling velocity of aerosol particles by quasi-elastic heterodyne light scattering is a major part of the operation of the PDSS. The use of heterodyning in this application differs from that of most of the other chapters in this volume (the other exception is the electrophoretic studies outlined by B. R. Ware et al. in Chapter 9) in that uniform, rather than diffusive motion, is the essential effect observed. This chapter is intended as a summary and review of the work that has been done on the PDSS in both development and in application of this technique in calibration of instruments and studies of aerosol aerodynamics and droplet growth in a cloud chamber.

II.   CONCEPT, DESIGN, AND PROCEDURE

The discussion of the PDSS instrument is separated into four parts: instrument design; heterodyne velocity measurement; calibration by Mie scattering pattern; and aerosol handling aspects.

Figure 1 is a schematic diagram of the optics and electronics. An argon ion laser operating on a single longitudinal mode at 514.5 nm is the light source. The unfocused beam, 3 mm in diameter, passes horizontally through the scattering chamber. Forward-scattered light from within the chamber is collected at one angle in the vertical plane. Collection lens, pinhole, and photomultiplier are mounted on a plat-

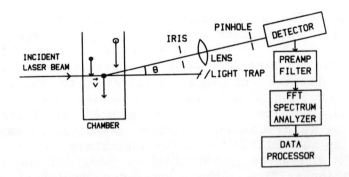

FIGURE 1.   Schematic diagram of the particle Doppler
shift spectrometer.   The vector **v** refers
to the velocity of the particle in the
beam.   (From Fletcher et al. [2].)

form that rotates vertically about the center of the
scattering volume. The diameter of the scattering
chamber is 1 cm. For collection of the scattered
light in a narrow range of angles from the width of
the entire chamber, a lens is used at its focal
length from a pinhole [2,3]. The scattering volume,
defined by the incident beam and the effective aperture of the lens-pinhole arrangement, includes the
beam spot on the entrance window of the scattering
chamber.

The beam spot on the flat glass window through
which the beam enters is used as the stationary local
oscillator source for heterodyning with the light
scattered by the moving particles in the chamber.
The signals from the photomultiplier pass through a
bandpass filter to a spectrum analyzer. A 400-point
hard-wired fast Fourier transform (FFT) instrument
was used in the experiments described here.

Subsequently, a microcomputer further processes,
analyzes, and stores the data generated. The data
are displayed as relative signal voltage versus
heterodyne beat frequency. An audio amplifier and a
loudspeaker are connected to the photomultiplier

output.   The audio output is an aid to alignment.
The beat frequencies due to falling of larger particles in the chamber give characteristic pitches
that can be used in conjunction with the background
noise due to the local oscillator to facilitate
alignment and proper operation of the chamber [4].
In addition, the "Music of the Spheres" or "Raindrops
Falling in My Chamber" provides a musical accompani-
ment to the activities in the laboratory.

A diffusion cloud chamber, Berglund-Liu (B-L)
vibrating-orifice aerosol generator, medicinal aerosol nebulizers, spray cans, and fine powder sources
have been used to produce particles for measurement
in the PDSS.   The particles, once generated, are
introduced into the top of a 1 m-long vertical glass
tube.   At the bottom end of the long tube is a narrower glass chamber, about 1 cm in diameter, through
which the laser beam passes.   The upper tube has an
inlet near the top and an outlet vent near the bottom
at the side, just below the opening to the scattering
chamber.   The scattering chamber is open only at the
top, where it receives particles from the long upper
tube.   Experiments are begun by flowing particles in
an airstream through the upper tube from the top
inlet to the lower vent.   When the upper tube is full
of a stream of particle-laden gas, the upper tube is
sealed off and the particles settle toward the scattering chamber at the bottom.

As a particle passes through the beam, a portion
of the scattered light is collected on the photomultiplier.   The light intensity at the detector
surface is proportional to the square of the field
amplitudes that reach the detector.   Fields with
different characteristic frequencies reach the detector simultaneously.   One field has the frequency of
the incident laser beam.   This originates in the
elastic scattering from the stationary window.   The
other frequency dependence is that due to the quasi-
elastic or Doppler-shifted scattering from the moving
particle.   When fields with both frequency characteristics are present simultaneously on the photodetec-

tor, a "beat" frequency (the difference in frequency between the fields at the detector) is imposed on the photocurrent [5]. The sum frequency is beyond the frequency response of the system. Several frequency components can be present in the scattered light, corresponding to several particles in the scattering volume with different velocities. In this case, provided the elastically scattered "local oscillator" field amplitude at the detector is much greater than any of the field amplitudes scattered by the particles, all frequency components will appear in the photocurrent. The complete frequency content of the photocurrent is then displayed by the spectrum analyzer. The field scattered by the particle is shifted and broadened in frequency in comparison to the field scattered by the stationary glass window. Brownian motion of the particle contributes a linewidth proportional to the square of the scattering vector (the vector difference between the incident and scattered wavevectors; see Fig. 1), $F = DK^2$. The slow drift velocity of the particle settling due to gravity imparts a frequency shift to the scattered field that is linearly proportional to the scattering vector. If the beat frequency is F due solely to the settling motion of a particle with downward velocity v, it is given by $F = \mathbf{v} \cdot \mathbf{K}$.

To achieve the best resolution of the settling velocities of the particles, the frequency shift should be maximized, whereas the broadening of the line is minimized. Small forward-angle scattering satisfies the criterion for resolution and maximizes the scattered intensity from the particles since scattering is strongly peaked in the forward direction for particles in the appropriate size range. For unequivocal resolution of the sizes of a collection of particles without reliance on any assumptions regarding their size distribution, the frequency shift must be somewhat greater than the linewidth. This imposes a lower size limit of about 4 µm diameter.

Measurement of the heterodyne beat frequency spectrum is the basis for determining the particle settling velocity, and that, in turn, is used to deduce particle size. Stokes law relates the settling velocity v of a particle to its radius r, density ρ, the viscosity of the medium through which it falls η, the density of that medium ρ', and the gravitational acceleration g. The expression is $v(r) = 2(\rho - \rho')gr^2/9\eta$ [6].

An example of the application of Stokes law to particle size determination with the Doppler shift instrument is shown in Fig. 2. Four particles were present in the scattering volume simultaneously. They gave rise to four frequency signatures since they fell at different velocities. The sizes deduced from application of Stokes law are indicated on the spectrum.

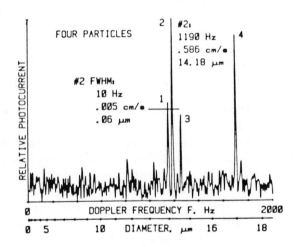

FIGURE 2.   A typical spectrum showing the frequency shifts and corresponding size scale based on Stokes law for four particles that were in the beam during the 0.2 sec time window. Analysis of the size, error limits, and linewidth for one particle are indicated. (From Bright et al. [4].)

An important question arises from examination of Fig. 2. Is the relationship between the frequency or velocity and the particle size given by Stokes law unique, and can the measurements be calibrated in an absolute sense? The presence of convection currents, static charge effects, photophoresis, and other variables clearly could disrupt the predicted relationship between size and velocity. The answer is affirmative since the presence of interfering effects can be checked and the size determination calibrated directly in the PDSS instrument. This can be done easily with samples of unknown size and size distribution. What is necessary (and easily obtained) is a sample of an aerosol of known index of refraction and density that forms a broad size distribution of spherical particles.

For particles of size comparable to the wavelength of the incident light, a plot of the elastic scattering (Mie) amplitude as a function of the particle size at one angle of observation has large, irregular peaks and valleys [7]. The characteristic peaks and valleys (elastic scattering resonances and antiresonances) of the Mie scattering pattern at one scattering angle for particles of varying size provides the basis for calibration. A typical example of some data, the appropriate calculated Mie scattering plot, and the comparison between them is shown in Fig. 3. Since the total amount of time that a particle of a certain size spends in the beam is inversely proportional to its velocity, the total amount of light scattered at the appropriate frequency for that particle is also inversely proportional to the velocity and thus to frequency. Therefore, the calculated Mie scattering curve must be weighted by $1/F$. This has been done for the calculated values shown in Fig. 3.

What is important in Fig. 3 is the match in position on the horizontal axis of the calculated curve and the data. The procedure for estimation of the error in size measurement is indicated in the inset to Fig. 3. A parabolic curve is fit to a

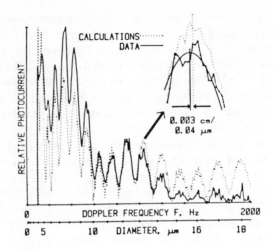

FIGURE 3.   Mie scattering intensity as a function of particle size and weighted by $1/F$. The inset illustrates the analysis of the spectrum and calculated plot for determination of small shifts in velocity. Only the parabola used to fit the data is shown in this example. (From Bright et al. [4].)

feature of the data and calculated curves, a local maximum or minimum, and the deviation of the center of the two parabolas from each other determined. This procedure is done for up to 20 extrema over the breadth of the experimental curve to find the best fit for each data set. The fact that the data do not follow the ordinate values of the calculated plot, especially for larger particle sizes, simply indicates that the size distribution of the real aerosol sample is not uniform. The deviation along the vertical axis of the data from the calculated value is, in fact, directly related to the deviation of the actual size distribution of the sample aerosol from a uniform one.

The data shown in Fig. 3 were collected for a very broad size distribution of dioctylphthalate (DOP) oil droplets produced by an aerosol nebulizer. Use of a broad size distribution is helpful in calibrating the PDSS because the breadth ensures that several extrema of the scattering curve will be visible in the PDSS spectrum, thus allowing unambiguous correlation between measured velocity and size. The sensitive correlation between size and velocity permitted us to determine the appropriateness of the slip correction to Stokes law behavior [8]. In the range of 5-15 µm at 1 atm of pressure in air, the slip correction gave increasingly poor results compared to the use of Stokes law alone [4]. The slip correction predicts settling velocities that are too large under the conditions of these experiments.

III. EXPERIMENTS AND APPLICATIONS

Since the PDSS can be accurately calibrated by reference to calculations based on basic electrodynamic theory, the instrument is an excellent means for calibration of other types of particle sizing devices. This is particularly useful since it is not feasible to produce a bottle of a standard aerosol of a desired size distribution. With other size measurement techniques, it is usually necessary to generate an aerosol of approximately known size distribution and relate measurements *in situ* to those made on similar particles in another medium (such as for polystyrene latex spheres in solution) or on a substrate (such as by optical or electron microscopy of particles or impact marks on a coated slide). These indirect methods are not needed with the PDSS.

A comparison was made between the particle size distribution measured by PDSS, that indicated by a commercial optical particle counter (OPC), and the expected size generated by the B-L device [2]. The aerosol generated by the B-L consisted of DOP drop-

lets dispersed in filtered air. Particles of 4-8 μm
in diameter were used. The aerosol was sampled in
parallel by the PDSS and the OPC. Care was taken to
ensure that the sampling procedure did not bias the
results due to the flow pattern or dynamics or time-
dependent effects. The conclusions of the study were
that (1) the B-L generator produced particles that
were within about 3% of the size predicted by use of
the operating parameters of the generator, although
several particle sizes were produced simultaneously
by the same generator conditions in some cases and
(2) the apparent size distribution obtained with the
OPC deviated significantly from the correct distribution in mean size and width.

Generally rather narrow distributions of particle size are produced by the B-L generator, but
some problems were revealed with the PDSS. The mean
size may drift by 2% or so over a few minutes and,
under some operating conditions, the generator produces several different particle sizes simultaneously. Since the data on the PDSS and the OPC were
taken in parallel and the same stream of aerosol
particles was sampled within seconds by each instrument, the problems with the Berglund-Liu method were
enlightening, but not a hindrance. The OPC, which
collects light scattered at 8-20° in a forward cone,
uses a multichannel analyzer to store the signal from
the particles passing singly through the scattering
volume. The scattered light intensity is approximately proportional to the square of the particle
diameter within the size range studied. On the basis
of the curve relating signal voltage to particle
size, the size distribution (dN/dD, where N is the
number of particles and D is their diameter) determined by the OPC contained artifacts. Significant
errors in mean diameters of distributions were
evident when OPC data based on the manufacturer's
calibration curve were compared to PDSS results. The
most pronounced effect was observed for the particles
with mean diameter of about 8 μm. Particle Doppler
shift spectrometer data indicated a rather narrow

distribution around 8.2 μm (standard deviation of 0.07 μm), whereas the OPC erroneously indicated a broad distribution from 8-10 μm.

Changes in the size distribution, as well as the distribution itself, have been measured with an early version of the PDSS. The growth of water droplets in a steady-state diffusion cloud chamber was studied by observing the Doppler-shifted light scattering from droplets at several heights within the cloud chamber [9,10]. A slow flow of water-saturated air moved from the upper, warm plate to the lower, cool surface. Data were taken at several heights in the chamber and for several combinations of temperatures on the upper and lower plates. The latter values determine the supersaturation profile in the chamber. Under conditions for which the maximum supersaturation was 1.047 and 1.024, the observed growth rate fit theoretical predictions well with reasonable values of the parameters. However, for lesser values of the supersaturation (~ 1.014), the observed growth was substantially slower than predicted. Further studies would be necessary to determine the source of the discrepancy.

Measurement of the velocity of aerosol particles is interesting not only as a means of determining size, but also as a way of investigating the aerodynamics of particles at very low Reynolds numbers [11]. The airflow generated by the motion of particles of a few micrometers in diameter was studied by PDSS. The effect of a stream of falling particles on a second stream in another portion of the chamber could be seen from the shift in velocity of both streams of particles. Figure 4 contains four examples. In each case the geometry of the particle stream and the position probed by the laser in the scattering chamber is indicated next to the appropriate curve. A single stream of particles falling in one part of the chamber falls faster than the Stokes law velocity, as is the case for the top part of Fig. 4. The ordinate is the observed velocity minus the Stokes law velocity as a function of par-

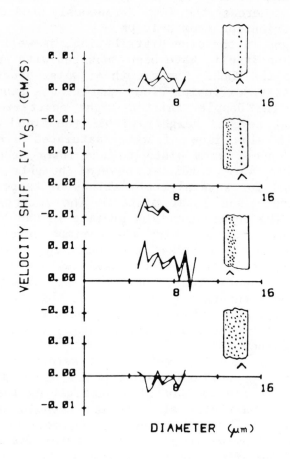

FIGURE 4. Summary of four experiments that probed the nature of particle-induced fluid flow. The carat beneath each sketch of the chamber cross section indicates the position of the laser beam. The ordinate is the difference in velocity between that measured and that predicted on the basis of Stokes law for each size particle. (From Fletcher et al. [11].)

ticle diameter. The other three cases illustrated in
Fig. 4 demonstrate the effects of one stream on
another, with the larger stream dominating the flow
effect in the chamber. In the case shown at the
bottom of Fig. 4 there is no apparent shift in
velocity because the particle stream fills the cross
section of the chamber. When this occurs, there is
no path devoid of particles that can be used as a
channel for return air circulation. The air
displaced by each particle is constrained by the air
displaced by all other particles. Therefore, all
particles fall at essentially the Stokes law velocity
without shift.

A broad size distribution of particle sizes was
used in all cases. The appropriate Mie scattering
pattern was compared to the data. The shift in the
pattern was computed extremum by extremum (up to 20
features of the curves were compared) and the measured deviation from the velocity calculated from
Stokes law for each size particle was determined. In
this manner, trends in the velocity that are dependent only on the pattern of particle streams in the
closed chamber were discerned. The effect is a
cumulative one in that the observed shifts are dependent on number density. The greater the particle
number density (100-1000 particles cm^3) the greater
the induced velocity shift. The shifts in velocity
are small, on the order of 10-100 μm/sec. The sensitivity of the PDSS and the ability to determine
accurately the size of the particles as well as their
velocities are crucial to the study of those motions
that occur with Reynolds numbers of 0.01-0.001.

IV. SUMMARY AND PROGNOSIS

The PDSS has been shown to be a valuable tool
for calibration of other instruments and for basic
studies of aerosol growth and aerosol aerodynamics.
The size range is limited by the Brownian motion
imposed linewidth to particles larger than about 4 μm

374   I. Chabay

in diameter. The method is accurate when used with
spherical particles of known index and density but
can also be used to measure relative size for unknown
particles with some constraints.

ACKNOWLEDGMENT

A great many of the results described in this
chapter were produced through the excellent efforts
of Robert A. Fletcher and David S. Bright, with whom
I was fortunate enough to collaborate at NBS.

REFERENCES

1. D. S. Bright and I. Chabay, Chemtech 9, 694 (1979).

2. R. A. Fletcher, G. W. Mulholland, I. Chabay and D. S. Bright, J. Aerosol Sci. 11, 53 (1980).

3. I. Chabay and D. S. Bright, J. Colloid Interface Sci. 63, 304 (1978).

4. D. S. Bright, R. A. Fletcher and I. Chabay, J. Phys. Chem. 84, 1607 (1980).

5. H. Z. Cummins and H. L. Swinney, Progr. Opt. 8, 135 (1970).

6. N. A. Fuchs, The Mechanics of Aerosols, Macmillan, New York, 1964.

7. M. Kerker, The Scattering of Light and Other Electromagnetic Radiation, Academic Press, New York, 1969.

8. C. N. Davies, Proc. Ry. Soc. Lond. 57, 18 (1945).

9. J. P. Gollub, I. Chabay and W.H. Flygare, Appl. Opt. *12*, 2838 (1973).

10. J. P. Gollub, I. Chabay and W. H. Flygare, J. Chem. Phys. *61*, 2139 (1974).

11. R. A. Fletcher, D. S. Bright and I. Chabay, J. Phys. Chem. *84*, 1611 (1980).

CHAPTER 14

Dynamic Depolarized Light-Scattering Studies of Rodlike Particles

KARL ZERO
Stanford University
Stanford, California

CONTENTS

I. INTRODUCTION

Dynamic light scattering (DLS) is a well-established, although relatively new, technique for measuring the diffusion coefficients of particles in solution [1,2]. These diffusion coefficients, in turn, can be related to the particle dimensions [3-8]. In polarized DLS the incident and scattered light have the same direction of linear polarization. Polarized DLS is widely used to determine the size distribution of spherical particles or nonspherical particles of known shape by measuring the distribution of the translational diffusion coefficients. However, except for extremely large particles, where structure factors contribute to the polarized correlation function [1], the dimensions of an arbitrary particle are difficult to determine from the polarized scattering alone. Often another property with a different shape dependence, such as the bulk viscosity or the rotational diffusion coefficient, is measured and combined with the translational diffusion coefficient to yield the shape and size of the particle.

One technique for measurement of the rotational diffusion coefficient is depolarized dynamic light scattering (DDLS) [1], where the polarization of the incident light and the scattered light differ by 90°. Michielsen and Pecora's studies [9], for example, found the dimensions of gramicidin dimer by combining the results from DDLS and polarized DLS. Furthermore, because the rotational diffusion coefficient has a much greater dependence on the largest dimension ($\sim 1/R^3$) than does the translational diffusion coefficient ($\sim 1/R$), DDLS is much more sensitive to small variations in size. Like polarized DLS, DDLS is also a relatively nonperturbing experiment. Unfortunately, the depolarized signal is usually very weak, resulting in poor signal/noise ratios. Thus size distributions are not readily obtained from DDLS data [10].

For several reasons, the effects of concentration on the rates of diffusion must be considered. Often, in order to achieve a reasonable signal/noise ratio, relatively concentrated solutions of particles (1-10% by weight) must be used. In addition, some particles, such as certain micelles, appear to change size and shape with concentration [11]. For long rodlike particles, the rotational diffusion coefficient should have a considerable dependence on concentration in such concentrated solutions. Thus, to establish the behavior of the DLS data in this region, a study, the results of which are presented here, was made of the concentration dependence of the DLS time constants for semidilute solutions of rodlike particles. The systems studied were poly(γ-benzyl-L-glutamate) (PBLG) in 1,2-dichlorethane and light meromyosin (LMM) in an aqueous buffer.

Besides overall rotation, internal motions should also contribute to the DDLS correlation functions. For example, a rigid rod with a joint inserted at the center, that is, a once-broken rod, should have an additional component in the DDLS correlation functions due to bending at the joint. Here the DDLS correlation functions for myosin rod, which is thought to be a once-broken rod, were found and compared with the predictions of two different theories.

II.   MOTION OF RODS IN THE SEMIDILUTE REGION AND THEIR DDLS TIME CORRELATION FUNCTIONS

Many different relationships are used for calculating the diffusion coefficients of rods in dilute solution with a given length L and diameter d, based mainly on corrections to the Stokes-Einstein equations for spheres [3-8]. For long, thin rods, the expressions for the rotational diffusion coefficient Ξ and the translational diffusion coefficient D can be approximated by the Kirkwood-Riseman results [12]:

$$\Xi = \frac{3k_B T \, \ln(L/d)}{\pi\eta_s L^3} , \tag{1}$$

$$D = \frac{k_B T \, \ln(L/d)}{3\pi\eta_s L} , \tag{2}$$

where k_B is Boltzmann's constant, T is the absolute temperature, and η_s is the bulk viscosity of the solvent. These relationships assume that the rods are sufficiently large that stick boundary conditions hold; that is, the particles behave like macroscopic rods in a viscous continuum.

As the concentration is increased, one would expect the motion of each rod to be hindered by its neighbors, resulting in a decrease in the rate of translational and rotational diffusion. Doi and Edwards [13,14] have presented a theory for the diffusion coefficients of rods with number concentrations c much greater than $1/L^3$ but less than $2\pi/dL^2$; that is, solutions where the rods are still oriented randomly but where their motions are strongly hindered by steric repulsions between rods. In their treatment Doi and Edwards assume that translational motion parallel to the axis of the rod ($D_{||}$) is unhindered (that is, the rods are infinitely thin) whereas translational motion perpendicular to the rod's axis ($D_\perp$) is restricted to about the distance a_c to the nearest rod. Since a_c is much less than the length for these concentrations ($a_c \sim 1/cL^2$ and $c \gg 1/L^3$), $D_\perp \sim 0$ compared to $D_{||}$; thus $D = 1/3(D_{||} + 2D_\perp) = 1/3D_{||}$. Since $D_{||} \sim 2D_\perp$ at infinite dilution, the translational diffusion coefficient in this semidilute region is roughly half the value of infinite dilution. For a particular rod A to rotate a significant amount, the nearest neighboring rod B must translate a distance on the order of its length, thus temporarily releasing the constraint on rod A (see Fig. 1). From this model, Doi and Edwards obtain an expression for the concentration dependence

of the rotational diffusion coefficient in the semidilute region $(1/L^3 \ll c < 2\pi/dL^2)$ [13,14]:

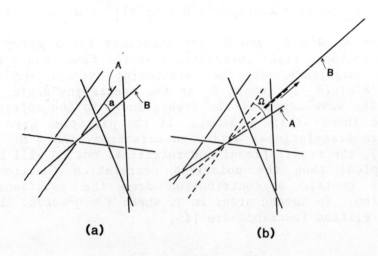

(a)   (b)

FIGURE 1. A schematic representation of a semidilute
      solution of rodlike macromolecules:
      (a) before rod A rotates; (b) after rod A
      rotates.

$$\Xi = \frac{\beta' k_B T \, \ln(L/d)}{\eta_s L^9 c^2} \tag{3}$$

where c is the number concentration of rods and β' is
a proportionality constant. Note that Ξ now has an
inverse c^2 dependence on concentration and an inverse
L^9 dependence on length.

  For dilute solutions of small rods with no
translational-rotational coupling, the polarized
$[I_{VV}(q,t)]$ and depolarized $[I_{VH}(q,t)]$ homodyne time
correlation functions are given by the expressions
[1]

$$I_{VV}(q,t) = A\{\exp[-q^2 Dt]\}^2 + B \ , \tag{4}$$

$$I_{VH}(q,t) = A'\{\exp[-(q^2 D + 6\Xi)t]\}^2 + B' \ , \tag{5}$$

where A, A', B, and B' are constants for a given q and incident light intensity, t is the time, and q is the magnitude of the scattering vector [$q = 4\pi n/\lambda \sin(\theta_s/2)$, where θ_s is the scattering angle, λ is the wavelength of the light, and n is the refractive index of the medium]. If the particles have a large translational diffusion anisotropy ($\Delta D = D_\parallel - D_\perp$), the translational and rotational motion will be coupled; thus the polarized correlation functions will contain a contribution from the rotational motion. To second order in γ, where $\gamma = q^2 \Delta D/\Xi$, the correlation functions are [15]

$$I_{VV}(q,t) = A\left[\sum_{i=1}^{2} a_i \exp[-\Gamma_i^0 t]\right]^2 + B \ , \tag{6}$$

$$I_{VH}(q,t) = A'\left(\left[\sum_{i=1}^{2} b_i \exp[-\Gamma_i^{21} t]\right]\cos^2 \frac{\theta_s}{2}\right.$$

$$\left. +\left[\sum_{i=1}^{2} c_i \exp[-\Gamma_i^{22} t]\right]\sin^2 \frac{\theta_s}{2}\right)^2 + B' \ , \tag{7}$$

where

$$\Gamma_1^0 = q^2 D - \frac{2}{135}\Xi\gamma^2$$

$$\Gamma_2^0 = q^2 D + \frac{4}{21} q^2 \Delta D + 6\Xi + \frac{2}{135}\Xi\gamma^2$$

$$\Gamma_1^{21} = q^2 D - \frac{2}{21} q^2 \Delta D + 6\Xi - \frac{4}{1029}\Xi\gamma^2$$

$$\Gamma_2^{21} = q^2 D - \frac{34}{231} q^2 \Delta D + 20\Xi + \frac{4}{1029} \Xi \gamma^2$$

$$\Gamma_1^{22} = q^2 D - \frac{4}{21} q^2 \Delta D + 6\Xi - \frac{2}{1029} \Xi \gamma^2$$

$$\Gamma_2^{22} = q^2 D + \frac{16}{231} q^2 \Delta D + 20\Xi + \frac{2}{1029} \Xi \gamma^2$$

$$a_2 = \frac{(2/135)\gamma^2}{1 + (2/63)\gamma + (4/135)\gamma^2} \, , \quad a_1 = 1 - a_2$$

$$b_2 = \frac{(4/1029)\gamma^2}{1 - (2/539)\gamma + (8/1029)\gamma^2} \, , \quad b_1 = 1 - b_2$$

$$c_2 = \frac{(2/1029)\gamma^2}{1 + (10/539)\gamma + (4/1029)\gamma^2} \, , \quad c_1 = 1 - c_2$$

Rods in semidilute solutions, where $\Delta D \sim D_{||}$, $D \sim 1/3(D_{||})$ and $\gamma = q^2 D_{||}/\Xi$, should have a significant amount of translation-rotation coupling, particularly when Ξ becomes small relative to $q^2 D_{||}$. For real systems, the faster components in the depolarized correlation function (that is, the $i = 2$ terms) would be difficult to observe experimentally, because of their low amplitude and their rapid decay. Thus Eq. (5) is still a good approximation to the depolarized correlation functions. However, the faster components should definitely make a contribution to the polarized DLS.

III. THE ONCE-BROKEN ROD

If a joint is inserted into the center of a rod, one would expect a contribution to the DDLS due to motion of the segment relative to the joint. For dilute solutions of once-broken rods with no coupling between translation, overall rotation and bending at the joint, the DDLS homodyne correlation functions can be approximated by [18]

$$I_{VH}(q,t) = A'\left(\sum_{i=1}^{2} C_i \exp[-\Gamma_i t] \right)^2 + B' , \qquad (8)$$

where

$$\Gamma_1 = q^2 D + 6\Xi + \nu_1 D_\beta ,$$

$$\Gamma_2 = q^2 D + 6\Xi + \nu_2 D_\beta ,$$

The amplitudes (C_i) and the factors ν_i depend on which model is chosen for the bending motion. The coefficient D_β is related to the rotational diffusion coefficient D_R^β of an unconnected segment. Because of hydrodynamic interaction between segments and differences in the center-of-mass positions, $D_\beta \sim 2/3(D_R)$ [16,17].

Most theories for the once-broken rod assume completely free motion at the joint [16,17]. For this special case $C_1 \sim \nu_1 \sim 0$ and $\nu_2 \sim 6$; thus almost all the contribution to the DDLS is from the independent motion of the segments. Therefore, a single exponential with a time constant close to that for a solution of unconnected segments would be expected. One correction to this model is to impose a maximum bending angle θ^0. For $(\theta^0/2) \leq 80°$, this model results in the following expressions [18]:

$$\nu_1 = 0 ,$$

$$\mu_0 = \cos \frac{\theta^0}{2} ,$$

$$C_1 = \frac{1}{4} \mu_0^2 (1 + \mu_0)^2 ,$$

$$c_2 = \frac{1}{20} (4 - \mu_0 - 6\mu_0^2 - \mu_0^3 + 4\mu_0^4) ,$$

$$v_2 = v_2^0(v_2^0 + 1) ,$$

where v_2^0 is a complicated function of $\theta^0/2$, given by Pal [19] and tabulated in Table 1. From Table 1 it can be seen that the bending mode decays faster and decreases in amplitude as the maximum bending angle is decreased.

TABLE 1. Parameters for the Once-Broken Rod with a Maximum Bending Angle θ^0

$\theta^0/2$	v_2^0	c_1	c_2	c_2/c_1	v_2
0°	∞	1	0	0	∞
15°	14.12	0.901	0.00084	0.00093	213.5
30°	6.83	0.653	0.0117	0.0179	53.5
45°	4.4	0.364	0.0470	0.129	24.4
60°	3.2	0.141	0.106	0.752	13.4
75°	2.4	0.0265	0.167	6.30	8.3
90°	2	0	0.2	∞	6

Another model assumes that any bending angle is possible, provided the segments do not pass through each other, but a cosine potential exists that influences motion of the segments. In other words, a force F_β is exerted on the segments proportional to the sine of the bending angle θ_β; thus

$$F_\beta = f \sin \theta_\beta ,$$

where f is a force constant. After solving the pertinent Smoluchowski equation, one obtains the values tabulated in Table 2, where the amplitudes and time decay constants are a function of $f/k_B T$ [20]. In this model, the bending motion contributes to both exponentials and decays faster with increasing $f/k_B T$.

TABLE 2.   Parameters for the Once-Broken Rod with a Cosine Potential

$f/k_B T$	ν_1	ν_2	c_1	c_2	c_2/c_1
0	0	6	0	1	∞
0.5	0.751	7.212	0.009	0.935	104
1.0	1.505	8.468	0.031	0.868	28
2.0	3.038	11.102	0.095	0.748	7.9
3.0	4.610	13.891	0.163	0.660	4.05
4.0	6.226	16.860	0.220	0.603	2.74
5.0	7.884	20.076	0.264	0.573	2.17
6.0	9.583	23.729	0.294	0.572	1.95

IV.   EXPERIMENTAL

A schematic of the dynamic light-scattering apparatus is shown in Fig. 2. The details of these experiments on PBLG and myosin rod are published elsewhere [15,18,21]. A 5-W argon-ion laser (488 nm) was used as the incident light source, and 10 mm^2 glass fluorimeter cells were used to hold the solutions of rods. Typically, suitable DDLS correlation functions were collected within 5-10 hr, whereas the polarized data were collected within a few minutes.

All measurements were taken at an angle of 90° and a
temperature of 20.0 ± 0.1°C.

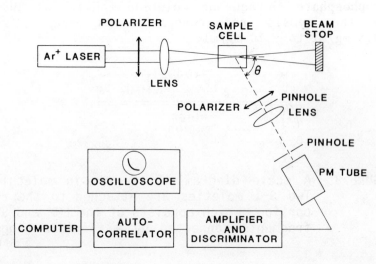

FIGURE 2.   Schematic diagram of the light-scattering
apparatus.

Three different molecular weights of PBLG
(150,000, 170,000, and 210,000 g/mol) were investigated at concentrations ranging from 4 mg/ml to
50 mg/ml. In the 1,2-dichloroethane solvent, PBLG
exists as an α-helix, with a diameter of about 15 Å
and a length per residue of ~ 1.5 Å. Thus PBLG is
essentially a rodlike particle with a persistence
length of ~ 1000 Å [24]. The three molecular weights
studied had lengths of 1030, 1160, and 1440 Å,
respectively.

The myosin molecule is shown in Fig. 3. The
myosin rod is the myosin molecule with the head
groups (S-1) removed, leaving the S-2 and light
meromyosin (LMM) fragments connected by a "hinge"
region. Both electric birefringence and electron
microscopy experiments [25-27] indicate that a considerable amount of flexibility exists at the joint
between the LMM and S-2 fragments, with a possible

maximum bending angle (θ^0) of 145° [25].   In the present study the solvent was 0.6 M KCl plus 0.010 M pyrophosphate in aqueous solution (pH 9.5 at 20°C) and the myosin rod concentrations ranged from ~ 0.5 mg/ml to ~ 10 mg/ml.

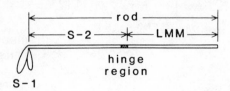

FIGURE 3.   A scale diagram of the myosin molecule. Two S-1 moieties are attached to the rod portion, which consists of the LMM and S-2 fragments connected by a flexible hinge.

V.   RESULTS

The time correlation functions $C(t)$ were fit by nonlinear least-squares analysis to two different forms:

$$C(t) = A \exp[-2\Gamma t] + B , \qquad (9)$$

$$C(t) = \left(\sum_{i=1}^{2} A_i \exp[-\Gamma_i t] \right)^2 + B , \qquad (10)$$

where the Γ_i are the time decay constants, B is the basline, and the A_i are the amplitudes.   Because of the poor signal/noise ratio of the data, the best fit to the data was obtained by allowing all the parameters to float, including the baseline, which, in principle, can be found experimentally.

The results for the PBLG solutions are shown in Figs. 4 and 5.   The rotational diffusion coefficient, found from Eq. (5), is plotted against $1/c^2$ in Fig. 4, whereas the reciprocal decay time for the

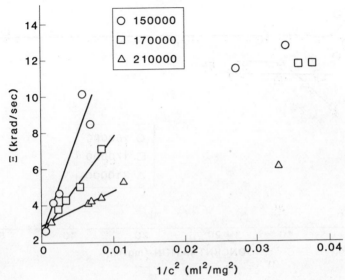

FIGURE 4.   Plot of the rotational diffusion coefficient Ξ versus reciprocal concentration
squared for three different molecular
weights of PBLG in 1,2-dichloroethane
solutions. The lines are linear least-
square fits of Ξ versus $1/c^2$ for $1/c^2 <$
0.02.

polarized correlated intensity $[1/\tau_p = \Gamma$; Eq. (9)] is
plotted against concentration in Fig. 5. The size of
the symbols does not indicate the experimental precision; the actual precision is a few percent for the
polarized data and about 10% for the depolarized
data.

The region defined by Doi and Edwards would
range from about 0.1 mg/ml to 10 mg/ml; however, the
actual onset of the region appears to be a little
greater than 5 mg/ml ($c \gg 1/L^3$ is expected). Beyond
this concentration, the plots of Ξ versus $1/c^2$ are
linear. For the molecular weights 150,000, 170,000,
and 210,000, comparison of the slopes with the Doi

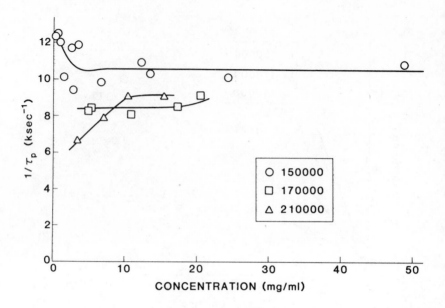

FIGURE 5.   Plot of reciprocal polarized decay time ($1/\tau_p$) versus concentration for three different molecular weights of PBLG in 1,2-dichloroethane solutions. The lines are not fits, but merely emphasize the trend of the data.

and Edwards relationship yields β' values of 1070, 1170, and 1768, respectively. Thus β' is several orders of magnitude larger than predicted. From the Doi-Edwards theory, the ratio ($R_{2:1}$) of the slopes (A_2, A_1) for two different lengths should fit the relationship (assuming that c is in units of milligrams per milliliter)

$$R_{2:1} = \frac{A_2}{A_1} = \left(\frac{L_1}{L_2}\right)^x \frac{\ln(L_2/d_2)}{\ln(L_1/d_1)} \left(\frac{MW_2}{MW_1}\right)^2 ,$$

where x is predicted to be 9 and L_i, d_i, and MW_i are, respectively, the length, diameter, and molecular weight for molecule i. Experimentally, the values found for x are 8.2, 7.5, and 7.1, where the value 8.2 is obtained from comparing the shortest PBLG rods. The average for a comparison of all three lengths is 7.6 ± 0.6. Considering the precision of the data points, this value, although slightly lower than predicted, is reasonably close to 9. One possible explanation, other than the precision of the data, for the increase in β' value (or decrease in x) with length may be flexibility, resulting in a smaller hydrodynamic length for the longer rods; estimates of the persistence length for PBLG range from 900 to 1400 Å [24]. Transient-electric-birefringence experiments on long, rodlike viruses were performed by Maguire et al. [22,23]. They also observed a $1/c^2$ dependence for Ξ and obtained a value of 7.7 for x, using only two different viruses. Depolarized dynamic light-scattering studies on LMM also yield a $1/c^2$ dependence for semidilute solutions [21].

Below about 5 mg/ml, the values for Ξ plateau and become concentration independent. For the molecular weights 150,000, 170,000, and 210,000, the values obtained for Ξ below 5 mg/ml are, respectively, 13 ± 3, 12 ± 2, and 6.1 ± 0.2 krad/sec. Within experimental error, these values correspond well with the values predicted from Broersma's relations for infinite dilution (15.5, 11.2, and 6.25 krad/sec, respectively).

The $1/\tau_p$ values, on the other hand, show relatively little concentration dependence in this region, as expected from the Doi-Edwards theory [13,14] (the lines in Fig. 5 are meant merely to guide the eyes; they are not fits). However, the values are about two-thirds the infinite dilution value rather than the predicted one-half. This discrepancy can be qualitatively explained by the two-exponential nature of the curves (as observed experimentally and predicted by the theory presented in Section II). Unfortunately, the quality of the data is insufficient to give a consistent fit to

Eq. (6). Although a double-exponential fit can yield a reasonably good value for the slower exponential (that is, corresponding to a diffusion coefficient of about $(1/3)D_{\parallel}$), the small difference in time scales between the two exponentials and the low amplitude of the fast exponential make a quantitative determination of the value of the time constants impossible for the lower end of the concentration range. The higher concentrations do show a reasonably strong contribution from the rotational time and fit Eq. (6) at least semiquantitatively. Qualitatively, the contribution from the faster mode could cause the apparent rise in the polarized decay time (as determined from a single exponential fit) seen in Fig. 5.

For the Doi-Edwards initial theory, one would expect the intercept of these plots in Fig. 4 to be zero (i.e., no rotation at infinite concentration). However, the actual intercepts are fairly substantial (2-3 krad/sec). This discrepancy arises from the finite thickness of the rods. At the higher concentrations, translational diffusion parallel to the axis would be expected to no longer be completely unhindered as the distance between rods becomes comparable to their diameters. Consequently, the rate of rotational diffusion should begin to drop faster than the $1/c^2$ dependence, and a fit to $1/c^2$ would yield a false intercept. A simple, tentative treatment [13,14] for this effect is to simply correct the size of a_c for the diameter; that is, let $a_c = \langle r_c \rangle - d$ and $\langle r_c \rangle = \alpha_c/cL^2$, where α_c is some numerical factor. The new equation obtained for Ξ is

$$\Xi = \frac{\beta' k_B T \ln(L/d)(\alpha_c - cdL^2)^2}{\alpha_c^2 \eta_s L^9 c^2} .$$
(11)

Although the data are not sufficiently precise to fit to this form, the magnitude of the intercepts is consistent with this equation.

The results for the myosin rod are tabulated in Tables 3 and 4. At the concentrations used, there was no significant concentration dependence for the decay constants. The values for Ξ in Table 3 were calculated from Eq. (5); that is, it was assumed that only overall rotation was present. Clearly, the Ξ value for myosin rod found by this assumption is larger than one would predict for a stiff, unbroken rod. From polarized DLS measurements [Eq. (4)] [21], q^2D was found to be 4.8 ± 0.8 ksec^{-1}, somewhat lower than the infinite dilution value predicted by Broersma's relationships for an unbroken rod (8.0 ksec^{-1}) [4-6]. By comparing the fits to the depolarized data at different time scales, it was determined that Eq. (2) (that is, a two-exponential fit) gave the most consistent results [21]. A single-exponential fit to the tail of the DDLS correlation functions ($t > 8$ µsec) was done to yield a more reliable value for the slower time constant (Γ_2).

The ratio A_1/A_2 (where Γ_1 is the faster time) in Eq. (2) should correspond to the ratio C_2/C_1 in Eq. (8). From experiment, $A_1/A_2 = 1.2 \pm 0.4$. This corresponds to an average angle θ^0 of 128° (in the range of 121-132°). From Pal's equations [19], the value of v_2^0 at 128° was found to be $2.97 \sim 3$. Using this value for v_2^0 and the values found for the decay constants, one obtains a value for D_β of 24 ± 6 krad/sec. The rotational diffusion constants for LMM and S-2 free in solution have been measured (Table 1) [10,13]; the average of their values is about 35 krad/sec. Thus the value obtained for D_β is about 69% of this average value. For the case where the hinge is a universal joint between two identical segments, the theoretical effect of the joint on the rotational diffusion coefficients of the segments has been calculated [16,17]. There are two major distinctions between a dilute solution of once-broken rods and one of unconnected segments. First, the relationships between center-of-mass (translational) motions and the orientational motions of the segments

TABLE 3. Rotational Diffusion Coefficients for
Myosin Rod and Its Fragments, Assuming One
Decay Time

| Fragment | Contour Length (nm)[a] | $\Xi_{\perp 20°,w}$ (krad/sec) | |
		Theory[b]	Experiment[c,d]
LMM	78.5	23.2	21.7 ± 0.8,[c] 21.8 ± 0.3[d]
S-2	65.0	38.2	47.5 ± 1.6[d]
Rod	136.0	5.3	7.7 ± 0.8,[c] 6.9 ± 0.1[d]

[a]The contour length is based on electron micrograph data [28].

[b]The theoretical Ξ value is based on Broersma's relations for a dilute solution of rods, using the contour length and a diameter of 2 nm [4-6].

[c]From light-scattering measurements [21].

[d]From electric birefringence data [26], correcting the results for viscosity and temperature (3-20°C).

TABLE 4. Depolarized Decay Constants for Myosin Rod,
Assuming Two Decay Times

Type of Fit	Γ_1 (ksec^{-1})	Γ_2 (ksec^{-1})
Eq. (1)[a]	———	34 ± 5
Eq. (2)	319 ± 84	28 ± 6

[a]The fits were done for time $t > 8$ μsec.

are different. Second, there exists some hydrodynamic interaction between connected segments that
is not present for the free segments. As a result,
the rotational diffusion coefficient for a connected
segment is expected to be smaller than the value for
an unconnected segment. Wegener et al. [17] predict
that D_β for a connected segment would be 78% of the
value for an unconnected segment, whereas Fujiwara et
al. [16], from a more detailed account of the hydrodynamic interactions, predict that D_β is 53% of that
for an unconnected segment. Thus the value obtained
for D_β is consistent with the current theories.
    The slow time in the two-exponential fit roughly
corresponds to the overall rotational diffusion of
the myosin rod. Because of the difficulties involved
in fitting two exponentials to noisy data, the value
for $\Xi_\perp$ obtained from the second exponential (3.7 ±
0.8 krad/sec) may be somewhat low. The value
obtained by fitting a single exponential to the long
time tail of the correlation function ($t > 8$ μsec) is
4.8 ± 0.8 krad/sec. Considering the theoretical
approximations, both values agree well with the value
predicted if myosin were a stiff, unbroken rod
(5.3 krad/sec).
    Some theories for muscle contraction postulate
that elasticity may be a property of the hinge region
in the myosin rod [29-31]. Thus a restoring force
may be generated as the rod bends, producing tension.
A reasonable approximation of such a force should be
the cosine potential theory, whose results are presented in Section III and tabulated in Table 2.
Again, two exponentials are predicted. However, the
time constants predicted are clearly much faster than
those observed experimentally when theoretical relative amplitudes (C_2/C_1) of the same magnitude as
experiment (A_1/A_2) are compared. Consequently, a
substantial restoring force does not appear to be
responsible for the DDLS correlation functions. A
combination of the two models may be used; that is,
both a maximum bending angle and a weak restoring
force could be used in the theory. However, the

assumption of a maximum bending angle alone gives results accurate to within experimental error. Furthermore, the maximum angle found (121-132°) is reasonably close to the angle found from electron microscopy (145°). Addition of a restoring force would not improve the results significantly.

VI. CONCLUSIONS

For semidilute solutions of rodlike particles, the rotational diffusion coefficient is proportional to $\sim 1/c^2$ and $\sim 1/L^8$. Considering experimental errors and the theoretical approximations, the trends obtained are consistent with the Doi-Edwards theory, which also predicts a $1/c^2$ dependence but expects a larger length dependence ($1/L^9$). The proportionality constant β' is about two orders of magnitude higher than was predicted, and thus the value of the rotational diffusion coefficient is much higher than that predicted. Further work on the motions of rodlike particles is needed to test the theory more fully and the limits of its applicability. In particular, it would be useful to be able to predict values for β' and the intercept for any rodlike system, either semiempirically or purely theoretically.

The myosin rod has two contributions to its DDLS time correlation functions. The slower time is comparable to the value expected if the myosin rod were a stiff, unbroken rod. The faster time is attributed to bending of the myosin rod at the hinge region. A substantial restoring force does not appear to be present; instead, the myosin rod seems to bend freely within a maximum bending angle of 128°. Additional experiments on other once-broken rod systems, particularly ones where a restoring force is known to exist, may help to verify the tentative conclusions reached here.

ACKNOWLEDGMENTS

The DDLS experiments were performed in the laboratories of, and under the guidance of, Professor R. Pecora at Stanford University and were funded by NIH Grant No. 5R01 GM 22517 and NSF Grant No. CHE 79-01070. The myosin rod was supplied by Professor Stefan Highsmith (University of the Pacific). The DDLS experiments on myosin rod were performed by Dr. Chun-Chen Wang (currently at the Naval Research Laboratories, Washington, D.C.).

REFERENCES

1. B. J. Berne and R. Pecora, *Dynamic Light Scattering* , Wiley, New York, 1976.

2. B. Chu, *Laser Light Scattering*, Academic Press, New York, 1974.

3. S. H. Koenig, Biopolymers *14*, 2421 (1975).

4. S. Broersma, J. Chem. Phys. *32*, 1626 (1960).

5. S. Broersma, J. Chem. Phys. *32*, 1632 (1960).

6. J. Newman, H. L. Swinney and L. A. Day, J. Molec. Biol. *116*, 593 (1977).

7. M. M. Tirado and J. G. de la Torre, J. Chem. Phys. *71*, 2581 (1979); *73*, 1986 (1980).

8. J. E. Hearst, J. Chem. Phys. *38*, 1062 (1963).

9. S. Michielsen and R. Pecora, Biochemistry *20*, 6994 (1981).

10. C. R. Crosby, N. C. Ford Jr., F. E. Karasz and K. H. Langley, J. Chem. Phys. *75*, 4298 (1981).

11. P.J. Missel, N. A. Mazer, G. B. Benedek and C. Y. Young, J. Phys. Chem. *84*, 1044 (1980).

12. J. Riseman and J. G. Kirkwood, J. Chem. Phys. *18*, 512 (1950).

13. M. Doi and S. F. Edwards, J. Chem. Soc. Faraday II *74*, 560 (1978).

14. M. Doi and S. F. Edwards, J. Chem. Soc. Faraday II *74*, 918 (1978).

15. K. Zero and R. Pecora, Macromolecules *15*, 87 (1982).

16. M. Fujiwara, N. Numasawa and N. Saito, Rep. Progr. Polym. Phys. Jap. *23*, 531 (1980).

17. W. Wegener, R. Dowben and V. Koester, J. Chem. Phys. *73*, 4086 (1980).

18. K. Zero and R. Pecora, Macromolecules *15*, 1023 (1982).

19. B. Pal, Bull. Calcutta Math. Soc. *10*, 3 (1919).

20. K. Zero, Stanford University, Chemistry Dept., Ph.D. Thesis (Aug. 1982).

21. S. Highsmith, C.-C. Wang, K. Zero and R. Pecora, Biochemistry *21*, 1192 (1982).

22. J. F. Maguire, J. P. McTague, and F. Rondelez, Phys. Rev. Lett. *45*, 1891 (1980).

23. J. F. Maguire, J. P. McTague and F. Rondelez, Phys. Rev. Lett. *47*, 148 (1981).

24. N. Ookubo, M. Komatsubara, H. Nakajima, and Y. Wada, Biopolymers *15*, 929 (1976).

25. K. Takahashi, J. Biochem. *83*, 905 (1978).

26. S. Highsmith, K. M. Kretzschmar, C. T. O'Konski and M. F. Morales, Proc. Natl. Acad. Sci. (USA) *74*, 4986 (1977).

27. A. Elliott and G. Offer, J. Molec. Biol. *123*, 505 (1978).

28. S. Lowey, H. S. Slayter, A. G. Weeds and H. Baker, J. Molec. Biol. *42*, 1 (1969).

29. A. F. Huxley and R. Simmons, Nature *233*, 533 (1971).

30. W. F. Harrington, Proc. Natl. Acad. Sci. (USA) *68*, 685 (1971).

31. S. Highsmith, Biochim. Biophys. Acta *639*, 31 (1981).

CHAPTER 15

Transient Electric Birefringence of DNA Restriction Fragments and the Filamentous Virus Pf3

DON EDEN*and JOHN G. ELIAS[†]
Yale University
New Haven, Connecticut

CONTENTS

*Present address: Chemistry Department, San Francisco State University, San Francisco, California 94132.

†Present address: E. I. DuPont de Nemours and Company, Wilmington, Delaware 19898.

I. INTRODUCTION

When a pulsed electric field is applied to a solution of macromolecules, the macromolecules orient and the solution exhibits a macroscopic optical anisotropy. The amplitude and temporal response of this transient electric birefringence is a function of the intrinsic optical, electrical, and hydrodynamic properties of the macromolecules in the solution. Recent advances in instrumentation have made it easy to study the transient electric birefringence (TEB) of rodlike molecules at low concentrations and low electric fields. Information about the hydrodynamic properties and electrical polarizability of the molecules can be obtained in a straightforward manner. By combining the rotational diffusion constant determined by TEB and the translational diffusion constant obtained by use of quasielastic light scattering (QELS), it is possible, in principle, to obtain both the length and diameter of a rodlike molecule. For flexible molecules, deviations from rodlike behavior can be characterized statistically in terms of a persistence length.

Several recent monographs [1,2] review the theory and experimental techniques of transient electric birefringence and give examples of applications to biopolymer systems. Recent conference proceedings [3,4] contain numerous papers reporting on theoretical advances and experiments on a wide variety of systems.

In this chapter we demonstrate the sensitivity of TEB and give examples of recent work in our laboratory on restriction fragments of DNA and the filamentous virus Pf3.

The investigations result in excellent estimations of the hydrodynamic dimensions of the virus and short double-helical DNA fragments, in the force constant for flexing DNA molecules, and in an understanding of how flexibility affects the electrical polarizability of DNA.

Many of the optical and electronic components used in TEB are the same as those used in QELS. Therefore, it is straightforward and reasonably inexpensive to set up TEB apparatus in a laboratory already equipped for QELS.

II. THEORY

A. Steady-State Electrical Birefringence

In the electric birefringence technique, an electric field is used to orient molecules in solution. The orientation results from the torque exerted on the molecules' permanent and induced dipole moments. There is thus a tendency for the overall moment to align along the field direction. The optical properties of the system change with the ordering, and the steady-state magnitude of the change yields information on the electric moments. When the applied field is removed, the orientational order is randomized by rotational diffusion. As the molecules diffuse and the macroscopic order becomes smaller, the optical properties of the system change correspondingly. The time dependence of this change provides information on the size and shape of the molecules. In a dilute solution a system of isotropically oriented molecules, which may individually be optically anisotropic, does not affect the polarization of the incident light. However, when net order is induced, the refractive properties of the system become anisotropic and birefringence results. The amplitude of the birefringence depends on the degree of orientation. The birefringence of a medium subjected to a uniform electric field is defined as [5]

$$\Delta n = n_{||} - n_{\perp} , \tag{1}$$

where $n_{||}$ and $n_{\perp}$ are the refractive indices of the solution parallel and perpendicular to the applied field, respectively. In the Rayleigh-Gans approximation Δn is directly proportional to the difference in the molecular optical anisotropy. Linearly polarized light that is parallel to any of the principal axes of the refractive index tensor will be refracted in the normal manner. However, if the polarization of the incident light is not parallel or perpendicular to the field, the two components of the optical electric vector propagate through the medium at different velocities. One component will lag behind the other by an amount determined by the distance the light travels through the medium and by the difference in the two indices of refraction. The retardation or phase shift between the two electric components is given by

$$\delta = \frac{2\pi l \ \Delta n}{\lambda_0} , \tag{2}$$

where l is the distance the light travels through the medium and λ_0 is the wavelength of light *in vacuo*. Unless the phase shift is a multiple of $\pi/2$, the emerging light will be elliptically polarized.

A saturation value for the birefringence Δn_{sat} is observed when the undistorted molecules are fully oriented in the external field. For solutions comprised of highly symmetrical molecules, the saturation value of the birefringence is directly related to the difference in the optical polarizabilities $\Delta\alpha_0$ along the directions parallel and perpendicular to the molecular symmetry axis

$$\Delta n_{sat} = \frac{N(\alpha_{0,3} - \alpha_{0,1})}{2n\varepsilon} \;, \tag{3}$$

where N is the number of molecules, n is the refractive index of the isotropic solution, and ε is the permittivity *in vacuo*; Δn_{sat} is related to the optical anisotropy factors g_3 and g_1 by [6],

$$\Delta n_{sat} = \frac{2\pi C_v(g_3 - g_1)}{n} \;, \tag{4}$$

where C_v is the volume fraction of the particles. The angular orientational distribution function $W(\theta)$, where θ is the angle between the symmetry axis of the cylindrical molecule and the applied electric field, depends on the potential energy of each molecule in the electric field and is calculated from the Boltzmann equation

$$W(\theta) = \frac{\exp\left[-U/k_B T\right]}{2\pi \displaystyle\int_0^\pi \exp\left[-U/k_B T\right] \sin\theta \; d\theta} \;, \tag{5}$$

where U is the potential energy of each molecule in the electric field and is given by

$$U = \mu E \cos\theta - \frac{1}{2}(\alpha_{E,3} - \alpha_{E,1})E^2 \cos^2\theta \;, \tag{6}$$

where E is the magnitude of the electric field at the molecule, μ is the magnitude of the permanent dipole moment along the symmetry axis, and $\alpha_{E,3} - \alpha_{E,1}$ is the difference between the electrical polarizability parallel and perpendicular to the molecular symmetry axis.

The orientation factor [7] $\Phi(\beta,\gamma)$ that results from a spatial integration of $W(\theta)$ is given by

$$\Phi(\beta,\gamma) = \int_0^\pi W(\theta) \; \frac{3 \cos^2 \theta - 1}{2} \; 2\pi \sin \theta \; d\theta \; , \qquad (7)$$

and depends on the interaction energies relative to $k_B T$.

$$\beta = \frac{\mu E}{k_B T} \; , \qquad (8)$$

$$\gamma = \frac{(\alpha_{E,3} - \alpha_{E,1})E^2}{2 k_B T} \; . \qquad (9)$$

Therefore, $\Phi(\beta,\gamma)$ will be zero for an isotropic solution and one for a solution with all molecules oriented. For low electric fields

$$\Phi(\beta,\gamma) = \frac{\beta^2 + 2\gamma}{15} \; . \qquad (10)$$

The steady-state birefringence, which for arbitrary fields can be written as

$$\Delta n = \Delta n_{sat} \Phi(\beta,\gamma) \; , \qquad (11)$$

is proportional to the square of the applied electric field for low voltages:

$$\lim_{E \to 0} \frac{\Delta n}{E^2} = C_v n K_{sp} \; , \qquad (12)$$

where K_{sp}, the specific Kerr constant, is given by

$$K_{sp} = \frac{2\pi(g_3 - g_1)}{15n^2} \left\{ \frac{\mu^2}{(k_B T)^2} + \frac{\alpha_{E,3} - \alpha_{E,1}}{k_B T} \right\} . \quad (13)$$

Therefore, when the optical anisotropy $g_3 - g_1$ of a molecule and the relative contributions of the birefringence arising from permanent and induced moments are known, it is possible to evaluate the magnitude of the electrical terms.

B. Rotational Diffusion and Time Dependence of Birefringence

The time dependence of the electric field free decay of the birefringence of a dilute solution of macromolecules is determined by the rotational diffusion constants of the molecules. Wegener et al. [8] have shown that for a monodisperse solution containing arbitrarily shaped molecules, there can be as many as five exponentials. The decay is less complicated for molecules of higher symmetry. For cylindrically symmetric molecules with a large axial ratio only one exponential is observed and

$$\Delta n(t) = \Delta n_0 \exp[-6D_r t] . \quad (14)$$

In this case D_r can be related to the length L and diameter d of the molecule through

$$D_r = \frac{3k_B T}{\pi \eta_0 L^3} \left\{ \ln \frac{L}{d} - \gamma_r \right\} \quad (15)$$

where η_0 is the solvent viscosity, $k_B T$ is the thermal energy, and γ_r is a correction term for end effects that depends on L and d. There are numerous theoretical and empirical determinations of γ_r. Many have been presented both analytically and graphically by Elias and Eden [9].

The analogous expression for the translational diffusion of a cylinder is given by

$$D_t = \frac{k_B T}{3\pi\eta_0 L} \left\{ \ln \frac{L}{d} - \gamma_t \right\}, \tag{16}$$

where the correction term γ_t has been evaluated by numerous authors [10-13].

A comparison of Eqs. (15) and (16) shows that rotational diffusion is much more sensitive to the length of a cylinder than is translational diffusion. A measurement of both D_r and D_t on the same system permits the unambiguous determination of both the length and diameter of a cylinder when the correct forms for γ_r and γ_t are known. This powerful approach has been used to determine the dimensions of double-helical DNA and the filamentous virus Pf3 in the work described below.

Deoxyribonucleic acid and other linear polymers may behave as right circular cylinders only if the molecules are short enough. Very long molecules exhibit the characteristics of a random coil. A statistical treatment, which addresses these limiting cases as well as that of the more mathematically difficult intermediate length regime, is the wormlike coil model given by Kratky and Porod [14]. In this model the molecules are characterized by an isotropic chain flexibility parameter (the persistence length), by the distance between frictional elements along the chain, and by the diameter of each element. Hearst [15] derived equations for the rotational diffusion of wormlike and weakly bending rod molecules by using the methods developed by Riseman and Kirkwood [16] and Hermans and Ullman [17] for cases where excluded-volume effects are negligible. His treatment considered only the rotation of the entire molecule in which changes in internal coordinates were forbidden.

The resulting form of the rotational diffusion constant for a weakly bending rod about its minor axis is

$$D_r = \frac{k_B T}{\pi \eta_0 L^3} \left\{ 3 \ln \frac{L}{b} - 7.0 + 4 \frac{b}{a} + \lambda L \left[2.25 \ln \frac{L}{b} - 6.66 + 2 \frac{b}{a} \right] \right\} \tag{17}$$

where b is the distance between frictional elements, a is the Stokes diameter of each element, L is the contour length of the chain, and λ^{-1} is twice the persistence length.

To avoid the errors that can arise in the calculation of the hydrodynamic properties of flexible polymers using "configurational preaveraging" of the hydrodynamic interactions, Hagerman and Zimm [18] have recently performed a Monte Carlo analysis to relate the flexibility of short wormlike chains to their smallest rotational diffusion constant. Their analysis gives results that are qualitatively similar to those of the analysis by Hearst [15].

In general, the rise of the birefringence will have a temporal dependence that is more complicated than the decay [1,2,19,20] if the molecules have a permanent dipole moment or a moment that is induced slowly. Tinoco and Yamaoka [20] have treated the time dependence not only of the birefringence growth, but also the response on field reversal for low electric fields. Their analysis is particularly useful in determining the orientational mechanism of a macromolecule.

III.   EXPERIMENTAL APPARATUS

The birefringence apparatus was designed to measure moderately fast and relatively weak birefringence signals. Use of laser illumination

410   D. Eden and J. G. Elias

combined with fast analog-to-digital conversion and
signal averaging resulted in a birefringence amplitude resolution of 4×10^{-12} and time resolution of
50 nsec. The orientating electric field can be used
as an alternating unipolar pulse or as a reversing
bipolar pulse, and its amplitude is variable up to
7.5 kV/cm, with fast rise and decay times. The
design philosophy is similar to that described by
Newman and Swinney [21], but several changes have
been made to improve the sensitivity and response
time. A brief description of the apparatus, which is
shown schematically in Fig. 1, follows.

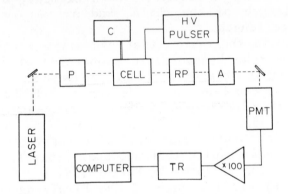

FIGURE 1.   Schematic diagram of birefringence apparatus. LASER, argon-ion Lexel Model 95; C,
Lauda K2/R bath and circulator; P and A,
Glan-Thompson polarizers; CELL, birefringence sample cell and thermostatic
container; HV PULSER, Cober Model 605P and
Velonex Model 350 high-voltage pulse
generators; RP, quarter-wave plate; PMT,
1P28 phototube; ×100, Textronix Model 1121
amplifier; TR, LeCroy 2256 transient
recorder; COMPUTER, LSI-11 computer.
[Reprinted with permission from J. G.
Elias and D. Eden, Macromolecules *14*, 410
(1981).]

The light source is a Lexel Model 95 argon-ion laser operating in the light-regulated mode at 514.5 nm. The laser beam passes first through a high-quality Glan-Thompson polarizer and then is reduced to a diameter of 1 mm with a pinhole before it enters the sample cell, which is shown schematically in Fig. 2.

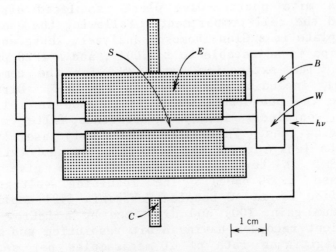

FIGURE 2.   Cross-sectional side view of Kerr cell (B, polycarbonate cell body; W, window; E, top electrode; C, bottom electrode connection; S, sample volume).

The cell was machined from a solid block of black polycarbonate and has an optical path length of 30 mm and an electrode gap of 2.7 mm. The sample volume is 0.7 cm^3. The cell windows are 8.1 mm thick optical glass and are attached, with minimal strain, to the cell body with cover plates. Strain birefringence is negligible. The massive electrodes are gold-plated stainless steel. They have a large heat capacity and will not flex, even in a very large externally applied electric field. The top electrode is removable to allow access to the sample chamber for filling the cell. A calibrated thermistor is embedded in the cell body 1.5 mm from the sample volume. The cell is completely encased by a thermostated aluminum block that has 6 mm diameter optical entrance and exit holes. The temperature is constantly monitored during an experiment and is controlled to ± 0.02°C.

A mica quarter-wave plate is placed directly behind the cell compartment. Following the quarter-wave plate is a Glan-Thompson analyzer, whose angular position is adjustable to ± 0.001°, and a 1P28 phototube. The extinction ratio K_{ex} for the complete system, including the cell, is 1 × 10^{-6}. Directly before the phototube, which is encased in a light-tight container, is an interference filter and a diverging lens. The filter passes only laser light, and the lens spreads the beam onto the photocathode. The PMT is terminated with 340 Ω to give fast response (~ 50 nsec). The resulting voltage is amplified by an AC coupled Tektronix 1121 amplifier (maximum gain: 100) and digitized by a LeCroy 2256 transient recorder having 8-bit resolution and maximum digitizing rate of 20 megasamples per second. Each birefringence signal is stored in 1024 words of digitizer memory and later transferred, between high-voltage pulses, to an on-line DEC LSI-11 computer by means of a laboratory-designed interface. The computer stores the birefringence signals, and after the requested number of accumulations it calculates the amplitude, fits the rise, reversal, or decay portions of the curve; plots the raw data; and stores the data on the disk. During the experiment the accumulated birefringence signal is displayed on a large screen oscilloscope so that the buildup of the signal can be followed visually.

The high-voltage exciting pulse is produced by coupling, in parallel, Cober Model 605P and Velonex Model 350 high-power pulse generators. The Velonex produces the negative pulse and the Cober, the positive pulse. When a unipolar exciting pulse is desired, the Cober generator is used, and the pulse polarity is alternated with mercury wetted switches to prevent electrode polarization and net electrophoresis. The pulse length is continuously adjustable from several hundred nanoseconds to 10 milliseconds. When pulse widths greater than 2 msec were desired, extra capacitance (1000 μF) was added to the Cober's floating supply to reduce pulse droop to an acceptable level (< 2%/msec). The electric field is continuously variable over the range 0-7.5 kV/cm, and it is accurately measured for each run by a high-voltage pulse amplitude meter. The high voltage switching time is approximately 100 nsec.

As a check of the capability of the system we measured the birefringence of water at low electric fields. The Kerr law was obeyed at all field values accessible to this apparatus and resulted in a Kerr constant of 3.44×10^{-12} cm/V^2 at room temperature. The sensitivity of the system is demonstrated by the fact that the birefringence of water can be measured at fields as low as 150 V/cm, which is equivalent to a birefringence of 4×10^{-12}.

With the apparatus described above, the intensity of the light that passes through the Glan-Thompson analyzer is

$$I(t) = I_0 K \left[\sin^2(\beta - \delta/2) + K_{ex} \right] , \qquad (18)$$

where I_0 is the intensity of the light incident on the sample, K is a light reflection and absorption factor (which is < 1), β is the angle between the analyzer and its crossed position, and K_{ex} is the overall extinction ratio:

$$K_{ex} = \frac{I(0)_{\beta=0}}{I(0)_{\beta=\pi/2}} . \qquad (19)$$

It is an indication of the overall quality of the optical components. Substitution of Eq. (2) into Eq. (18) and normalization to the steady-state light intensity present in the absence of a field gives the relative change in the intensity due to the birefringent nature of the ordered system:

$$\frac{\Delta I(t)}{I_0} = \frac{\sin^2[\beta - (\pi l/\lambda_0)\Delta n(t)] - \sin^2 \beta}{\sin^2 \beta + K_{ex}} . \tag{20}$$

When β is less than $10°$ and when $\beta > 5\delta_0$, where δ_0 is the steady-state retardation, Eq. (20) reduces to a simpler form for which the change in intensity is linearly, within 1%, proportional to the birefringence. Under these conditions the decay of the light intensity for a solution containing cylindrical molecules is

$$I(t) = \Delta I(0)\exp[-t/\tau_r] , \tag{21}$$

where the relaxation time $\tau_r = (6D_r)^{-1}$ can be determined directly.

Further details of the transient electric birefringence apparatus have been given by Elias [22].

IV.   SAMPLES

A.   DNA Restriction Fragments

The monodisperse restriction fragments were prepared by digesting the plasmids pBR322 and pMB9 with the restriction fragments *Hae* III and *Eco* RI. The details of the preparation and isolation have been given by Elias and Eden [19] and Elias [22]. For all the experiments described below, the sample temperature was 3.8°C and the DNA concentrations were the same (~ 5 µg of DNA/cm^3). Experiments were performed on fragments from 64 to 5000 base pairs in a sodium phosphate buffer adjusted to a sodium ion concentration of 1.0 mM and pH 7.2 [19]. In addition, five fragments, 64 to 124 base pairs, were studied as a function of ionic strength (0.2-2.5 mM in sodium) in a phosphate buffer at pH 7.6 [9].

B. Pf3 Virus

The filamentous virus Pf3 was propagated on its host strain of *Pseudomonas aeruginosa* PA01 (RP1) bearing the RP1 plasmid that determines, in part, its host range. Its isolation, which has been described by Newman et al. [23], is very similar to that used by Chen et al. [13] in their study of another filamentous virus, Xf. The samples were run in sodium phosphate buffer (pH 7.0) at 24.75°C at concentrations of 2 and 8 $\mu g/cm^3$ and ionic strengths of 1 and 10 mM.

V. RESULTS AND DISCUSSION

A. DNA Restriction Fragments

A typical birefringence response to an alternating unipolar pulse for a 124 base pair fragment in 1 mM Na phosphate buffer is given in Fig. 3. The inset to Fig. 3 shows the decay of the birefringence as log I versus t. Note that the intensity change is negative. This results because the optical anisotropy g_3 - g_1 for DNA is negative (-0.0215 [24]) and since the DNA tends to align with its long axis parallel to the applied electric field, the birefringence [Eqs. (12) and (13)] is negative.

The birefringence resulting from a field reversal experiment for a solution of 192 base pair fragments in 1 mM sodium phosphate is given in Fig. 4. Note that there is a transient that results on field reversal. Analysis of this transient [22] assists in the determination of the electrical mechanism by which orientation is achieved. As has been discussed elsewhere [22,25], the orientation arises from a slow induced moment resulting from the net displacement of the counterions relative to their equilibrium position near the phosphates on the DNA double helix.

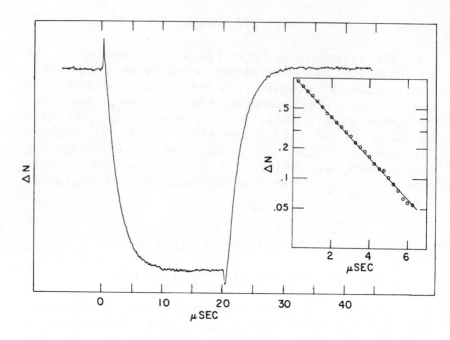

FIGURE 3.   Typical example of an electric birefringence signal resulting from alternating
unipolar pulses. The signal results from
averaging 600 sweeps arising from 20 μsec
pulses with field strength of 1800 V/cm
applied to a solution of 124-BP DNA at a
concentration of 5.2 μg/cm³ in 1 mM sodium
phosphate buffer at pH 7.2 at 3.8°C. The
spikes on the rising and falling edges are
due to the positive birefringence of the
buffer. The semilog inset shows that the
decay is a single exponential with a
characteristic time of 2.17 μsec.

Figures 3 and 4 demonstrate the high signal/
noise that can be achieved for low DNA concentrations
at low electric fields. For a 124 base pair (BP)
fragment with a concentration of 5 μg/cm³, Δn_{sat} as
calculated from Eq. (3) would be -2.54×10^{-7}. This

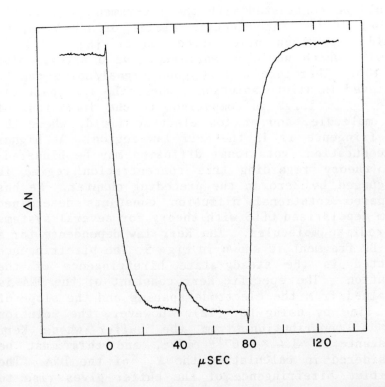

FIGURE 4. Typical example of a reversing low-field electric birefringence signal. The maximum in the center occurs when the applied electric field is suddenly reversed in polarity and the DNA molecules partially reorient. The spikes on the rising, reversal, and falling edges are due to the positive birefringence of water. The signal results from averaging 400 field-reversing pulses with field strengths of 2200 V/cm. The positive and negative positions of the pulse are each 40 μsec wide. The sample is 188-BP DNA at a concentration of 5.1 μg/cm³. The other conditions are as in Fig. 3.

should be contrasted with the instrument sensitivity of 4×10^{-12}. At electric fields of 600 V/cm, we would expect an orientation factor [Eq. (11)] of 0.0016, which would potentially give a signal/noise of 100. Therefore a very good signal/noise can be achieved in dilute solutions, where the interparticle spacing is large in comparison to the dimensions of the molecule, and at low electric field, where the birefringence is in the Kerr law region. At higher concentration, rotational diffusion may be hindered. The theory regarding this concentration regime is discussed by Zero in the preceding chapter. He has compared rotational diffusion constants determined from depolarized QELS with theory for several systems of rodlike molecules. The Kerr law dependence for a 124-BP fragment is shown in Fig. 5. The birefringence plotted is the steady-state birefringence of the solution. The specific Kerr constant of the DNA is obtained from the electrode spacing and the slope of the line by using Eq. (12). However, the solution has a contribution from the buffer whose Kerr constant is 3.4×10^{-12} cm/V^2, and this must be considered in calculating the K_{sp} of the DNA. The positive birefringence of the buffer gives rise to the sharp transients in Figs. 3 and 4 that occur when the field is turned on, reversed, or turned off.

B.   Short-Fragment Ionic Strength Study

The rotational diffusion constant of five fragments (64, 80, 89, 104, and 124-BP) have been studied as a function of ionic strength (0.2-2.5 mM) at 3.8°C [9]. The diffusion constants were obtained from the decay of the birefringence, and all fragments exhibited single exponential decay as determined from both a least-squares fit to a single exponential and from a cumulants analysis [26]. No ionic strength dependence of the rotational relaxation time was observed. The average experimentally determined rotational relaxation times for the five fragments

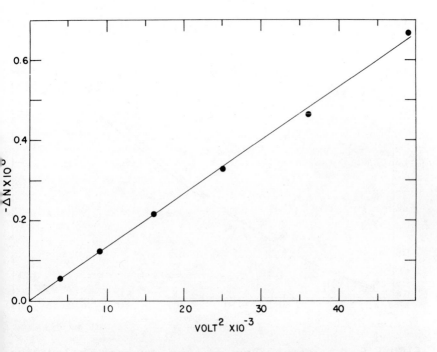

FIGURE 5.   The steady-state birefringence of 124-BP
DNA as a function of applied voltage
squared. The solution conditions are the
same as in Fig. 3. The specific Kerr
constant is determined from the slope and
Eq. (12).

are shown in Fig. 6, along with two best-fit theoretical curves for the four shortest fragments, using
Eq. (15) with the γ_r functions reported by Mandelkern
and Crothers [9,27] and Yoshizaki and Yamakawa [28].
In this analysis we were striving to obtain a structural constant, the rise per base pair (RPBP), for
rodlike DNA in solution by use of a reasonable choice
for the hydrodynamic diameter and five expressions
for γ_r. We fitted the theory, with RPBP as the
adjustable parameter, to experiment by using two data
sets: one containing all five fragments and one that

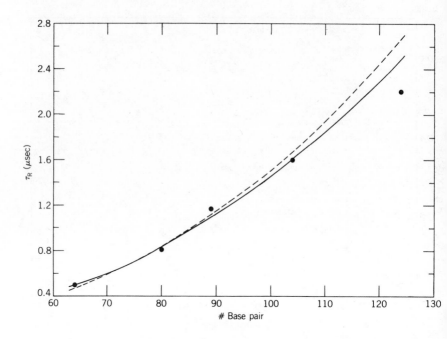

FIGURE 6.   A comparison of the average relaxation time of the pBR322 fragments at 3.8°C in phosphate buffer (pH 7.6) with the theoretical predictions for γ_r by Mandelkern and Crothers [27] (solid line) and Yoshizaki and Yamakawa [28] (dashed line) for a hydrodynamic diameter of 26 Å. Only the four shortest fragments have been used in the fits. [Reprinted with permission from J. G. Elias and D. Eden, Biopolymers 20, 2369 (1981).]

excluded the 124-BP fragment. The error in the four fragment fit was less by a factor of 2 or more than the error in the five fragment fit for all theories considered. As a consequence of this analysis and the analysis of larger fragments, which is described below, we believe that the four fragments ($\leqq$ 104 base pairs) behave as straight right circular

cylinders to a very good approximation, but that
significant deviations from rodlike behavior exist
for 124-BP and longer fragments. For a hydrodynamic
diameter of 26 Å, we obtain an RPBP between 3.32-
3.45 Å, which is consistent with the range for RPBP
(3.04-3.41 Å) obtained from X-ray diffraction results
from B-genus DNA in fibers. The variation in RPBP
for the fiber results has been attributed to the
difference in the base composition and sequence [29].

    In this analysis we had to choose a diameter to
obtain the RPBP. However, by measuring both D_r and
D_t, it is possible to obtain both of the structural
parameters, RPBP and d. This was accomplished [30]
by combining Mandelkern's data [31] for the translational diffusion constant, obtained by using QELS,
with D_r from our birefringence and Mandelkern's [31]
dichroism data. We find that the birefringence decay
times and the D_t are consistent with a diameter of
26 Å when Broersma's original expressions [10] for D_r
and D_t are used. Subsequent analysis [30] gives an
RPBP value of (3.34 ± 0.1) Å.

C. Length Dependence of D_r Yields the Flexibility

    The deviation of the rotational relaxation time
obtained from the field free birefringence decay from
that expected for a rigid cylinder becomes greater as
the size of the DNA increases. The results of measurements in 1 mM sodium phosphate buffer at 3.8°C
are shown in Fig. 7. For the 410- and 587-BP fragments the decay curves were the sum of two exponentials, one long and one very short with a small
amplitude. For these fragments, the longer times
were used in the hydrodynamic analysis. For the
410-BP fragment, τ_R is 60% that expected for a rigid
cylinder. The data were analyzed by using Hearst's
[15] expression for the rotational diffusion of a
weakly bending rod [Eq. (17)]. The persistence
length was determined to be 495 Å. This persistence
length is considerably smaller than that obtained

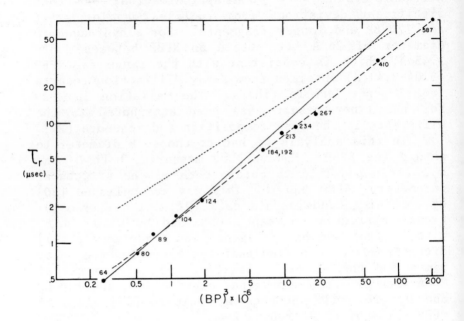

FIGURE 7.   Log-log plot of the rotational times versus the number of DNA base pairs cubed. The circles represent the experimental points. The solid line was calculated from Broersma's original equation [Eq. (15) and γ_r from reference 10] using 3.15 Å for the RPBP and 27 Å for the diameter. The dashed line was calculated from Hearst's weakly bending rod model [Eq. (17)], using 495 Å for the persistence length, 3.15 Å for the RPBP, 30.4 Å for the distance between frictional elements, and 27 Å for the diameter. The dotted line was calculated from Hearst's [15] wormlike model, using the same parameters as above. All the experiments were performed at 3.8°C in 1 mM Na⁺. [Reprinted with permission from J. G. Elias and D. Eden, Macromolecules *14*, 410 (1981)].

from light-scattering and flow birefringence experiments at this ionic strength [32]. In those experiments on large DNA, significant corrections, which
arise from excluded volume effects, must be made.
Excluded volume corrections should be insignificant
for the short restriction fragments used in our
electric birefringence experiment. In large DNA the
persistence length is observed to be strongly dependent on ionic strength [32] because of its effect on
the electrostatic interactions between phosphates on
the helix. To date, the persistence length has not
been studied systematically as a function of ionic
strength by use of rotational diffusion constants
determined by electrooptical techniques. This work
is planned for the future in our laboratory.

It is worthwhile to consider the implication of
our measured persistence length on the organization
of DNA in chromatin. According to the treatment of
Landau and Lifshitz [33], the Hooke's law spring
constant α for bending an isotropic spring can be
related to the persistence length Q of a Kratky-Porod
wormlike chain.

The free energy required to flex a short segment
of contour length L of a long chain is

$$\Delta G = \frac{\alpha \theta_L^2}{2L} ,$$
(22)

where θ_L is the angle between the ends of the segment. When placed in a thermal bath, bending occurs
in two mutually perpendicular directions and

$$\langle \theta_L^2 \rangle = \frac{2Lk_BT}{\alpha} .$$
(23)

The spring constant can be related to the persistence
length by

$$\alpha = Qk_BT .$$
(24)

Our persistence length of 495 Å at 4°C corresponds to a spring constant α of 1.89×10^{-19} erg·cm/rad^2. For a contour length $L = \varrho$, $\langle\theta_L^2\rangle = 2$ radians^2. In the chromatin structure, DNA is wrapped around histone cores to form nucleosomes, and the nucleosomes are connected by "linker" DNA. A 145-BP length of DNA wraps around the histone core 1 3/4 times and then proceeds to the next nucleosome [34]. Therefore, one turn of double-helical DNA requires about 83 base pairs. Given an RPBP of 3.34 Å, an 83-BP fragment will have a contour length of 277 Å, slightly more than half a persistence length. As calculated from Eq. (23), this fragment free in solution will have $\langle\theta_L^2\rangle = 1.12$ radians^2, which is much less than that for a fragment of this size in nucleosomal DNA, where $\langle\theta_L^2\rangle = 4\pi^2$ radians^2. Therefore, the energy associated with wrapping the DNA around the histones to form a nucleosome is from Eq. (22), 35 $k_B T$, or 80.6 kJ/mole. The origin of the extra energy must come from strong protein-DNA interactions, which are probably many times the necessary 81 kJ/mole.

D.   Length Dependence of the Electrical Polarizability and Kerr Constant

There are many theoretical treatments of the length dependence of the electrical polarizability of a cylindrical polyelectrolytes [19,22]. One of the the earliest and simplest is the treatment by Mandel [45], who used a model that emphasized the discrete distribution of polyion charges to derive the polarizability and the relaxation time for polarization of cylindrical, highly charged polyions due to the longitudinal motion of bound counterions. In the model the bound counterions were constrained to a region close to the polyion where the electrostatic potential was greater than $k_B T$. Potential-energy minima were located at the charged sites attached to the polyion, and the dipole moment induced by an external electric field was calculated from the probability of locating counterions a certain distance away from their minimum energy positions. The polarizability $\Delta\alpha = \alpha_{E,3} - \alpha_{E,1}$ was given by

$$\Delta\alpha = \frac{\gamma z e_0^2 N L^2}{12 k_B T}, \tag{25}$$

where L is the length of the rod, N is the number of charged sites, γ is the fraction of bound counterions, e_0 is the elementary electrical charge, and z is the valence of the counterions. For a polymer, such as DNA, with a linear charge distribution the electrical polarizability should then depend on the cube of the length of a straight cylindrical polyion. We have analyzed our Kerr constant data to obtain $\Delta\alpha$ by using Eqs. (12) and (13), taking the optical anisotropy $g_3 - g_1 = -0.0215$ and calculating C_v, assuming that the partial specific volume of DNA is 0.50 cm^3/g. From symmetry considerations and from the temporal course of the birefringence on field reversal, we have set the permanent dipole moment μ in Eq. (13) equal to zero and can therefore explicitly calculate the electrical polarizability. The electrical polarizability of our five shortest fragments in 1 mM sodium phosphate buffer is presented in Fig. 8 as a function of the length cubed. The solid line represents a least-squares fit to the four shortest fragments. Clearly, the length dependence of the electrical polarizabilities for the four shortest fragments is cubic, as predicted by Eq. (25), and the fit goes through the origin. Although alternative length dependencies could also be used to describe the data, other best-fit expressions would not go through the origin. Also included in Fig. 8 are Stellwagen's [35] data on 80- and 98-BP fragments in Tris buffer (pH 8) at a temperature near 20°C. Note that the largest fragment (124-BP) plotted in Fig. 8 has a polarizability less than that predicted by a cubic length dependence. Stellwagen's 129-BP fragment also has a polarizability significantly below the line representing a cubic length

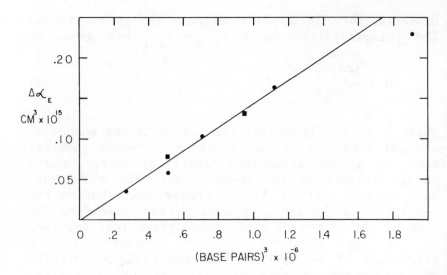

FIGURE 8.   Electrical polarizability of short DNA restriction fragments versus $(BP)^3$. The circles represent the polarizability in 1 mM sodium phosphate buffer at 3.8°C for 64-, 80-, 89-, 104- and 124-BP fragments. The squares are the results of Stellwagen [35] in Tris buffer near 20°C.

dependence. We believe that in the larger fragments we are observing the effects of flexibility on the Kerr constant and hence on the electrical polar-izability. The deviations from cylindrical behavior begin to occur for fragments of the same size as they did for the rotational relaxation time (see Fig. 6).

For fragments larger than 124-BP, we observe a greater deviation of our experimental polarizabil-ities from that of a cubic dependence. We expect the electrical polarizability to have a weaker length dependence since there will be a contribution to the polarizability perpendicular to its principal axis, which results from the overall curvature of the molecule. The perpendicular component reduces the effective electrical polarization. For these longer molecules, the optical anisotropy will also depend on the length and stiffness. This problem has been analyzed for a wormlike chain by Arpin et al. [36]. Because both the electrical and optical terms in Eq. (13) are in general length dependent, we have presented the specific Kerr constant versus the size of the DNA in Fig. 9. For small fragments, the Kerr constant is cubic with length, and this length dependence decreases to a linear or weaker dependence for large fragments.

In a sufficiently high electric field, flexible polyelectrolytes will stiffen and straighten, making it possible in principle to determine the optical anisotropy by extrapolation to infinite field. This is the approach used by Stellwagen [35]. However, the optical properties for a flexible molecule in low field, where we have made our measurements, will be different from those in high field.

We would like to use the length dependence of the Kerr constant as shown in Fig. 9 to characterize the flexibility of a polyelectrolyte. This is a particularly difficult problem since the theory for the electrical polarizability of a straight cylindrical polyelectrolyte is still being developed. Jernigan and Thompson [37] have shown how flexibility should affect the length dependence of the Kerr constant for a flexible polymer, but they have not considered polyelectrolytes explicitly. Van der Touw and Mandel [38] have treated the effect of flexibility on the electrical polarizability of flexible polyelectrolytes. Elias [22] has considered a variety of ways of combining the predictions for the variations in the optical anisotropy and the electrical polarization for prediction of the length dependence of the Kerr constant. There is a great need for further theoretical work on this problem since its solution will permit the evaluation of the persistence length of more flexible electrolytes from birefringence amplitude measurements.

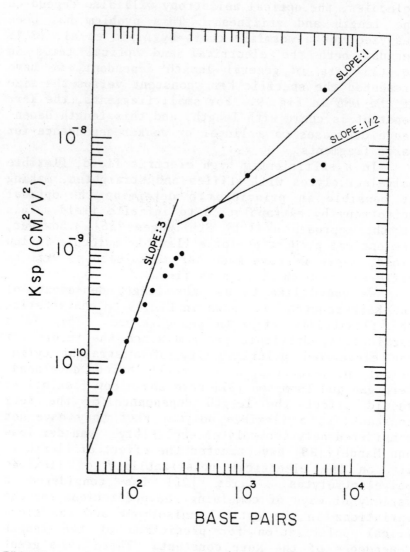

FIGURE 9. Log-log plot of the specific Kerr constant (conditions as in Fig. 3) versus the size of the fragments.

E. Pf3 Virus

The transient electric birefringence experiments performed on the filamentous virus Pf3 were designed to determine its rotational diffusion constant. When combined with other hydrodynamic measurements, it is possible to develop a structural picture of the virus. This approach had already been used by Newman et al. [39] for fd virus and by Chen et al. [13] for Xf virus. All three filamentous viruses have gram-negative bacteria as hosts and have a circular single-stranded DNA molecule packed in a helical sheath of protein subunits.

Figure 10 is an electron micrograph (provided by L. A. Day) of fd and Pf3 viruses. The smaller particles are the Pf3: they have a length of 0.773 that of fd as determined from the analysis of many particles [40]. The preparative conditions for the electron microscopy have been given by Berkowitz and Day [40].

Large particles such as Pf3 orient very easily in an electric field. However, the time needed to achieve steady-state orientation can be quite long, resulting in heating of the solution. For these reasons, and since we were not interested in the Kerr constant of the virus, we used pulses 2.5 msec long and electric field strengths of 185 V/cm. The pulse polarity was alternated to prevent net electrophoresis. The resulting birefringence response is shown in Fig. 11. Note that the steady state is not obtained before the field is turned off. The field free decay was analyzed by using Eq. (26), which is analogous to Eq. (21):

$$I(t) = A\exp[-D_r t] + B . \tag{26}$$

A second-order cumulant analysis was used with the background B taken as the experimental average of the long time intensity ($t \gg 6D_r^{-1}$). Measurements were made on three preparations at concentrations of 2 and 8 $\mu g/cm^3$ and ionic strengths of 1 and 10 mM at

FIGURE 10. Electron micrograph of the filamentous viruses Pf3 and fd. Of the 12 particles present, the five shorter ones are Pf3.

24.75°C. At these concentrations D_r was independent of concentrations, as would be expected since the interparticle spacing is at least twice the length of the virus. The rotational diffusion constant in 1 mM buffer is approximately 10% less than in 10 mM buffer. Since the electrostatic persistence length increases with decreasing salt concentration, we may be observing a limited amount of flexibility of the Pf3 virus at the higher ionic strength. Therefore, we have chosen to use the low salt value. Analysis of all our results gives $D_{r,25°C} = 47.4 \pm 1.4$ sec^{-1}. Without an estimate for the diameter, the determination of D_r is not of much utility. However, the measurements of the translational diffusion constant of the same particle and the substitution of D_r and

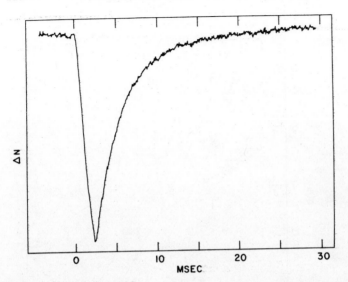

FIGURE 11. Transient electric birefringence resulting from alternating unipolar pulses
applied to a sample of Pf3 in a 1 m*M*
sodium phosphate buffer at pH 7.0 and a
temperature of 24.75°C. The sample concentration was 8 μg/cm³. Nine hundred
2.5 msec long pulses of field strength
185 V/cm were used to achieve the signal.
The decay time was 3.53 msec.

D_t into Eqs. (15) and (16) permits the determination
of both L and d once the expressions for γ_r and γ_t
are given. This is the approach we have taken.
Newman has deterined $D_{t, 25°C} = (3.30 \pm 0.10) \times 10^{-8}$ cm²/sec from low-angle polarized QELS measurements [23].

For a given D_r and γ_r, or alternatively, for a
given D_t and γ_t, a length that would be associated
with a given diameter can be generated. The results
of this analysis obtained from our experimental
values of D_r and D_t [Eqs. (15) and (16)] and
Broersma's expressions for γ_r and γ_t [11,13,39] are
represented by the curves given in Fig. 12. As Chen

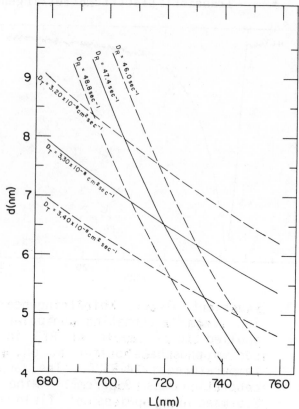

FIGURE 12. The simultaneous solution of the Broersma relationships [see Eqs. (15), (16), and reference 13 for γ_r and γ_t] for the lengths and diameters of rigid rods, using $D_{t,25} = (3.30 \pm 0.10) \times 10^{-8}$ cm²/ sec and $D_{r,25} = 47.4 \pm 1.4$ sec⁻¹. The results for the hydrodynamic length and diameter are 720 ± 25 nm and 6.5 ± 1.5 nm, respectively. A solution similar to that from the Tirado and Garcia de la Torre equations [Eqs. (27) and (28)] results in a hydrodynamic length and diameter of 707 ± 25 nm and 8.3 ± 1.9 nm, respectively. [Reprinted with permission from J. Newman, L. A. Day, G. W. Dalack, and D. Eden, Biochemistry *21*, 3352 (1982)].

et al. [13] have pointed out, the intersection of the two solid lines in Fig. 12 gives the values of L and d, which simultaneously satisfy both Eqs. (15) and (16), and the intersections of the dashed lines indicate the variation in L and d that can arise from the uncertainties in D_r and D_t.

From this analysis [23] we determine $L = 720 \pm 25$ nm and $d = 6.5 \pm 1.5$ nm. The value for the hydrodynamic length is 5% greater than the length determined by electron microscopy by Berkowitz and Day [40] and 6% less than an earlier electron microscopic determination [41].

Tirado and Garcia de la Torre [12,42] have calculated the length and diameter dependence of γ_t and γ_r by modeling rods as stacked rings of spheres. By performing linear fits to the large axial ratio data in Table 1 of reference 12 and Table 2 of reference 42, we have obtained the following expressions for γ_t and γ_r.

$$\gamma_t = 0.313 - 0.601 \frac{d}{L} , \tag{27}$$

$$\gamma_r = 0.662 - 0.92 \frac{d}{L} . \tag{28}$$

Use of their model results in a value of $L = 707 \pm 25$ nm, 1.8% smaller than the value obtained from the Broersma equations and a value of $d = 8.3 \pm 1.9$ nm, considerably greater than that obtained by use of the Broersma expressions.

In previous studies on fd [39] and Xf [13] the hydrodynamic diameter determined from the Broersma relationships was in excellent agreement with that obtained by electron microscopy and X-ray unit cell dimensions. The principal impact of the larger hydrodynamic diameter that results from the Tirado and Garcia de la Torre expressions is a significantly thicker layer of hydrodynamically bound water: 0.96 cm^3 solvent per gram of virus versus 0.31cm^3/g [23].

VI.   CONCLUSIONS

We have demonstrated the utility of transient electric birefringence for determination of the rotational diffusion constant of long cylindrical molecules in dilute solutions. When this method is combined with QELS measurements of the translational diffusion constant, the length and diameter of the particles can be obtained. Although the length can be determined with a high degree of precision, the uncertainty in the diameter will be relatively large because the diameter enters only logarithmically in the expressions for D_r and D_t. The deviations of D_r from that expected for rodlike behavior can be used to determine a measure of the molecular flexibility, the persistence length. It has been observed that B-form DNA has a persistence length of 495 Å in 1 mM sodium phosphate buffer (pH 7) at 3.8°C. The length dependence of the Kerr constant is a strong function of the ratio of the contour length relative to the persistence length. In the future this may become a convenient way to determine the flexibility of poly-electrolytes.

The experiments described above have been performed in the authors' laboratory. Other recent transient electric birefringence experiments on DNA restrictions fragments have been reported by Stellwagen [35] and Hagerman [43]. Allen [44] has recently reviewed electrooptic measurements of viruses and bacteriophages.

ACKNOWLEDGMENTS

We wish to thank our colleagues at Yale University for providing some of the DNA samples, for experimental hints, and for useful discussions. Particular thanks go to Gregory W. Dalack, who performed the initial measurements on Pf3, and to D. M. Crothers, N. Dattagupta, M. Fried, and M. Mandelkern. We thank Loren Day of the Public Health Research

Institute of the City of New York who provided the Pf3 samples and Jay Newman of Union College who measured the translational diffusion constant of Pf3. The research was supported by NIH Grant RR 07015.

REFERENCES

1. E. Fredericq and C. Houssier, *Electric Dichroism and Electric Birefringence*, Clarendon Press, Oxford, 1973.

2. C. T. O'Konski, Ed. *Molecular Electro-Optics*, Dekker, New York, 1976.

3. B. R. Jennings, Ed., *Electro-Optics and Dielectrics of Macromolecules and Colloids*, Plenum Press, New York, 1978.

4. S. Krause, Ed., *Molecular Eelctro-Optics: Electro-Optic Properties of Macromolecules and Colloids in Solution*, Plenum Press, New York, 1981.

5. H. Benoit, Ann. Phys. 6, 561 (1951).

6. V. Peterlin and H. A. Stuart, Z. Physik *112*, 129 (1939).

7. C. T. O'Konski, K. Yoshioka, and W. H. Orttung, J. Phys. Chem. *63*, 1558 (1959).

8. W. A. Wegener, R. M. Dowben, and V. J. Koester, J. Chem. Phys. *70*, 622 (1979).

9. J. G. Elias and D. Eden, Biopolymers *20*, 2369 (1981).

10. S. Broersma, J. Chem. Phys. 32, 1626, 1632 (1960).

436   D. Eden and J. G. Elias

11. S. Broersma, J. Chem. Phys. *74*, 6989 (1981).

12. M. M. Tirado and J. Garcia de la Torre, J. Chem. Phys. *71*, 1986 (1979).

13. F. C. Chen, G. Koopmans, R. L. Wiseman, L. A. Day, and H. L. Swinney, Biochemistry *19*, 1373 (1980).

14. O. Kratky and G. Porod, Rec. Trav. Chim. *68*, 1106 (1949).

15. J. E. Hearst, J. Chem. Phys. *38*, 1062 (1963).

16. J. Riseman and J. G. Kirkwood, *Rheology*, Vol. 1, Academic Press, New York, 1956, p. 495.

17. J. J. Hermans and R. Ullman, Physica *18*, 951 (1952).

18. P. J. Hagerman and B. H. Zimm, Biopolymers *20*, 1481 (1981).

19. J. G. Elias and D. Eden, Macromolecules *14*, 410 (1981).

20. I. Tinoco and K. Yamaoka, J. Phys. Chem. *63*, 423 (1959).

21. J. Newman and H. L. Swinney, Biopolymers *15*, 301 (1976).

22. J. G. Elias, Ph.D. thesis, Yale University, New Haven, Conn. 1981.

23. J. Newman, L. A. Day, G. W. Dalack, and D. Eden, Biochemistry *21*, 3352 (1982).

24. S. Sokerov and G. Weill, Biophys. Chem. *10*, 161 (1979).

25. D. Eden and J. G. Elias, Biophys. J. *37*, 330a (1982).

26. D. E. Koppel, J. Chem. Phys. *57*, 4814 (1972).

27. M. Mandelkern and D. M. Crothers, private communication.

28. T. Yoshizaki and H. Yamakawa, J. Chem. Phys. *73*, 1986 (1980).

29. S. Arnott, R. Chandrasekaran, and E. Selsing, in *Structure and Conformation of Nucleic Acids and Protein-Nucleic Acid Interactions*, M. Sundaralengam and S. T. Rao, Eds., University Park Press, Baltimore, 1975.

30. M. Mandelkern, J. G. Elias, D. Eden, and D. M. Crothers, J. Molec. Biol. *152*, 153 (1981).

31. M. Mandelkern, Ph. D. thesis, Yale University, 1980.

32. K. L. Cairney and R. E. Harrington, Biopolymers *21*, 923 (1982).

33. E. M. Lifshitz and L. P. Pitaevskii, *Landau and Lifshitz Statistical Physics*, 3rd ed., Pergamon Press, Oxford, 1980.

34. R. D. Kornberg and A. Klug, Sci. Am. *244* (2), 52 (1981).

35. N. Stellwagen, Biopolymers *20*, 399 (1981).

36. M. Arpin, G. Strazielle, G. Weill, and H. Benoit, Polymer *18*, 262 (1977).

37. R. L. Jernigan and D. S. Thompson, in *Molecular Electro-Optics*, C. T. O'Konski, Ed., Dekker, New York, 1976.

438   D. Eden and J. G. Elias

38.   F. van der Touw and M. Mandel, Biophys. Chem. 2,
       218 (1974).

39.   J. Newman, H. L. Swinney, and L. A. Day, J.
       Molec. Biol. *116*, 593 (1977).

40.   S. A. Berkowitz and L. A. Day, Biochemistry *19*,
       2696 (1980).

41.   D. E. Bradley, Biochem. Biophys. Res. Commun.
       57, 893 (1974).

42.   M. M. Tirado and J. Garcia de la Torre, J. Chem.
       Phys. 73, 1986 (1980).

43.   P. J. Hagerman, Biopolymers *20*, 1503 (1981).

44.   F. S. Allen, in *Molecular Electro-Optics:
       Electro-Optic Properties of Macromolecules and
       Colloids in Solution*, S. Krause, Ed., Plenum
       Press, New York, 1981.

45.   M. Mandel, Mol. Phys. *4*, 489 (1961).

CHAPTER 16

Particle Size Distribution
by Optical Transients and
Quasi-Elastic Light Scattering

V. NOVOTNY*
XEROX Research Centre of Canada
Mississauga, Ontario

CONTENTS

+Present address: EXXON Enterprises, 328 Gibraltar Drive, Sunnyvale, California 94086.

INTRODUCTION

Microscopic sizes and shapes of powders, colloids, and aerosols can be measured by a myriad of experimental techniques, including microscopic, light-scattering, and miscellaneous methods such as sedimentation, centrifugation, electrical resistance, and sieving. The selection is particularly wide and applications are more straightforward when particles are monodisperse and spherical. Most systems that are technologically interesting are polydisperse and nonspherical. Characterization of these systems is more difficult, and it is preferable to use several complementary techniques.

Optical microscopy can be used for the evaluation of particles down to ~ 0.1 µm, especially when dark-field illumination is employed. Scanning and transmission electron microscopies are applicable to objects down to 10 Å in size. When microscopies are combined with computer image analysis, accurate size distribution and shape information can be obtained. The application of microscopies to colloids is sometimes limited because of particle agglomeration or other artifacts that may occur during sample preparation. Microscopies are the most direct methods of size analysis, but they are not convenient for *in situ*, real-time measurements.

Light-scattering techniques can be subdivided into two categories, according to whether the light is scattered from a single particle or from a number of widely separated particles. The most notable single scatterer techniques include angular measurements [2] and low-angle light scattering [3], where the intensity of light scattered is directly proportional to the size. These techniques yield accurate size distributions but are applicable to scatterers typically above ~ 0.5 µm. Smaller particle sizes can be measured by techniques that collect light from many scatterers. Quasi-elastic light scattering (QELS) [4] can detect scatterers down to 10 Å. This technique is based on the detection of the fluctuations of light scattered from scatterers undergoing diffusional and/or translational motion. At a given scattering angle, QELS gives only the average diffusion coefficient and the width of its distribution, which are related to the particle sizes. Because particles of a given size contribute more or less to the overall signal according to the scattering angle, the particle size distribution should be reconstructed with measurements taken at several scattering angles; this is referred to as *angular QELS*.

When scatterers settle gravitationally, other light-scattering methods can be employed. Optical transients (OT) [5] are based on the detection of time dependent changes in the intensity of the light scattered from the particles during their translational motion. Both QELS and OT can be used to measure particle mobilities, and OT can also determine particle-substrate interaction forces.

In this chapter, optical transients are applied to sedimenting systems to determine their particle size distributions, and the results are compared with angular QELS measurements.

II. METHODS

The principle on which OTs are based is the temporal detection of scattered light intensities from particles during their translational motion (Fig. 1). Optical transients can be applied to size measurements when the scatterers settle gravitationally. The scattering medium is suddenly injected into the cell with at least one window. The scatterers start in the dispersed state and then settle continuously toward the planar detection boundary and terminate their motion in the nondispersed state. This experiment is referred to as "sweepout." Monochromatic light impinges on the cell, and scattered or transmitted light is detected. Integral and differential techniques can be distinguished. In the differential technique, only scatterers at or near

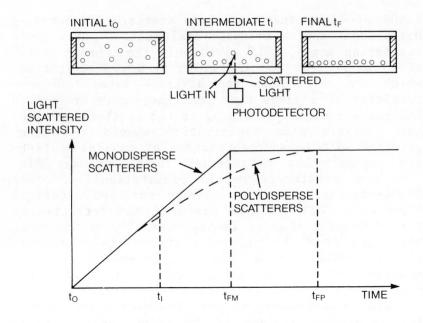

FIGURE 1. Principle of optical transients as applied to particle size measurements and scattered light intensity for monodisperse and polydisperse scatterers in a medium with infinitely high extinction in a backscattering configuration.

the observation boundary are detected. This can be accomplished in two backscattering arrangements. Total internal reflection (TIR) or an optical waveguide can be employed to sample only those scatterers that reach the evanescent wave. Alternatively, absorption of the suspending medium in ultraviolet, visible, or infrared regions can be utilized to minimize the contribution of scatterers away from the observation boundary. In the integral technique, transmission or small-angle forward scattering arrangements are used. In this case all scatterers contribute to the signal, which relies on small multiple light-scattering differences between the dispersed and nondispersed states. The interpretation of differential transients is simpler than that of integral transients. Moreover, the differential technique has the potential of single scatterer detection and, therefore, is emphasized here.

Differential optical transients [5] can be described relatively easily. Assuming initially that all scatterers are identical and very dilute, and that the medium has infinite extinction at the detection wavelength, the optical signal would be proportional to the number of scatterers at the detection interface, and thus the light intensity will increase linearly with time until all scatterers reach the detection boundary as indicated in Fig. 1. When the extinction coefficient of the medium is finite, the detected light intensity can be derived in the following manner: N scatterers with identical scattering power i are initially distributed uniformly in the cell of thickness d. The concentration of scatterers is so low that multiple light scattering is negligible. If the problem is considered to be one dimensional, the x axis is chosen to be perpendicular to the cell surface and its origin is placed at the interface opposite to the detection side. Monochromatic light incident on the cell at an angle θ_{i1} propagates into the cell at an angle θ_{i2} with respect to the x axis so that $\sin \theta_{i2} = (n_1 \sin \theta_{i1})/n_2$, where n_2 and n_1 are refractive indices of the medium and air, respectively. Only light scattered from points at distances $(d - x)$ from the detection boundary and parallel to the x axis is monitored by a photodetector $(\theta_f = 180°)$. Light travels the total distance $a(d - x)$ through the medium, where

$$A = 1 + \left[1 - \left(\frac{n_1}{n_2} \sin \theta_{i1} \right)^2 \right]^{-1/2} . \qquad (1)$$

The light intensity I from the single scatterer at $(d - x)$ is

$$I = I_0 i \exp[-\alpha a(d - x)] \ , \tag{2}$$

where I_0 is the incident light intensity corrected for the reflective losses at the cell interfaces and α is the product of the extinction coefficient and the concentration of the absorbing medium.

When all particles have identical settling velocity v, the total detected light intensity $I_s(t)$ between $t = 0$ and $t = d/v$ is given by the summation of all contributions from Eq. (2):

$$I_s(t) = \frac{NI_0 i}{d} \left\{ \int_{vt}^{d} \exp[-\alpha a(d - x)]dx + vt \right\}$$

$$= \frac{NI_0 i}{d} \left\{ \frac{1}{\alpha a} \{1-\exp[-\alpha a(d - vt)]\} + vt \right\}. \tag{3}$$

The first term is contributed by particles in the dispersed state, whereas the second term corresponds to particles at the detection boundary. Equation (3) can be generalized when the scatterers exhibit a settling velocity distribution $n(v)$ that satisfies the normalization condition $\int_0^\infty n(v)dv = N$.

The sweepout from the initial uniform distribution can be described with the contributions of scatterers that originated at x' from section dx' and that have at time t the spatial distribution $n(x - x',t)$ corresponding to the velocity distribution $n(v)$. The optical signal from scatterers that started at a particular x' is

$$J(x',t) = I_0 i \left\{ \int_{x'}^{d} \exp[-\alpha a(d - x)] \, n(x - x',t)dx \right.$$

$$\left. + \int_{d}^{\infty} n(x - x',t)dx \right\}. \tag{4}$$

Scatterers that reached the detection boundary at $x' = d$ are represented by the distribution from d to $+\infty$. The total signal is the summation of all $J(x',t)$ contributions, which is

$$I_s(t) = \frac{1}{d} \int_0^d J(x',t) \, dx' \; . \tag{5}$$

Different velocities and corresponding spatial distributions, including Gaussian, Maxwellian, gamma, and log-normal distributions, can be used in the modeling of $I_s(t)$. The gamma distribution is given here for illustration:

$$n(v) = \frac{Nb^p}{\Gamma(p)} \, v^{p-1} \, \exp[-bv] \; , \tag{6}$$

where $v = (x - x')/t$, $\Gamma(p)$ is the gamma function, $p = (v_a/v_s)^2$ and $b = v_a/v_s^2$, v_a is the average velocity, and v_s is the standard deviation of the velocity distribution.

Scattered light intensities $I_s(t)$ from Eqs. (4) and (5) can be evaluated either analytically or numerically. The velocity distribution is varied until there is good agreement between the experimental and predicted $I_s(t)$. The settling velocities v can be related to the radius r of spherical particles with

$$r = \left(\frac{9\eta}{2 \, \Delta\rho \, g} \right)^{1/2} v^{1/2} \tag{7}$$

where η, $\Delta\rho$, and g are the medium viscosity, the density difference between particles and the medium, and the gravitational constant, respectively. The velocity distribution $n(v)$ is converted into the size distribution $n(r)$ by relating r_i to v_i with Eq. (7) and setting $n(r_i) = n(v_i)$.

Finally, the dependence of light-scattered intensity on particle size can be taken into account. Mie scattering theory [6] is used to calculate the scattered light intensities $P(r,\theta)$ as a function of the particle radius r for a given scattering angle $\theta = (\theta_f - \theta_{i2})$. Equations (4) and (5) are generalized by multiplying $n(x - x',t)$ by $P\{r[v = (x - x')/t], \theta\}$:

$$I_s(t) = \frac{I_0 i}{d} \int_0^d \left\{ \int_{x'}^d \exp[-\alpha a(d - x)] \right.$$

$$\times \, n(x - x',t) \, P[r(v = \frac{x - x'}{t}),\theta]dx$$

$$\left. + \int_d^\infty n(x - x',t) \, P[r(v = \frac{x - x'}{t}),\theta]dx \right\} dx' \quad (8)$$

The calculated $I_s(t)$ is compared with the experimental $I_s(t)$, and $n(v)$ is repeatedly modified to provide better agreement with the experimental light-scattered intensities.

Our quasi-elastic measurements were performed in the time regime by detecting the autocorrelation function of the scattered light. The autocorrelation function of the photocurrent $C(\tau)$ for identical, dilute, spherical scatterers in the homodyne experiment is [4]

$$C(\tau) = B + A \, \exp[-2D(r)K^2\tau] \, , \quad (9)$$

where B and A are experimental constants, D is the diffusional coefficient, and K is the scattering K vector given by

$$D(r) = \frac{k_B T}{6\pi\eta r} \, , \quad (10)$$

$$K = \frac{4\pi n_2 \sin(\theta/2)}{\lambda} , \tag{11}$$

where T is the absolute temperature, k_B is Boltzmann's constant, λ is the wavelength, and the other
parameters are as defined above. For a system of
polydisperse spherical scatterers, $C(\tau)$ can be generalized

$$C_p(\tau)$$

$$= B + A \left| \int_0^\infty n(r) \, P(r,\theta) \, \exp[-D(r)K^2\tau] dr \right|^2 . \tag{12}$$

A single measurement of $C_n(\tau)$ is not sufficient
for definition of the particle size distribution
$n(r)$. A measurement at a single-scattering angle θ
can yield the average particle size $2r_a$ and the width
of the distribution $2r_s$ by the method of cumulants
[7]. The determination of $n(r)$ requires measurements
over a wide range of scattering angles and consistently good fit of experimental $C_n(\tau)$ according to
Eq. (12). As the scattering angle is varied, different parts of the distribution contribute more significantly to $C_p(\tau)$.

An experimental OT arrangement suitable for size
measurements is shown schematically in Fig. 2.
Monochromatic light is directed into the scattering
cell along optical paths 1, 2, or 3, which correspond
to backscattering, total internal reflection, or
small-angle forward scattering, respectively. The
scattered light is detected by a photomultiplier and
then amplified and digitized with a recorder such as
a multichannel analyzer. For slowly settling particles (> 1000 sec), the light scattering system
itself must be extremely stable. To handle drift of
the light source, the incident light beam is split
and recorded and the ratio of the scattered light to
the reference is taken. Finally, data are transferred to a computer for analysis and plotting.

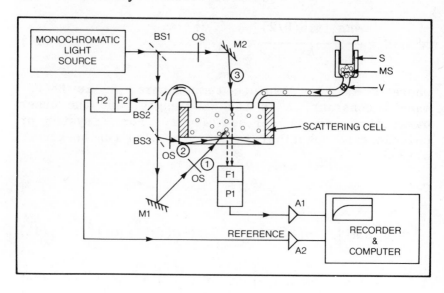

FIGURE 2.   Optical transient setup for particle size
measurements. Optical path 1 is used for
backscattering differential measurements;
path 2, for total internal reflection
arrangement; and path 3, for low-angle
forward scattering, integral measurements
(BS1, 2, and 3, beamsplitters; M1 and M2,
mirrors; OS, optical stops; F1 and F2,
filters; P1 and P2, photodetectors; A1 and
A2, amplifiers; S, syringe; MS, magnetic
stirrer; and V, valve).

Two QELS systems were utilized in this study.
The first featured a single-clipped autocorrelator
designed and built in this laboratory, an argon-ion
laser operating at a wavelength of 5145 Å, and a
Dewar-type cell holder for temperature-dependent
measurements between -30 and +150°C. The optical
system for the homodyne detection was standard,
whereas the heterodyne detection for electrophoretic
and sedimentation experiments used beating of the
scattered light with the reference light provided by

the optical path B1-M-B2 as indicated in Fig. 3.
Additional experimental details were outlined in
previous papers [8]. The second system was based on
a Malvern light-scattering spectrometer, full autocorrelator, and a helium-neon laser.

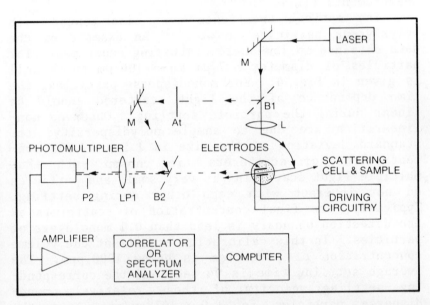

FIGURE 3. Quasi-elastic light-scattering system for
heterodyne settling or electrophoretic
measurements and homodyne experiments
(without electrodes, driving circuitry and
with a beam stop at the attenuator At) (P1
and P2, pinholes; L, lenses; M, mirrors;
B1 and B2, beamsplitters).

III. RESULTS AND DISCUSSION

The performance of both QELS systems was tested
by measuring the diffusion coefficients of monodisperse polystyrene spheres in water. Agreement
within 1% was obtained between particle sizes determined from Eq. (10) and those measured by electron

microscopy for sizes 0.110, 0.482, 1.01, and 5.7 µm. Moreover, the experimental value of B did not exceed the theoretical value of B by more than 0.1%. For scattering angles of 30-145°, the diffusion coefficient did not vary by more than the accuracy of the measurements (1%).

The optical transients were also tested with polystyrene particles in water. An example of the data obtained in low-angle scattering experiments for particles of diameter 5.7 µm in a 100 µm thick cell is given in Fig. 4. For monodisperse particles, the time dependence of the light scattered should be linear during the particle settling. Observed nonlinearities are due to sample polydispersity; the standard deviation of the size is 1.5 µm. The concentrations of scatterers are chosen such that multiple light scattering is very small and particles do not interact with each other during settling. Typically, the final concentration of scatterers at the detection boundary is less than 0.1 monolayers of particles. In this calibration experiment the volume concentration of particles is below 1000 ppm. The average settling time is 94 sec, and the corresponding settling velocity of these relatively monodisperse particles is 1.0×10^{-6} m/sec, and the particle size is 5.5 µm. The calibration result indicates that Brownian motion [9] and charged particle repulsion effects are negligible for this sample.

Subsequently, the optical transient and quasielastic methods were applied to polydisperse systems based on titanium dioxide particles dispersed in aqueous and nonaqueous media. Electron microscopy of titanium dioxide powder shows that particle sizes vary between 0.05 and 0.5 µm. Many particles in the micrographs are not individually discernible as a result of clustering, which prevents a reliable analysis of powder size distribution. Although most of the work with aqueous colloids was performed at pH = 7.0, the variation of particle size with pH was also examined. A typical optical transient from a

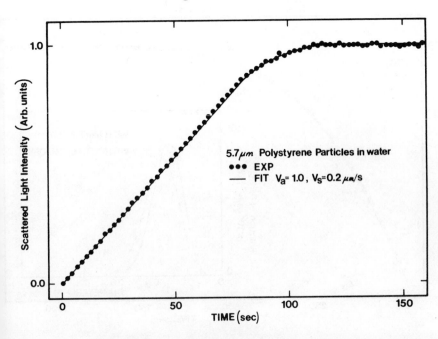

FIGURE 4.   Optical transient obtained from relatively monodisperse polystyrene colloid. Only every fiftieth averaged experimental data point is shown. The solid line corresponds to theoretical data.

polydisperse aqueous sample (pH = 7.0) is shown in Fig. 5. The experimental conditions were as follows: total internal reflection; cell thickness, 100 µm; extinction distance, $l = 1/\alpha \sim 0.1$ µm; and volume concentration of titanium dioxide, 50 ppm. The scattering function P was calculated from Mie theory for diameters from 0.05 µm to several µm. (The refractive indices of titanium dioxide and medium are 2.4 and 1.33, respectively.) The scattering function P varies considerably around the average diameter $2r_a \sim 0.3$ µm, and its dependence on size cannot be disregarded. The theoretical transient from Eq. (8) is compared with the experimental data, and the particle size distribution is shown in Fig. 5.

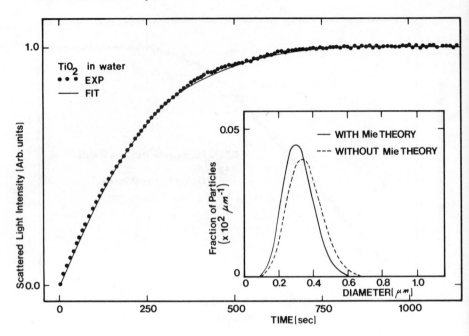

FIGURE 5.   Typical optical transient from the polydisperse system with every fiftieth averaged experimental data point plotted.
Theoretical fit and the corresponding
particle size distribution are also
included.

The autocorrelation function $C_p(\tau)$ taken from
the same polydisperse sample at $90°$ is exhibited in
Fig. 6.   The predicted $C_p(\tau)$, according to Eq. (12)
with Mie theory, corresponds to the size distribution
also shown in Fig. 6.   In addition, $C_p(\tau)$ was studied
as a function of scattering angle θ, and $P(r,\theta)$
values were again calculated with Mie theory.   The
particle size distributions were adjusted until
satisfactory fits to the experimental $C_p(\tau)$ were
obtained at all angles (θ = 45, 60, 75, $90°$, 105, and
$120°$).   Both visual and least-square methods were
used to judge the quality of fits.   The overall

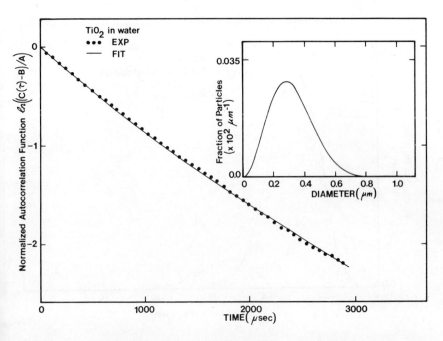

FIGURE 6.   Autocorrelation function obtained from the polydisperse system and theoretical fit based on Gaussian size distribution.

distribution, defined as the envelope of distributions evaluated at the above angles, is compared with optical transient results in Fig. 7. Quasi-elastic light scattering gives an average diameter $2r_a$ = 0.34 μm and an average width $2r_s$ = 0.28 μm, whereas OT yields $2r_a$ = 0.32 μm and $2r_s$ = 0.18 μm. Cumulants analysis without Mie corrections overestimates the size ($2r_s$ = 0.43 μm). The differences observed between QELS and OT distributions arise mainly from rapid general increases of scattering intensities with increasing particle size. Additional differences may be due to some nonsphericity and agglomeration within the sample.

    Titanium dioxide particles in water at pH = 7.0 are only very weakly charged; therefore, corrections

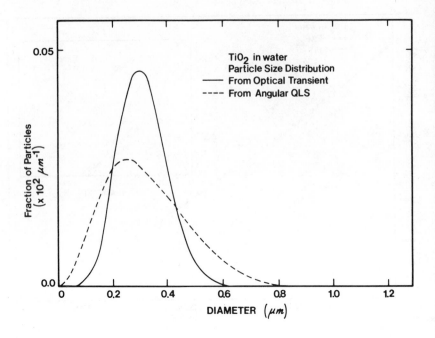

FIGURE 7.   Comparison of particle size distributions
determined by optical transients and by
angular QELS.

for electrostatic repulsion during settling are
negligible. The measurements of particle size as a
function of pH revealed that the size decreases
gradually to 0.23 µm at high and low values of pH.
This is expected because the particles become more
significantly charged in both regimes and electrostatic repulsion prevents agglomeration.

Particle size measurements were also applied to
nonaqueous colloids for which size distribution was
studied as a function of surfactant concentration.
The distributions resulting from optical transients
at zero, intermediate, and high surfactant concentrations are plotted in Fig. 8. At low concentrations
the colloid size is large because of insufficient
surfactant – particles are large agglomerates. With

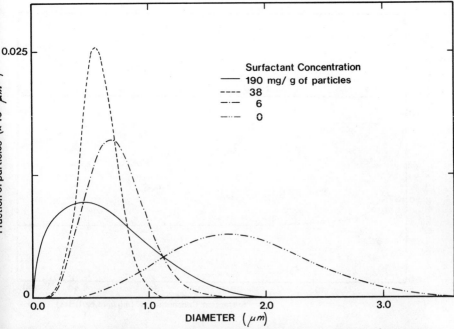

FIGURE 8.   Particle size distributions determined by optical transients at different surfactant concentrations.

increasing surfactant concentration, the average particle size decreases until it reaches a minimum and is independent of concentration. Surprisingly, the size distribution broadens considerably at high surfactant concentrations. The data are summarized in Fig. 9, where both the average size and the width of the distribution are plotted against surfactant concentration. It is interesting to point out that the adsorption isotherm of the surfactant on titanium dioxide coincides with the particle size dependence. Surfactant adsorption reaches a monolayer around the 4 mg of surfactant per gram of titanium dioxide, where the average size becomes independent of surfactant concentration.

Normally, QELS must be performed at low concentrations to avoid multiple light scattering. In our

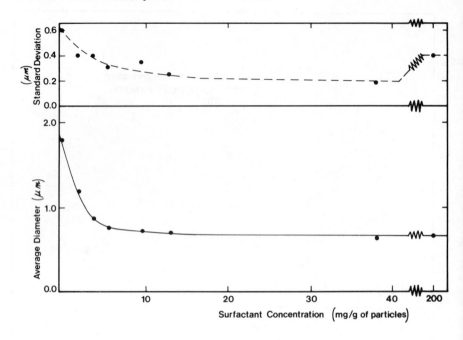

FIGURE 9.   Average particle size and the standard
            deviation plotted as a function of sur-
            factant concentration.

experiments, volume concentrations were below 1 ppm.
This restriction does not exist with OTs where particle concentrations at the detection boundary can be
calibrated against scattered light intensities.
Optical transients at different particle concentrations are shown in Fig. 10.   Particles settle independently at low particle concentrations; at higher
concentrations, interactive motion contributes to the
settling velocity and settling is affected by both
electrostatic repulsion of charged particles and
interactive motion at the highest concentrations.
    Quasi-elastic light scattering and optical
transient techniques were applied to a number of
different colloids, including other inorganic pigment
colloids, nonaqueous copolymer dispersions, carbon

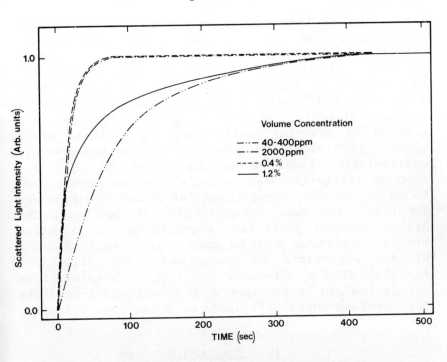

FIGURE 10.   Normalized optical transients obtained at different particle concentrations.

black colloids, and coal slurries. For very absorbing systems such as carbon black, optical transients exhibit signals that decrease as a function of settling time. In QELS and OT from highly absorbing particles, thermophoresis can occur. This is caused by absorption of light by particles, thus leading to a temperature increase of the particles and their surroundings and resulting in movement in addition to Brownian motion. An apparent decrease of particle size with increasing light power density is indicative of thermophoresis, which can be avoided by measurements at low power densities.

When optical transients are applied to colloids, extinction of the ultraviolet or infrared light by the suspending fluid itself may be adequate for limitation of the scattering signal from beyond the observation boundary. In some cases the colloidal properties in the presence of dissolved dye are of interest and visible detection is most convenient. This cannot be done for powders and aerosols, where differential OT with total internal reflection or integral OT must be employed. Optical transients are limited to gravitationally settling systems, which require particle sizes typically 0.1 µm and greater. Quasi-elastic light scattering relies on the diffusion of scatterers and is useful for scatterers down to 10 Å. In OTs, corrections for Brownian motion are important for small scatterers. In addition, for highly charged particles, corrections for electrostatic repulsions must be made. Also, both QELS and OT are applicable to measurements of mobilities. Finally, OT can determine particle-wall interaction forces and can be combined with electrical transients for complete characterization of nonaqueous colloids.

IV.   CONCLUSION

Optical transient and angular QELS techniques were used successfully to measure particle size distributions of polydisperse systems. Variations in light-scattering intensities with particle size and scattering angles were taken into account with Mie theory. Nonsphericity of scatterers was neglected. Satisfactory agreement was obtained between the two methods.

ACKNOWLEDGMENT

Valuable discussions with Dr. M. L. Hair and comments on the manuscript by Mr. W. Harter are gratefully acknowledged.

REFERENCES

1. M. Kerker, *The Scattering of Light*, Academic Press, New York, 1969; H. C. van de Hulst, *Light Scattering by Small Particles*, Wiley, New York, 1957.

2. W. Kaye and A. J. Havlik, Appl. Opt. *12*, 541 (1973).

3. D. D. Cooke and M. Kerker, J. Colloid Interface Sci. *42*, 150 (1973); T. R. Marshall, C. S. Parmenter, and M. Seaver, J. Colloid Interface Sci. *55*, 624 (1976).

4. H. Z. Cummins and H. L. Swinney, in *Progress in Optics*, Vol. 8, E. Wolf, Ed., North Holland, Amsterdam, 1970; H. Z. Cummins and E. R. Pike, Eds., *Photon Correlation and Light Beating Spectroscopy*, Plenum Press, New York, 1974; N. C. Ford, Chemica Scripta 2, 193 (1972); B. Chu, *Laser Light Scattering*, Academic Press, New York, 1974; B. J. Berne and R. Pecora, *Dynamic Light Scattering*, Wiley, New York, 1976.

5. V. Novotny, J. Appl. Phys. *50*, 324, 2787 (1979).

6. G. Mie, Ann. Physik *25*, 377 (1908); J. Stratton, *Electromagnetic Theory*, McGraw-Hill, New York, 1941.

7. D. E. Koppel, J. Chem. Phys. *57*, 4814 (1972).

8. V. Novotny, Chem. Phys. Lett. *70*, 321 (1980) and J. Chem. Phys. *78*, Jan. (1983); V. Novotny and M. L. Hair, in *Polymer Colloids II*, R. M. Fitch, Ed., Plenum Press, New York, 1979.

9. M. Mason and W. Weaver, Phys. Rev. *23*, 412 (1924); W. Weaver, Phys. Rev. *27*, 499 (1926).

10.   A. J. Goldman, R. G. Cox, and H. Brenner, Chem. Eng. Sci. *21*, 1151 (1966); M. Stimson and G. B. Jeffrey, Proc. Roy. Soc. Lond. Ser. A *111*, 110 (1926).

CHAPTER 17

A Photon Correlation Spectroscopy Study of Size Distributions of Polystyrene Initiated in a Microemulsion

HSIANG-IN TANG, PATRICIA L. JOHNSON, and ESIN GULARI
Wayne State University
Detroit, Michigan

CONTENTS

I. INTRODUCTION

The physical properties of a polymer depend on the average molecular weight and the molecular weight distribution of that polymer. Therefore, it is important to be able to predict and control the molecular weight and polydispersity of a polymer. In emulsion polymerization, it has been proposed [1,2] that initiation by a water-soluble free-radical initiator starts in the aqueous phase and follows one of two paths. In the first, radicals generated in the aqueous phase penetrate the monomer-swollen micelles and polymerize the solubilized monomer. In the second, radicals initiate the dissolved monomer molecules, forming oligomeric radicals that further grow into latex particles. Monomer droplets that contain most of the monomer do not compete in capturing radicals because the surface area of the droplets is smaller than that of micelles and latex particles. Monomer droplets act as reservoirs and supply monomer by diffusion for further growth of particles formed by either of the two mechanisms. However, if the dispersion of the monomer can be rendered finer, resulting in a much larger surface area, the monomer droplets can also be initiated. Ugelstad et al. [3] have initiated sytrene droplets using mixtures of sodium hexadecyl sulfate and hexadecanol as emulsifier that produce drops of 0.5-1 μm, in comparison to ~ 5 μm drops with surfactant as emulsifier only. The electron micrographs indicating bimodal dispersion of latex particles are evidence for a mechanism involving initiation and polymerization in monomer droplets.

Microemulsions are thermodynamically stable colloidal dispersions of water in an organic medium, water-in-oil (W/O) type, or an organic medium in water, oil-in-water (O/W) type. Polymerization of methyl acrylate in a W/O microemulsion has been reported by Stoffer and Bone [4]. They report a phase separation during the polymerization process but no change in molecular weight of the polymer due to phase separation.

In this study microemulsions formed by styrene, water, sodium dodecyl sulfate (SDS), and pentanol were initiated by potassium persulfate. The polymerizations were carried out at 60°C to completion. The isolated polystyrene products were studied by light scattering and gel-permeation chromatography (GPC). The molecular weight and polydispersity of these samples were determined from time-averaged intensity measurements and photon-correlation spectroscopy.

II. EXPERIMENTAL

The reagents used were of the following purities. Styrene from Matheson Coleman and Bell Chemical, Inc. of 98% purity was further purified by vacuum distillation to remove the inhibitor. Pentanol was from Fisher Scientific Company of boiling range 137.2-138.0°C. Sodium dodecyl sulfate from Bio-Rad. Laboratories was of 99% electrophoresis purity reagent and was used without further purification.

Microemulsions were prepared by titrating various sytrene, SDS, and pentanol mixtures with water containing initiator dissolved in it. Ampules filled with these microemulsions were sealed under nitrogen and kept in a constant temperature bath at 60°C for 2 days.

The polystyrene samples were isolated by standard procedures of filtration and washing with water and methanol followed by dissolving them in benzene and precipitating into methanol.

Gel-permeation chromatography measurements were made on a Waters Associates Model 440 liquid chromatograph with absorbance detector at 254 nm, using styragel columns for a molecular weight range of 10^5-10^6.

The components of the light-scattering spectrometer are essentially similar to that described elsewhere [5]. We used a Spectra-Physics model 165 argon-ion laser operating at about 300 mW with λ_0 = 488.0 or 514.5 nm. The time-averaged intensity measurements were counted on a Hewlett-Packard Model 5316A universal counter. The correlation functions were computed by a Malvern Scientific Model K7025 real-time multibit correlator. The reference intensity was monitored by using a power meter coupled with a digital voltmeter. The photometer was controlled by a Hewlett-Packard Model 85 calculator.

III.   METHODS OF DATA ANALYSIS

The methods employed in the analysis of the intensity and correlation function data are described by use of the same notation as in a previous publication [5].

A.   Intensity Data

The Rayleigh ratio R_v for vertically polarized light at finite concentrations has the approximate form

$$\frac{HC}{R_v} = (\frac{1}{M_w} + 2A_2C)[1 + (1 + 2A_2CM_w)^{-1} \frac{16\pi^2 n_0^2}{3\lambda_0^2}$$

$$\times <r_g^2(C)>_z \sin^2 \frac{\theta}{2}] , \qquad (1)$$

where $H = 4\pi^2 n_0^2 (\partial n/\partial C)^2/N_A\lambda_0^4$, with n_0, C, N_A, and λ_0 as, respectively, the refractive index, concentration, Avogadro's number, and wavelength of light in vacuo and $R_v = r^2 i_v/I_{ref}$, with r, i_v, and I_{ref} as the distance between the scattering center and the point of observation, the excess vertically polarized scattered intensity, and the reference incident intensity, respectively. In a plot of HC/R_v versus $\sin^2(\theta/2)$,

$$\langle r_g^2(C)\rangle_z^* = \frac{\langle r_g^2(C)\rangle_z}{1 + 2A_2CM_W}$$

$$= \frac{\text{initial slope}}{\text{intercept}} \; \frac{3\lambda_0^2}{16\pi^2 n_0^2} \; , \qquad (2)$$

where $\langle r_g^2(C)\rangle_z [\equiv \Sigma_i M_i C_i (r_g^2)_i / \Sigma_i M_i C_i]$ at infinite dilution is the square of radius of gyration. The terms A_2 and M_W are determined from a plot of $\lim_{\theta \to 0}(HC/R_v)$ versus C, where

$$\lim_{\theta \to 0} \frac{HC}{R_v} = \frac{1}{M_W} + 2A_2 C \; . \qquad (3)$$

B. Correlation Function Data

The measured homodyne, intensity autocorrelation function has the form

$$G_k^{(2)}(I\Delta\tau) = A(1 + \beta|g^{(1)}(I\Delta\tau)|^2) \; , \qquad (4)$$

where $|g^{(1)}(\tau)|$ is the normalized correlation function of the scattered electric field; k, A, β, I, and $\Delta\tau$ are, respectively, the clipping level, the background, an adjustable parameter in the fitting procedure, the delay channel number, and the delay time increment; and $\tau = I\Delta\tau$. In the cumulants method [6],

$$\ln(A\beta)^{1/2}|g^{(1)}(\tau)| = \ln(A\beta)^{1/2}$$

$$- \bar{\Gamma}\tau + \frac{1}{2!}\mu_2\tau^2 - \frac{1}{3!}\mu_3\tau^3 + \cdots \; , \qquad (5)$$

where

$$\bar{\Gamma} = \int_{\Gamma=0}^{\infty} \Gamma G(\Gamma) d\Gamma \qquad (6)$$

and

$$\mu_i = \int_{\Gamma=0}^{\infty} (\Gamma - \bar{\Gamma})^i \, G(\Gamma) d\Gamma \ , \qquad (7)$$

where $G(\Gamma)$ is the normalized distribution of decay rates. For $Kr_g < 1$

$$\langle D \rangle_z = \frac{\Sigma C_i M_i D_i}{\Sigma C_i M_i} = \frac{\Gamma}{K^2} \ , \qquad (8)$$

where $K[\equiv(4\pi n_0/\lambda_0) \sin(\theta/2)]$ is the magnitude of the momentum transfer vector. At infinite dilution,

$$\langle D \rangle_z^0 = \frac{k_B T}{6\pi\eta\langle r_h\rangle} \ , \qquad (9)$$

where $\langle r_h \rangle$ $[=\lim_{c \to 0} \Sigma_i C_i M_i / \Sigma_i (C_i M_i / r_{h,i})]$ is the weighted equivalent hydrodynamic radius and η is the solvent viscosity.

C. Histogram Method

In the histogram method [7], a discrete step function (a histogram) is used to approximate $G(\Gamma)$, such that

$$\sum_{j=1}^{n} G(\Gamma_j)\Delta\Gamma = 1 \qquad (10)$$

and

$$|g^{(1)}(\tau)| = \sum_{j=1}^{n} G(\Gamma_j) \int_{\Gamma_j - \Delta\Gamma/2}^{\Gamma_j + \Delta\Gamma/2} \exp[-\Gamma I \Delta\tau] d\Gamma \ , \quad (11)$$

where $G(\Gamma_j)$ represents the weighting factor based on integrated scattered intensity of molecules having linewidths from $\Gamma_j - \Delta\Gamma/2$ to $\Gamma_j + \Delta\Gamma/2$. Here n is the number of j steps in the histogram and $\Delta\Gamma[=(\Gamma_{max} - \Gamma_{min})/n]$ is the width of each step. Values for Γ_{min}, Γ_{max}, and n are determined relative to the precision of experimental data and to the average strength of $G(\Gamma_j)$. The histogram method is essentially a nonlinear least-squares technique that minimizes the sum of squared errors χ^2 with respect to each $G(\Gamma_j)$, simultaneously, such that

$$\frac{\partial}{\partial G(\Gamma_j)} \chi^2$$

$$= \frac{\partial}{\partial G(\Gamma_j)} \sum \frac{1}{\sigma_I^2}[Y_m(I\Delta\tau) - Y(I\Delta\tau)]^2 = 0, \quad (12)$$

where σ_I is the uncertainty of $Y_m(I\Delta\tau)$, with subscript m denoting the measured value. The net signal correlation function $Y(I\Delta\tau)[=A\beta|g^{(1)}(I\Delta\tau)|^2]$ has the form

$$Y(I\Delta\tau) = A\beta \left(\sum_{j=1}^{n} G(\Gamma_j) - \frac{1}{I\Delta\tau} \{\exp[-(\Gamma_j + \frac{\Delta\Gamma}{2})I\Delta\tau] \right.$$

$$\left. - \exp[-(\Gamma_j - \frac{\Delta\Gamma}{2})I\Delta\tau]\} \right)^2 \quad (13)$$

IV.   RESULTS AND DISCUSSION

A.   Phase Diagrams

The phase regions of a four-component system containing water, SDS, pentanol, and styrene may be illustrated by a tetrahedron. The determination of the details of the phase behavior of this sytem at various temperatures is a research project in itself. In this investigation we obtained the microemulsion regions on selected planes in which the ratio of styrene to (SDS + pentanol) was constant. In Fig. 1 microemulsion regions at 50°C are shown on a triangular plane, intersecting the SDS-pentanol-styrene face at 27.3 wt.% sytrene and extending to the apex representing pure water. The boundaries of the microemulsion regions were obtained by titrating with water various SDS-pentanol-styrene mixtures to the point of dissolution of SDS, determining the right-hand limit of the solubility region and then the clouding point defining the left-hand limit. There are two regions of solubility. The larger region, W/O region, corresponds to dispersion of water, ranging from 10 to 50 wt.% water, in a continuous phase containing styrene and pentanol. In the smaller region, O/W region, droplets containing styrene are dispersed in water ranging from 80 to 90 wt.%.

Two solutions, a W/O microemulsion and an O/W microemulsion, were initiated by using potassium persulfate as initiator. The compositions of the two solutions are listed in Table 1. The concentration of the initiator was such that the styrene/initiator ratio corresponded to 150. The polymerizations were carried out to completion at 60°C, and the poly-styrene samples were precipitated and purified by standard procedures [8]. The molecular weight and the radius of gyration of these samples were determined from measurements of excess scattered intensities at various angles and concentrations. The polydispersity of these samples were studied by use of the histogram analysis of the time correlation function profiles.

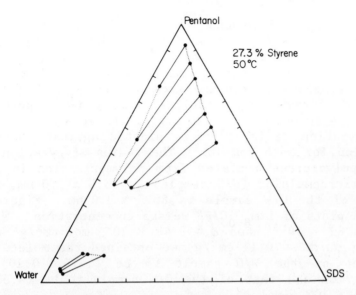

FIGURE 1.   Microemulsion regions containing 27.3 wt.% styrene based on (SDS + pentanol + sytrene) at 50°C.

TABLE 1.   Composition of O/W and W/O Microemulsions: Concentrations (Percent by weight)

	Water	Styrene	SDS	Pentanol
O/W microemulsion	83.3	4.56	7.80	4.34
W/O microemulsion	19.5	22.0	15.6	42.9

B.   Intensity Results

We used benzene as reference [8] for computing the Rayleigh ratio and took R_v ($\theta = 90°$) = 2.32 × 10^{-5} cm^{-1} at 25°C for λ_0 = 514.5 nm.   Corresponding to the benzene reference, we calculated an instrument constant $Q = R_v/R_v^* = 4.82 \times 10^{-10}$. Excess scattered intensities from the two polystyrene samples dissolved in toluene were measured at 14 different scattering angles ranging from $\theta = 44$ to $104°$ and at five different concentrations ranging from approximately 4 to 0.6 mg/ml. In Fig. 2 the apparent radius of gyration $\langle r_g^2(c) \rangle_z^{*1/2}$ is plotted against concentration for both samples. The radius of gyration of the polystyrene isolated from polymerization in the W/O microemulsion (W/O sample) is 75.5 ± 5.0 nm, and that of the O/W sample is 86.5 ± 3.0 nm. Figure 3 shows plots of $\lim_{\theta \to 0} C/R_v^*$ versus concentration. With $Q = 4.82 \times 10^{-10}$ and $H = 2.48 \times 10^{-7}$ mol·cm²/g² and using $dn/dC = 0.11$ cm³/g, we obtained the molecular weight of the W/O sample to be $(1.15 \pm 0.10) \times 10^6$ g/mol and that of the O/W sample to be $(4.53 \pm 0.40) \times 10^6$ g/mol.

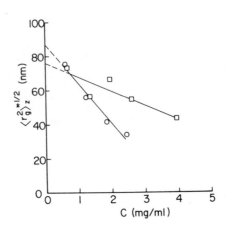

FIGURE 2.   Plot of $\langle r_g^2 \rangle_z^{*1/2}$ versus concentration at 25°C in toluene. Circles refer to the O/W sample and squares to the W/O sample.

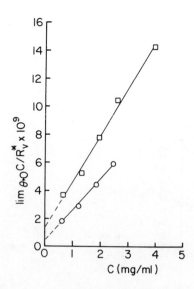

FIGURE 3.   Plot of $\lim_{\theta \to 0} (C/R_v^*)$ versus C at 25°C in toluene. Circles refer to the O/W sample and squares to the W/O sample.

C.   Analysis of Correlation Function Profiles

Correlation function profiles measured at the lowest concentration, $C = 0.66$ mg/ml, and $\theta = 90°$ were analyzed by the histogram method and the cumulants method. The distribution of decay rates obtained by the histogram method are shown in Fig. 4 for W/O and O/W samples. In the histogram analysis, for a step size $\Delta\Gamma = 500$ sec^{-1}, the number of steps determined was 7 for the choice of Γ_{min} and Γ_{max} corresponding to a value of χ^2 in Eq. (12) that was within experimental noise. Since Γ_{min} and Γ_{max} were varied parametrically and the measure was χ^2 within experimental uncertainties, we repeated the histogram analysis for sliding values of Γ-range over a step size. The results of five such fits, the quality of which were all acceptable according to values of χ^2, were averaged and are displayed in Fig. 4. We

TABLE 2. Physical Parameters Computed by the Histogram and Cumulants Methods for W/O Sample in Toluene at $C = 0.655$ mg/ml

	A. $\theta = 90°$; $\Delta\tau = 2.5$ μsec		B. $\theta = 60°$; $\Delta\tau = 6$ μsec
	Histogram	Best Cumulants (Fifth Order)	Best Cumulants (Fifth Order)
$\bar{\Gamma}$ (sec^{-1})	$(1.57 \pm 0.05) \times 10^4$	$(1.63 \pm 0.01) \times 10^4$	$(7.18 \pm 0.40) \times 10^3$
$\bar{D}$ (cm^2/sec)	2.12×10^{-7}	2.20×10^{-7}	1.93×10^{-7}
$\bar{r}_h$ (nm)	18.4	18.0	20.5
$\mu_2/\bar{\Gamma}^2$	0.56	0.74	0.92
$\bar{\Gamma}_L$ (sec^{-1})	2330		
$\bar{r}_{hL}$ (nm)	124.		
$\bar{\Gamma}_S$ (sec^{-1})	2.59×10^4		
$\bar{r}_{hS}$ (nm)	11.2		
A_L/A_S	43/57		
Red SSE[a]	6.7×10^6	2.5×10^6	1.0×10^7

[a] Red SSE $= \sqrt{(SSE/[N-M])}$, where SSE is the sum of squared errors when the fit converges and N and M are the number of data points and number of parameters, respectively.

TABLE 3. Physical Parameters Computed by the Histogram and Cumulants Methods for O/W Sample in Toluene at C = 0.66 mg/ml

	A. $\theta = 90°$; $\Delta\tau = 2.5$ μsec		B. $\theta = 60°$; $\Delta\tau = 6$ μsec
	Histogram	Best Cumulants (Fifth Order)	Best Cumulants (Fifth Order)
$\bar{\Gamma}$ (sec^{-1})	$(1.38 \pm 0.04) \times 10^4$	$(1.42 \pm 0.10) \times 10^4$	$(6.45 \pm 0.35) \times 10^3$
$\bar{D}$ (cm^2/sec)	1.86×10^{-7}	1.91×10^{-7}	1.74×10^{-7}
$\bar{r}_h$ (nm)	21.0	20.7	22.7
$\mu_2/\bar{\Gamma}^2$	0.28	0.45	0.50
$\bar{\Gamma}_L$ (sec^{-1})	3950		
$\bar{r}_{hL}$ (nm)	73.3		
$\bar{\Gamma}_S$ (sec^{-1})	1.92×10^4		
$\bar{r}_{hS}$ (nm)	15.1		
A_L/A_S	36/64		
Red SSE[a]	4.5×10^7	1.4×10^7	7.4×10^6

[a]Red SSE $= \sqrt{(\text{SSE}/[N-M])}$, where SSE is the sum of squared errors when the fit converges and N and M are the number of data points and number of parameters, respectively.

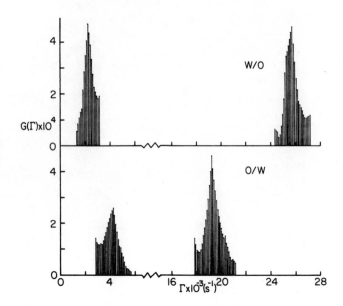

FIGURE 4.   Plots of linewidth distributions of W/O
and O/W samples measured in toluene at
25°C and at $c = 0.655$ mg/ml for the W/O
sample and at $c = 0.66$ mg/ml for the O/W
sample. The computed means and the variances are listed in Tables 2 (column A)
and 3 (column A). The measurements were
made at $\theta = 90°$ and $\Delta\tau = 2.5$ μsec.

observed bimodal distributions for both samples and
noted that the two different size fractions for the
W/O sample were further apart in size than the O/W
sample; namely $\bar{\Gamma}_2/\bar{\Gamma}_1$ (W/O sample) was 11 and $\bar{\Gamma}_2/\bar{\Gamma}_1$
(O/W sample) was 4.8. The means and variances of the
distributions shown in Fig. 4 are compared to those
obtained by cumulants analysis in Tables 2 and 3.

It must be noted that the histogram method is a
viable approximation to an otherwise difficult inversion for data existing over a finite time range.
When the samples are highly polydisperse, the distributions obtained by the histogram method depend on

the correlation time range. The simple reason for this is that present-day correlators make measurements at equally spaced delay times and information on smaller sizes is limited to only the early part of the correlation function profile. This effect is illustrated in Fig. 5, where two distributions of decay rates for the same sample measured at two different delay time settings are displayed. For the upper distribution, τ_{max} was 96 μsec and the hydrodynamic radii of the two fractions calculated by use of Eq. (8) and (9) were 96.1 and 11.7 nm, respectively. For the lower distribution, τ_{max} was 240 μsec and the hydrodynamic radii were 124 and 11.2 nm. Although the mean values of the sizes remain similar for both distributions, the width depends on the correlation time and changes appreciably as indicated by r_{hL}/r_{hS} = 8.2 for τ_{max} = 96 μsec and r_{hL}/r_{hS} = 11 for τ_{max} = 240 μsec.

D. Nature of the Bimodal Distributions

The bimodal nature of the size distributions for both the O/W and W/O samples were also detected by GPC. The GPC results are displayed in Figs. 6 and 7. In both cases the absorbance of the samples during elution are compared to the absorbance of the NBS-705 polystyrene standard, which has a molecular weight of 1.79×10^5. We found that the O/W sample shown in Fig. 6 contained molecular weights larger than 1.79×10^5 whereas the absorbance trace for the W/O sample in Fig. 7 indicated the presence of three different fractions with molecular weights larger and smaller than 1.79×10^5. The O/W sample had a narrower size distribution than the W/O sample did. This compared favorably with the values of μ_2/Γ^2 equal to 0.28 for O/W and 0.56 for W/O samples obtained from the histogram method.

For various monodisperse polystyrene standards of differing molecular weights, the ratio of hydrodynamic radius to radius of gyration r_h/r_g has been measured [9] in benzene at 25°C to be 0.60 $\pm$ 0.02. In

476   H.-I. Tang et al.

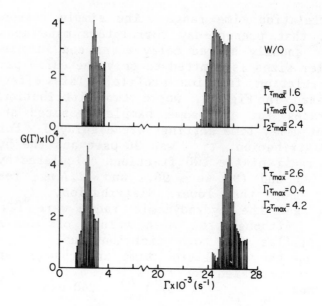

FIGURE 5.   Plots of the linewidth distribution of
the W/O sample measured at two different
delay time settings. For the upper
distribution τ_{max} = 96 μsec and for the
lower distribution τ_{max} = 240 μsec. The
measurements were made at 25°C for θ =
90°.

this study, at $C = 0.66$ mg/ml for the O/W sample in
toluene, we obtained $r_h = 21.0$ nm and could estimate
$\langle r^2 \rangle_z^{1/2} = 75.5$ nm, yielding $r_h/r_g = 0.28$. The corresponding value for the W/O sample was 0.21. For a
bimodal distribution, such a comparison becomes
meaningless. Instead, we need to compare both the
large and the small fractions separately. From the
analysis of the correlation function profiles
(Table 2, column A) for the W/O sample and θ = 90°,
we had $A_L/A_S = 43/57$ and $r_{hs} = 11.2$ nm. If we take
$r_h/r_g = 0.6$ for the small fraction, then $r_{gs} =$
18.6 nm. At θ = 90°, the ratio of the intensities of
the two fractions is proportional to the ratio of

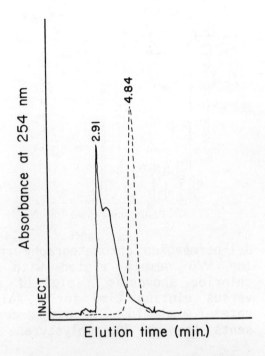

FIGURE 6.   Gel-permeation chromatography results for the O/W sample eluted with methylene chloride shown as a plot of absorbance versus elution time for a solvent flow rate of 2 ml/min. The dashed curve is the NBS 705 polystyrene standard.

areas in $G(\Gamma)$. With $I_{total}(\theta = 90°) = 8.34 \times 10^4$ counts, we could construct an intensity contribution due to the small fraction to be

$$\left(\frac{1}{R_v^*}\right)_S = 1.97 \times 10^{-5} + 0.256 \times 10^{-5} \sin^2 \frac{\theta}{2} . \quad (14)$$

Then the intensity contribution of the small fraction was substracted from the total measured intensity to obtain the intensity contribution of the large fraction at each scattering angle, as shown in Fig. 8.

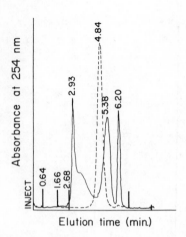

FIGURE 7. Gel-permeation chromatography results for the W/O sample eluted with methylene chloride shown as a plot of absorbance versus elution time for a solvent flow rate of 2 ml/min. The dashed curve represents the NBS 705 polystyrene standard.

From $(1/R_v^*)_L$ versus $\sin^2(\theta/2)$, we obtained $\bar{r}_{gL} = 196 \pm 40$ nm, corresponding to $(r_h/r_g)_L = 0.63 \pm 0.10$, which is in agreement with Miyaki et al.'s [9] findings. In Fig. 9 the same analysis is presented for the O/W sample with $A_L/A_S = 36/64$, $\bar{r}_{hS} = 15.1$ nm corresponding to $\bar{r}_{gS} = 25.2$ nm. For I_{total} ($\theta = 90°$) $= 1.563 \times 10^5$ counts, the contribution of the small fraction was computed to be

$$\left(\frac{1}{R_v^*}\right)_S = 9.09 \times 10^{-6} + 1.82 \times 10^{-6} \sin^2\frac{\theta}{2} . \quad (15)$$

For the O/W sample, we obtained $\bar{r}_{gL} = 157 \pm 30$ nm, resulting in $(r_h/r_g)_L = 0.47 \pm 0.10$.

Miyaki et al. [9] gave the relationship between molecular weight and radius of gyration as

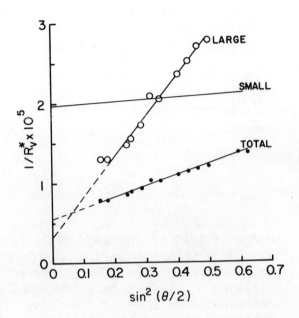

FIGURE 8.   Plots of $1/R_v^*$ of the W/O sample in toluene at $c = 0.655$ mg/ml versus $\sin^2(\theta/2)$. The solid circles are the measured values yielding $r_g = 86.5$ nm. The open circles represent the $1/R_v^*$ of the large fraction obtained after subtracting the intensity contribution of the small fraction represented by Eq. (14) and correspond to $r_{gL} = 196$ nm.

$$r_g(\text{Å}) = 0.13 \, M^{0.6} \, . \tag{16}$$

In Table 4 molecular weights calculated by use of Eq. (16) are listed together with values of r_h and r_g for both samples.

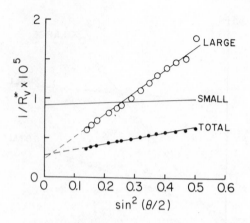

FIGURE 9.   Plots of $1/R_v^*$ of the W/O sample in toluene at $c = 0.66$ mg/ml versus $\sin^2(\theta/2)$. The solid circles are the measured values yielding $r_g = 75.5$ nm. The open circles represent the $1/R_v^*$ of the large fraction obtained after subtracting the intensity contribution of the small fraction represented by Eq. (15) and correspond to $r_{gL} = 157$ nm.

V.   CONCLUSIONS

Initiation of polymerization of styrene by potassium persulfate in both oil-in-water and water-in-oil microemulsions produced polystyrene containing at least two different size fractions. The size distributions of these polymers were measured by both photon correlation spectroscopy and GPC. For the O/W sample, the bimodal nature of the size distribution was evidence for the presence of at least two competing mechanisms for initiation. For the W/O sample, there could be three competing mechanisms based on GPC results.

The use of a combination of time-averaged intensity measurements and photon correlation spectroscopy provides an ideal nonintrusive method for characteri-

TABLE 4. Hydrodynamic Radii, Radii of Gyration, and Molecular Weights for the O/W and W/O Samples

	r_h (nm)	r_g (nm)	r_h/r_g	Molecular Weight
O/W small	15.1 ± 2.0	25.2	0.6	3.0×10^5
O/W large	73.3 ± 3.0	157 ± 30	0.47 ± 0.10	6.3×10^6
O/W overall	21.0 ± 0.6	75.5 ± 5.0	0.28	4.53×10^6
W/O small	11.2 ± 2.0	18.6	0.6	1.8×10^5
W/O large	124.1 ± 6.0	196 ± 40	0.63 ± 0.10	9.2×10^6
W/O overall	18.4 ± 0.6	86.5 ± 3.0	0.21	1.15×10^6

zation of suspended particles. However, in case of bimodal or very broad size distributions, the time range of the data becomes very crucial for obtaining the width of the distribution, as illustrated in Fig. 5. It should also be pointed out that the average molecular weight and radius of gyration determined from the time-averaged intensity measurements are mere averages differently weighted according to the details of the size distribution. However, the information on the hydrodynamic radii obtained from photon correlation spectroscopy can be used to resolve the intensity data into contributions of each fraction, thus yielding the molecular weight and radius of gyration of the two fractions.

REFERENCES

1. W. D. Harkins, J. Polym. Sci. *5*, 217 (1950).

2. R. M. Fitch, in *Polymer Colloids*, R. M. Fitch, Ed., Plenum Press, New York, 1971, p. 73.

3. J. Ugelstad, F. K. Hansen, and S. Lange, Die Makromolekulare Chem. *175*, 507 (1974).

4. J. O. Stoffer and T. Bone, J. Polym. Sci. *18*, 2641 (1980).

5. Es. Gulari and B. Chu, Biopolymers *18*, 2943 (1979).

6. D. E. Koppel, J. Chem. Phys. *57*, 4814 (1972).

7. Es. Gulari, Er. Gulari, Y. Tsunnashima, and B. Chu, J. Chem. Phys. *70*, 3965 (1979).

8. J. Ehl, C. Loucheux, C. Reiss, and H. Benoit, Makromol. Chem. *75*, 35 (1964).

9. Y. Miyaki, Y. Einaga, and H. H. Fujita, Macromolecules *11*, 6 (1978).

CHAPTER 18

Measurement of Ciliary Activity by Quasi-Elastic Light Scattering

WYLIE I. LEE
University of Washington, Seattle

CONTENTS

I. INTRODUCTION

One of the important aspects of aerosol research is the implication of these minute particles in the human respiratory system. For example, there is considerable evidence that cigarette smoking causes lung disease in animals. The common reaction of coughing or hawking is a natural mechanism for clearing these foreign objects from the body. However, there exists in mammals a physiological machinery that works continuously throughout the lifetime for the sole purpose of cleaning the airways. This is known as the *mucociliary transport*. There are three anatomic sites where mucociliary transport has been studied: the nasal passages; the airways of the lungs; and the trachea. Methods of these studies involve mainly the monitoring with external detectors the rate of disappearance or the change in distribution of radioactive particles deposited in the airways. Our interest in this problem is the detailed mechanism of this transport, which apparently is the interaction between the mucous layer and the cilia lining the epithelium of the airways. This chapter covers the recent studies on the measurement of ciliary activity.

Cilia are hairlike organelles that appear in many animals. It is the swimming apparatus of protozoa; it is the food collector of an oyster and a frog. But in mammals, ciliated epithelia are known to perform very important functions in both the respiratory and reproduction organs. It is estimated that ciliated epithelia cover an area in the human body of 5000 cm^2. The density of cilia is 600 million/cm^2. These cilia beat an average 15 times in every second. To date there are a sufficient number of medical records to correlate ciliary dysfunction with chronic headaches, mental depressions, reduced hearing and sense of smell, chronic bronchitis and bronchiectasis, infertility, and situs inversus (so-called mirror-person).

The ultrastructure of a cilium has been known for many decades. There are nine outer microtubular doublets and two central filaments. This is sometimes called "9 + 2 machinery." Afzelius [1] discovered two rows of arms on each of the nine outer filaments. These arms are now named *dynein arms*, which function the same way as the bridges on the myosin filaments in striated muscle. Experimental studies verified later the sliding filament theory of cilia and found the dynein arms have ATPase activity, just as the cross-bridge of muscle [2,3].

Despite the critical function that cilia perform in humans, a simple and objective method for study of ciliary activity has been developed only recently. The earlier methods used to assess ciliary activity were indirect. For example, the ciliary activity was related to the transport of charcoal powder or seeds over ciliated cells [4]. A better direct method was introduced in 1933 by Proetz [5], who used a combination of a microscope and a motion picture camera. To date, high-speed cinematography combined with phase-contrast microscopy is regarded as the most reliable method of measuring ciliary activity. However, the tedious procedure of data analysis discourages most investigators and also limits its abilities for on-line studies. A popular set up known as the *photometric method*, which detects the transilluminated light through the cilia [6], provides good information only for an *in vitro* study and is not reliable when the culture thickness increases or when the beating mode of cilia becomes asynchronic.

This chapter describes a new light-scattering technique which was developed originally for the studies of oviductal cilia [7] and some applications of this technique to the study of oviductal and respiratory ciliary activity.

II. MATERIALS AND METHOD

The tissue cultures of ciliated epithelia were grown in Rose chambers according to the method proposed by Rumery et al. [8]; this method yields cellular monolayers abundant in ciliated cells and with a few secretory cells. The overall design of the chamber is well suited for photomicroscopy, microcinematography, and also for laser light scattering.

The cultured cells in the chamber were first scanned under a phase-contrast microscope. After the clusters of ciliated cells were sighted, these locations were marked and the chamber was positioned onto the sample holder of the laser light-scattering spectrometer. The block diagram of this spectrometer is shown in Fig. 1. The beam of a helium-neon laser (Spectra Physics Model No. 124A) was focused on the preselected ciliated cells. The precise position was achieved by an *x-y* translation stage to which the chamber holder was mounted. When the ambient light was kept dark, a distinct speckle pattern consisting of some stationary spots and an oscillating pattern could be seen on the screen. The light scattered from both the ciliated cells and from the air-glass interface of the culture chamber was collected at near backscattering and imaged on the cathode of a photomultiplier (RCA C7164R) to achieve the heterodyne detection [9]. The spectral structure of scattered-light intensity fluctuations can be expressed by the autocorrelation function of photocurrent as

$$G^{(2)}(\tau) = ei_0\delta(\tau)$$

$$+ i_0^2 \left\{ 1 + \frac{i_s}{i_0} \varepsilon^2 \{g^{(1)}(\tau)\exp[-iw_L\tau] + \text{C.C.}\right\}, \quad (1)$$

where i_s and i_0 are the average photocurrent produced by the light scattered by the cilia and the elastic scattering at the air-glass interface, respectively. The heterodyne mixing efficiency ε represents the degree to which these two light waves have matched phase fronts over the detector's surface and w_L is the laser frequency. Therefore, the intensity autocorrelation function includes a term describing shot noise of the detector (first term), a DC term, and a term describing the fluctuation of the scattered electric field. The term $g^{(1)}(\tau)$ is the normalized field autocorrelation function while C.C. denotes the complex conjugate of the preceding term.

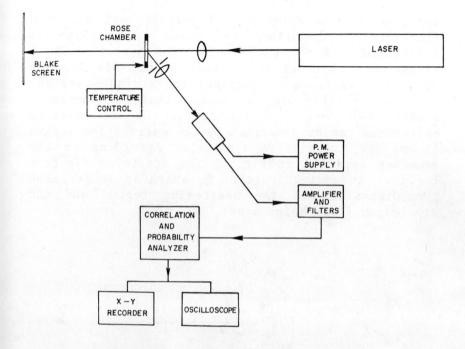

FIGURE 1. Schematic diagram of the laser light-scattering spectrometer.

$$g^{(1)}(\tau) = \frac{G^{(1)}(\tau)}{G^{(1)}(0)} = \frac{\langle E_s^*(R, t + \tau) E_s(R, t) \rangle}{\langle |E_s(R, t)|^2 \rangle} \ . \qquad (2)$$

A detailed derivation of the electric field autocorrelation function $G^{(1)}(\tau)$ was reported earlier [10]. In brief, the scattering of light from ciliated epithelium is treated as in the case of scattering from rough surfaces [11,12]. The surface of ciliated epithelium is characterized by a phase variable which includes two components: one comes from the periodic beating and the other from the random phase of the surface,

$$\phi(\mathbf{r'}, t) = \omega_c t + \phi_s(\mathbf{r'}, t) \ , \qquad (3)$$

488   W. I. Lee

where ω_c is the frequency of ciliary beat and $\phi_s = (2\pi/\lambda)h(\mathbf{r}',t)$ describes the phase shift which is associated with the surface function $h(\mathbf{r}',t)$. If we assume that $h(\mathbf{r}',t)$ is a Gaussian and its mean is zero, the variance σ describes the roughness of the surface. Furthermore, we assume that the coherency between two points on the surface of the ciliated epithelium can be described by the correlation length ξ and the correlation time τ_f. According to the geometry depicted in Fig. 2, the scattered electric field at the detector point $\mathbf{R}$, which is longer than the dimensions of the scattering region and the wavelength (Fraunhofer zone), is

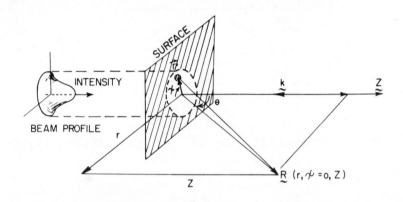

FIGURE 2.   Geometry of light scattering from the cultured cells of a ciliated epithelium. The intensity of the laser beam is assumed to have a Gaussian profile.

$$E_s(\mathbf{R},t) = A_0 \exp[-i\omega_L t] \int \exp[-k|\mathbf{R}-\mathbf{r}'|]$$
$$\times \exp[i\phi(\mathbf{r}',t)]\exp[-r'^2/\alpha^2]ds' , \qquad (4)$$

where k is the wave number $2\pi/\lambda$, A_0 is a constant that includes the amplitude of the incident light, and α is the width of intensity profile of the laser beam.

The autocorrelation function of the scattered field can be derived from its definition

$$G^{(1)}(\tau) = \langle E_s^*(R, t + \tau)E_s(R,t)\rangle \qquad (5)$$

and it becomes

$$
\begin{aligned}
G^{(1)}(\tau) &= \pi^2 \alpha^2 |A_0|^2 \exp[i(\omega_L + \omega_c)\tau] \\
&\quad \times \exp[-k^2\sigma^2\tau/\tau_f] \\
&\quad \times \int \exp[-u^2/4\gamma]\, J_0(ku \sin \theta)u\, du \\
&= \text{const} \times \exp[i(\omega_L + \omega_c)\tau] \\
&\quad \times \exp[-\gamma k^2 \sin^2 \theta]\, \exp[-k^2\sigma^2\tau/\tau_f] \;, \qquad (6)
\end{aligned}
$$

where J_0 is the zeroth-order Bessel function and $\gamma = [(2/\alpha^2) + (4k^2\sigma^2/\xi^2)]^{-1}$, which depends on the characteristic beam width α and the surface properties σ and ξ. In summary, the movement of the surface can be described as a superposition of two motions: the periodic oscillations of the cilia with an average frequency of ω_c and a random motion due to the metachronal coordination of the clusters of cilia. Consequently, the autocorrelation function of light scattered by cilia shows a frequency modulation as a result of the ciliary beat and a decay relaxation related to the space-time coherence of the ciliated epithelium. From Eqs. (1), (2) and (6), a photocurrent autocorrelation function such as follows can be expected:

$$
\begin{aligned}
G^{(2)}(\tau) &= \text{shot noise} + \text{DC} \\
&\quad + 2i_0 i_s \varepsilon^2 \exp[-k^2\sigma^2\tau/\tau_f]\cos \omega_c\tau \;. \qquad (7)
\end{aligned}
$$

We are interested in the third term, which is a damped cosine function whose frequency equals the frequency of ciliary beat ω_c and whose damping coefficient is related to the coherence of ciliary motions $\tau_c/k^2\sigma^2$.

In the experiment the DC component was filtered and the autocorrelation function was obtained by a correlator (SAICOR-43A or Langley-Ford digital correlator with analog input) as illustrated in Fig. 1. Figure 3 shows the result obtained from a cluster of cilia beating at a dominant frequency of 18 Hz. A more practical display of the results is to show the beating frequency in frequency space, which is the Fourier transformation of the autocorrelation function $G^{(2)}(\tau)$. A typical spectrum of such a measurement is shown in Fig. 4. The peak appearing near zero frequency indicates the loose cells and debris produced by secretory cells. The width of the peak at 18 Hz, the dominant ciliary beat frequency, is related to the coherence of ciliary movements just as is the damping coefficient of the autocorrelation function.

III.   RESULTS

Calcium has been implicated in excitation-contraction coupling [13], and our preliminary studies showed that ciliary beat frequency in mammalian ciliated cells could be modified by changing the extracellular calcium concentration. The effects of various concentrations of Ca^{2+} on oviductal cilia were assayed by measuring the ciliary beat frequency in cultures derived from fimbria of rabbits. In each of these experiments the cultures were first equilibrated in Hank's solution for at least 20 min at 37°C. After the control values of the frequency were measured and the chamber was perfused with 10 ml of calcium-free Hank's solution containing 2 mM EGTA, the ciliary beat frequency was monitored until ciliary arrest was observed. The chamber was drained and then refilled with Hank's solution containing increasing concentrations of calcium (10^{-6}-10^{-2} M). The cultures were equilibrated for 10 min, and then

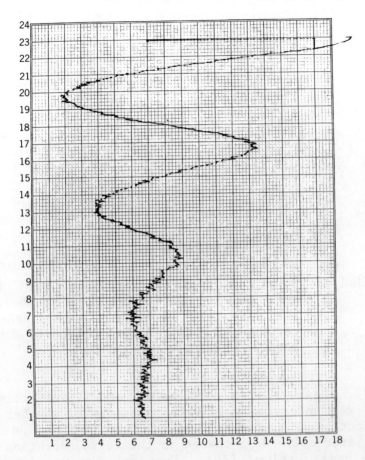

FIGURE 3.   Autocorrelation function obtained from light scattered from cultures of rabbit oviductal epithelium. The total time span of this trace is 200 msec. The frequency of cilicary beat calculated from this result is 18 Hz.

the frequencies of ciliary beat were measured. The results of measurements of calcium dependence of ciliary beat are shown in Fig. 5.

The mucociliary clearance in humans is improved by β-adrenergic stimulation such as isoproterenol

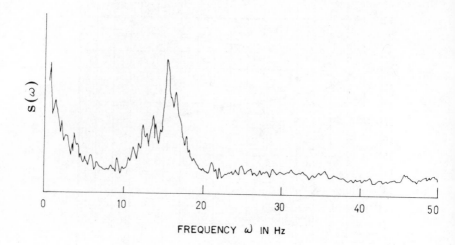

FIGURE 4. Power spectrum of scattered light from cultures of rabbit tracheal epithelium. The main peak shows 17 Hz. The peak at zero was the diffusional motion of debris produced by secretory cells.

[14]. However, the enhancement of ciliary activity was not understood since the question of whether the effect was a direct and specific action on the ciliated cells or a change in the mucus secretion has not been resolved [15]. This question was investigated recently by using a laser light-scattering technique, which is modified slightly from the one described here [16]. Figure 6 shows the results of measurements on cultures of rabbit tracheal epithelium by use of 0.15% hyaluronidase to obtain mucus-free preparations [8]. The dose response of ciliary beat frequency on isoproterenol indicates that the increase in mucociliary transport by β-adrenergic stimulation can be a direct stimulating effect on the ciliated cells of the respiratory tract. This conclusion was also supported by the fact that the presence of a β-adrenergic blocker, propranolol, which itself does not affect the ciliary activity

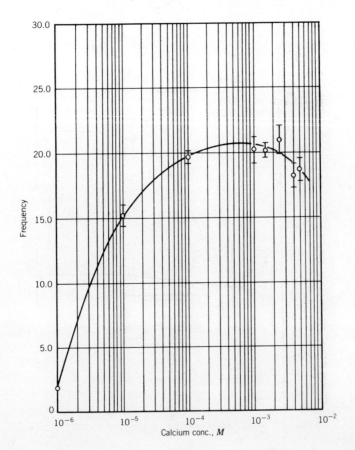

FIGURE 5. Dependence of rabbit oviductal ciliary beat on the extracellular concentration of Ca^{2+}.

unless its concentration exceeds 10^{-6} M, has nullified the stimulation due to isoproterenol.

IV. DISCUSSION

The frequency of ciliary beat is a useful and sensitive parameter for studying the ionic and hor-

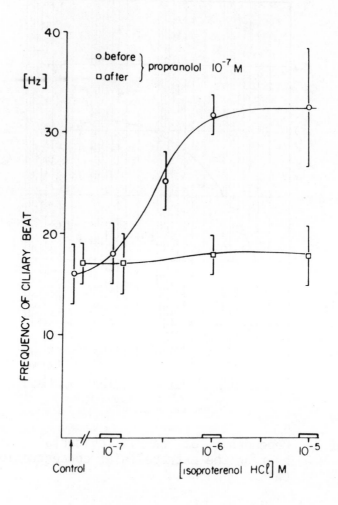

FIGURE 6. Dose-response relationship of effect of isoproterenol on the frequency of tracheal ciliary beat [Reprinted from P. Verdugo, T. J. Noel, and P. Y. Tam, J. Appl. Physiol.: Resp. Environ. Exercise Physiol. *48*, 868-871 (1980) with permission of the authors and editor of The Journal of Applied Physiology.]

monal control of ciliary activity. We have demonstrated that quasi-elastic light scattering (QELS) is
a reliable and convenient technique for measuring the
frequency of ciliary beat. Although it does not
detect the direction of the effective ciliary stroke,
it measures precisely the frequency of ciliary beat
without any of the ambiguities inherent in the
stroboscopic and photometric monitoring techniques.
With only a few modifications, the system can be
incorporated with a phase-contrast microscope to
provide an easy access for filming [17]. As we
suggested previously [10], a fiber-optic scattering
spectrometer can be developed for *in situ* studies of
ciliary activity and mucociliary transport. Our
experience with this system indicates that the main
obstacle in applying optical fibers is its ultra-
sensitivity to the noise caused by bending of the
fiber or by the fiber vibration due to minute
external disturbances. These noises may be greatly
reduced if a bundle of three fibers is used and a
differential spectral analysis is performed. The
block diagram of this spectrometer is shown in
Fig. 7. This new device was only recently introduced
in our laboratory, and the preliminary results are
very gratifying.

We hope that other investigators will be
inspired to use a QELS technique for the study of
mucociliary transport. Many challenges remain in its
development for new applications. A QELS technique
has been used to measure *in vitro* the rheological
properties of mucus [18]. Perhaps this is the best
technique for investigating the direct action of
irritants and pharmacological agents on the ciliary
activity, the rheological properties of mucus, and
the mucociliary transport in normal subjects and
patients with respiratory diseases.

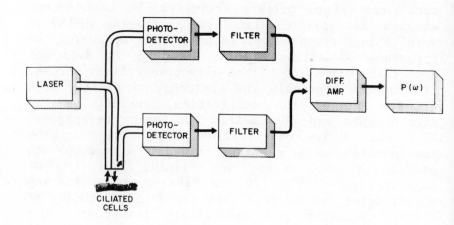

FIGURE 7.   Block diagram of the differential fiber-optic detection system.

ACKNOWLEDGMENTS

I would like to thank Dr. P. Verdugo for contributing his recent results. I would also like to acknowledge the support of Dr. B. Dahneke and Dr. N. C. Ford and the College of Engineering, University of Washington.

REFERENCES

1. B. A. Afzelius, J. Biophys. Biochem. Cytol. 5, 269-278 (1959).

2. B. H. Gibbons and I. R. Gibbons, J. Cell Sci. 13, 337-357 (1973).

3. P. Satir, in *Cilia and Flagella*, M. A. Sleigh, Ed., Academic Press, London, 1974, pp. 131-143.

4. W. Sharpey, in *Todd's Cyclopaedia of Anatomy and Physiology*, Longman, Brown, Green, Roberts, London, 1935, pp. 606-636.

5. A. W. Proetz, in *Transactions of the Section on Laryngology, Otology, and Rhinology. American Medical Association*, 1933, pp. 260-262.

6. T. Dalhamn and R. Rylander, Nature (Lond.) *196*, 592-593 (1962).

7. W. I. Lee and P. Verdugo, Biophys. J. *16*, 1115-1119 (1976).

8. R. E. Rumery, E. Phinney, and R. J. Blandau, in *Methods in Mammalian Embryology*, J. C. Daniel, Jr., Ed., Freeman, San Francisco, 1971, pp. 472-495.

9. H. Z. Cummins and H. L. Swinney, Progr. Opt. 8, 133-200 (1970).

10. W. I. Lee and P. Verdugo, Ann. Biomed. Eng. *5*, 248-259 (1977).

11. P. Beckmann and A. Spizzichino, *The Scattering of Electromagnetic Waves from Rough Surfaces*, Macmillan, New York, 1963.

12. A. S. Marathay, L. Heiko, and J. L. Zuckerman, Appl. Optics *9*, 2470-2476 (1970).

13. H. E. Huxley, Nature *243*, 445-449 (1973)

14. J. Iravani and G. N. Melville, Respiration *31*, 350-357 (1974).

15. A. Wanner, Am. Rev. Respir. Dis. *116*, 73-125 (1977).

498   W. I. Lee

16.   P. Verdugo, T. J. Noel, and P. Y. Tam, J. Appl.
       Physiol: Respi. Environ. Exercise Physiol. *48*,
       868-871 (1980).

17.   P. Verdugo, T. R. Hinds, and F. F. Vincenzi,
       Pediatr. Res. *13*, 131-135 (1979).

18.   W. I. Lee, P. Verdugo, R. J. Blandau, and P.
       Gaddum-Rosse, Gynecol. Invest. *8*, 254-266
       (1977).

PART FOUR
PARTICLE INTERACTION EFFECTS

CHAPTER 19

Effects of Interparticle Interactions on Particle Size Determination by QELS

D. F. NICOLI and R. B. DORSHOW
University of California at Santa Barbara

CONTENTS

I. INTRODUCTION

Quasi-elastic light scattering (QELS) is a powerful technique that measures the mutual diffusion coefficient (or diffusivity) D of particles suspended in solution [1-3]. From a measurement of D

501

one can, in principle, obtain the corresponding particle hydrodynamic radius R_h by use of the Stokes-Einstein relationship. The use of this procedure, however, assumes that the solution is sufficiently dilute such that the Brownian random-walk diffusion of a given macroparticle is due solely to solvent-particle collisions and is uninfluenced by the presence of other macroparticles nearby. In solutions of finite concentration, on the other hand, it is well known that the existence of repulsive and/or attractive interparticle forces may significantly affect D, and hence the inferred hydrodynamic radius R_h, of the scattering macroparticles [4,5]. The influence of interparticle interactions is particularly acute in the case of charged particles. Depending on the surrounding counterion concentration, the repulsive Coulombic interaction between like-charged species can produce large increases ($\geq 500\%$) in D at very low ionic strengths.

The importance of interparticle interactions for measurements of the scattered intensity have been recognized for some time [6]. Corrections to measured intensities have been made by using the so-called second virial coefficient, which is thermodynamic in origin. However, a cursory survey of the literature suggests that the effects of interactions on the diffusion coefficient have only recently been appreciated. As is the case with the scattered intensity, there is a perturbation to D of thermodynamic origin, described by the second virial coefficient; this is due to the variation of the osmotic pressure with particle concentration. In addition, however, there is a perturbation whose origin is hydrodynamic. Diffusivity D is affected by a friction factor between the particle and the solvent that is concentration dependent because there is a solvent velocity field generated in the vicinity of a given particle due to the motion of nearby particles. The hydrodynamic perturbation was calculated recently for spherical particles by both Batchelor [7,8] and Felderhof [9]. This formalism was used recently by

Corti and Degiorgio [10,11] and Finsy et al. [12] to interpret their diffusivity data for surfactant micelles (SDS + NaCl) and water-in-oil (W/O) micro-emulsions, respectively. Since there has been very limited application of the hydrodynamic theory to dynamic light-scattering results reported to date, it would appear useful to review linear interaction theory as it applies to diffusivity measurements.

We apply the theory to experimental data from three different kinds of diffusing particle systems: charged protein molecules (BSA) at very low ionic strengths, charged surfactant micelles in aqueous salt solutions, and uncharged emulsion particles consisting of droplets of a fluorocarbon solubilized by a nonionic surfactant. These systems differ greatly in physicochemical makeup and exhibit differing amounts of repulsive and attractive inter-actions. In the first two cases the particles are highly charged and perturbations to D are due mostly to electrostatic repulsions. In the latter system the particles are electrically neutral; the predomi-nant interactions are van der Waals attractions.

II. THEORY

The diffusion coefficient D of a semidilute monodisperse solution of particles of number concen-tration n and volume fraction ϕ can be described by the generalized Stokes-Einstein relationship [13],

$$D = \frac{(\partial\Pi/\partial n)_{P,T}}{f(n)} (1 - \phi) , \qquad (1)$$

where Π is the osmotic pressure and $f(n)$ is the friction factor of the particles. We restrict our discussion to spherical particles of radius a; no theory has been developed for spheres of mixed radii (that is, a polydisperse system) or for particle shapes different from a sphere. In this case ϕ is related to n by $\phi = 4/3(\pi a^3 n)$.

The factor $(\partial \Pi / \partial n)_{P,T}$ in Eq. (1) is usually referred to as the *thermodynamic term* because the osmotic pressure Π is closely related to the chemical potential of the solution. In a semidilute system, Π can be expanded in the concentration n,

$$\Pi = k_B T(n + B_2 n^2 + \cdots) , \qquad (2)$$

where k_B is Boltzmann's constant, T is the absolute temperature, and coefficient B_2 is the well-known second osmotic virial coefficient.

The influence of solvent flow on the diffusivity occurs through the particle mobility, which [Eq. (1)] equals $(1 - \phi)/f(n)$. This, in turn, can be expanded in the particle volume fraction ϕ:

$$\text{Mobility} = \frac{(1 - \phi)}{f(n)} = \frac{1}{6\pi \eta a} (1 + K_h \phi + \cdots) , \qquad (3)$$

where η is the solvent shear viscosity and K_h is the so-called hydrodynamic correction. Hence from Eq. (1) we obtain the linear perturbation of the diffusivity, valid at small volume fractions:

$$D = D_0 [1 + (K_t + K_h)\phi] , \qquad (4)$$

where coefficient K_t is the "thermodynamic" correction to the diffusivity, equivalent to $2B_2/[4/3(\pi a^3)]$. Prefactor D_0 is the diffusivity in the limit of infinite dilution, where interparticle interactions are nonexistent, and is given by the familiar Stokes-Einstein expression

$$D_0 = \frac{k_B T}{6\pi \eta a} . \qquad (5)$$

To the extent that interactions can be ignored, the particle hydrodynamic radius a can be obtained from the measured diffusivity by using Eq. (5).

Coefficient K_t can be related to the potential energy of interaction between two particles suspended in the solvent separated by distance r, $W_{12}(r)$, by using a "configuration" integral taken from thermodynamics [14],

$$K_t = \frac{3}{a^3} \int_0^\infty r^2 (1 - \exp[-W_{12}(r)/k_B T]) dr . \qquad (6)$$

The pair potential energy $W_{12}(r)$ in general consists of three contributions: hard-sphere repulsion W_{HS}, in which $W_{HS}(r) = 0$ for $r \geq 2a$ and $W_{HS}(r) = \infty$ for $r < 2a$; electrostatic repulsion W_R between macroparticles of like charge; and van der Waals attractions W_A. (The possibility of additional electrostatic forces associated with particle-solvent interactions is contained implicitly in W_R.) It is convenient to explicitly separate the hard-sphere exclusion term W_{HS} from the remaining interaction energy, which we call $W(r)$ [$\equiv W_R(r) + W_A(r)$]. Also, with Corti and Degiorgio [11] we use the dimensionless normalized separation parameter $x \equiv (r - 2a)/2a$. Correction coefficient K_t can then be expressed as

$$K_t = 8 + 24 \int_0^\infty (1 + x)^2 (1 - \exp[-W(x)/k_B T]) dx . \qquad (7)$$

The positive constant term (+8) is the contribution to the thermodynamic correction due to hard-sphere repulsions, which acts to increase the diffusivity.

Using different theoretical approaches, Batchelor [7,8] and Felderhof [9] independently obtained expressions for the hydrodynamic perturbation coefficient K_h due to solvent backstreaming effects. Corti and Degiorgio [11] expressed the Felderhof result in a convenient integral form that closely resembles the structure of K_t [Eq. (7)]:

$$K_h = -6.44 - \int_0^\infty F(x) \, (1 - \exp[-W(x)/k_B T]) dx \, . \qquad (8)$$

where $F(x) = 12(1 + x) - 15/8(1 + x)^{-2} + 27/64(1 + x)^{-4} + 75/64(1 + x)^{-5}$. The negative constant term (-6.44) is, again, due to the hard-sphere interaction. This contribution is *negative*, which acts to decrease the diffusivity. Interestingly, it nearly cancels the thermodynamic contribution K_t, so that the hard-sphere exclusion retains a relatively small influence on D; that is, $D = D_0[1 + (8-6.44)\phi] = D_0(1 + 1.56\phi)$. Felderhof also derived Eqs. (4), (7), and (8), using the Smoluchowski equation in the low-particle-density limit, including central potential and hydrodynamic interactions. [In Batchelor's calculation for K_h the hard-sphere constant is -6.55, with $F(x) = 11.89(1 + x) + 0.706 - 1.69(1 + x)^{-1}$, which results in nearly identical values for K_h as those obtained by use of the Felderhof expressions.]

To date there have been very few experimental tests of the above theory. Kops-Werkhoven and Fijnaut [15] performed both light-scattering and sedimentation measurements on neutral submicron silica particles that were stabilized against van der Waals attractions by a coating of polymer chains (to enforce a minimum separation). They obtained $K_h = -6 \pm 1$ and $K_t + K_h = (1.3 - 1.4) \pm 0.2$, in essential agreement with the formalism above, assuming the existence of only hard-sphere repulsions.

More recently, Finsy et al. [12] utilized a modified version of the Batchelor hydrodynamic theory to analyze their diffusion coefficient data for a W/O microemulsion. They interpreted the linear decrease in D with increasing ϕ in terms of hard-sphere repulsion plus a substantial attractive potential energy for a pair of neutral microemulsion droplets. Corti and Degiorgio [11] obtained qualitative fits to their

SDS micellar diffusivity data over a large range of surfactant and NaCl concentrations. In their system they observed diffusivities that either increased or decreased linearly with increasing surfactant concentration, depending on the added NaCl concentration (that is, positive slopes for D vs. ϕ for [NaCl] $\leqq$ 0.40 M at 25°C, negative slopes otherwise).

The application of Eqs. (4), (7), and (8) to a particular suspension of diffusing particles requires that we determine an appropriate pair interaction energy $W(x)$. We characterize the system as consisting of rigid, spherical particles of radius a, carrying a net charge Q smeared uniformly over the surface. Such was the idealized model originally used by Verwey and Overbeek [16] to develop a theory of colloidal stability. The so-called DLVO theory that subsequently evolved is based on the competition between a repulsive potential energy W_R due to electrostatics, and an attractive energy W_A due to van der Waals forces [17]. We first discuss the repulsive contribution.

The detailed functional form of the repulsive energy $W_R(x)$ depends on the relative sizes of the particle radius a and the range of the interaction. The latter depends on the ionic strength of the solvent, which is characterized by the Debye-Hückel inverse screening length κ;

$$\kappa = \left(\frac{8\pi I e^2}{\varepsilon(T)kT} \right)^{1/2} , \tag{9}$$

where, I is the solution ionic strength (number of monovalent ions per cubic centimeter), e is the electronic charge, and $\varepsilon(T)$ is the solvent dielectric constant. In DLVO theory the analytical form of $W_R(x)$ varies with the value of κa; different solutions exist for $\kappa a < 1$ and $\kappa a > 1$.

For $\kappa a < 1$, Verwey and Overbeek obtained

$$W_R(x) = \psi_0^2 \varepsilon a \, \frac{\exp[-2\kappa a x]}{2(1 + x)} \, \gamma \qquad (10)$$

where

$$\psi_0 = \frac{Qe}{a\varepsilon(1 + \kappa a)} \, f \, .$$

Here ψ_0 is the surface potential of the charged spherical particle; f and γ are functions of κa and x [given by Eqs. (79) and (83) on pp. 149-150 in reference 16]. In the case in which $\kappa a \ll 1$, functions f and γ both reduce to unity, so that

$$W_R(x) = \frac{Q^2 e^2}{2\varepsilon a(1 + \kappa a)^2} \, \frac{\exp[-2\kappa a x]}{1 + x} \quad (\kappa a \ll 1) \, . \quad (11)$$

As may be seen shortly, for micellar systems the salt concentrations of interest typically lie in the range 0.1-1 M, whereas the "minimum-sphere" radius is 20-35 Å, depending on the surfactant and temperature; hence κa is close to unity, so the $\kappa a \ll 1$ approximation [Eq. (11)] is not appropriate. Although several approximation methods have been suggested for evaluating Eq. (10), it is more straightforward to simply calculate $W_R(x)$ exactly from the Verwey-Overbeek theory by use of a digital computer.

For the case $\kappa a > 1$, Verwey and Overbeek derived an analytical function for W_R that is valid for small values of ψ_0:

$$W_R(x) = \frac{\varepsilon a \psi_0^2}{2} \, \ln(1 + \exp[-2\kappa a x]) \, . \qquad (12)$$

There is a graphical representation of the solution for W_R, corresponding to Eqs. (10) and (12), in the original Verwey and Overbeek monograph. In the intermediate region $1 < \kappa a < 10$ there is a smooth curve that connects the two limiting expressions.

Finally, we require an attractive contribution to the pair potential energy W_A. In DLVO theory this is the London-van der Waals expression for the energy of two dielectric spheres of radius a, as derived by Hamaker [18].

$$V_A(x) = - \frac{A}{12} \left[(x^2 + 2x)^{-1} + (x^2 + 2x + 1)^{-1} \right.$$

$$\left. + 2\ln \frac{x^2 + 2x}{x^2 + 2x + 1} \right]. \tag{13}$$

The strength of the interaction is determined by the magnitude of A, the Hamaker coefficient. Since $V_A(x)$ diverges at $x = 0$ (i.e., the spheres just touch), it is necessary that a lower cutoff in the separation x_L be utilized in the integral expressions for K_t and K_h, as pointed out recently by Corti and Degiorgio. This is equivalent to approximating the van der Waals potential as a square well of finite depth.

III. RESULTS: THEORY VERSUS EXPERIMENT

A. Charged Protein (BSA) Molecules

First we apply the above theory to the experimental diffusivity results obtained several years ago by Doherty and Benedek [19] for charged bovine serum albumin (BSA) molecules in solution. They adjusted the protein charge by varying the solution pH and found that D increased monotonically with increasing mean protein charge Q and decreasing solution ionic strength I. Increases in D of more than twofold relative to the high-salt, fully screened value were observed at very low salt concentrations.

The Doherty-Benedek results were expressed in terms of the dimensionless parameter $(D - D_0)/D_0$, which according to the linear interaction theory stated above is equal to $K\phi = (K_t + K_h)\phi$. We [20] have calculated this from Eqs. (7) and (8). Since the BSA molecules were quite highly charged (relative to their size) and the solution ionic strength very low in those experiments, we can easily verify that the electrostatic repulsive forces greatly outweighed the van der Waals attractions [assuming a typical value for A in Eq. (13)]. Hence we included only the repulsive term $W_R(x)$ in the pair potential energy $W_{12}(x)$ in Eqs. (7) and (8). Furthermore, given the low solution ionic strengths utilized (typically 0.004-0.014 M), we were able to use the $\kappa a \ll 1$ approximation for $W_R(x)$ given by Eq. (11). Thus computation of $K_t + K_h$ for various values of Q and I is straightforward. We assumed that the protein radius a can be approximated by the hydrodynamic radius as measured in the absence of interactions (i.e., from $D_0 = 6.2 \times 10^{-7}$ cm^2/sec at 20°C at either $Q = 0$ or in the limit of high salt concentrations), $R_h = 34.5$ Å.

All that remains is a determination of the volume fraction ϕ. Surprisingly, herein lies the greatest ambiguity in these calculations. Doherty and Benedek performed their measurements at fixed BSA concentration: 5 g%, or 50 mg/ml, which is equivalent to $n = 4.3 \times 10^{17}$ molecules/cm^3 (BSA molecular weight = 69,000). However, to compute ϕ, we must know the effective volume per molecule; a range of estimates is possible.

First, we can assume the volume of the equivalent hydrodynamic sphere as measured by dynamic light scattering; it is simply $(4/3)\pi R_h^3$, or 1.72×10^{-19} cm^3. The product of this volume times n yields $\phi = 0.075$. We see later that this is an upper limit of the estimates of ϕ.

Second, we can consider the fact that BSA does not possess a spherical shape. Rather, previous measurements suggest that it can be modeled as a prolate ellipsoid of axial ratio $b/a = 3.5$. Using Perrin's formula [2,21] for R_h as a function of a and

b for prolates, we can solve for a and b. The resulting prolate volume is $(4/3)\pi b a^2 = 1.14 \times 10^{-19}$ cm^3, substantially smaller than the volume of the equivalent hydrodynamic sphere (above). Multiplying by n, we obtain a smaller estimate for the volume fraction $\phi = 0.052$. Although this is surely a better estimate of the actual fraction of solution volume occupied by the BSA particles, it remains to be seen whether this value is appropriate for the theory, given the assumed nonspherical nature of the molecules. The theory outlined above assumes strictly spherical particles.

Finally, yet another calculation of ϕ is possible [6], based on an estimate of the specific volume of BSA: $\bar{v} = 0.74$ ml/g. Given a protein concentration of 0.05 g/ml, we obtain $\phi = 0.037$. We regard this estimate as a lower limit for the possibilities for ϕ.

The results of our theoretical predictions for $(D - D_0)/D_0$ versus BSA charge Q and solution ionic strength I (for 5 g% BSA) are summarized [20] in Figs. 1-3. The relevant experimental data due to Doherty and Benedek are shown for comparison with theory. We confine ourselves here to the three largest values of BSA charge that they employed — $Q = 7$, 10, and 18. In each figure we have plotted the results of four different theoretical predictions, labeled curves A through D.

Curve A represents the results of calculations made recently by Schor and Serrallach [22]. Their theory, like ours, is a linear interaction approach with a DLVO-type potential (only electrostatic repulsions); however, it considers only the thermodynamic correction to D (that is, the second virial coefficient) and ignores the concentration dependence of the mobility. This results in the contribution $K_t \phi$, where K_t is given by Eq. (7), with $W_{12}(x) = W_R(x)$ given by Eq. (11). The value of volume fraction chosen was $\phi = 0.075$, which assumes the full volume of the effective hydrodynamic sphere. The theoretical calculations by Schor and Serrallach represent a

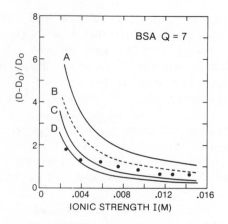

FIGURE 1.   Normalized correction to the diffusivity
$(D - D_0)/D_0$ versus solution ionic strength
I for 5% (w/v) BSA of charge $Q = 7$. The
solid dots represent the experimental data
due to Doherty and Benedek [19]. Curve A
is the theoretical prediction by Schor and
Serrallach [22], with $\phi = 0.075$. Curve B
is the Schor-Serrallach prediction for
$\phi = 0.052$. Curve C is our theoretical
result, which includes the hydrodynamic
correction to D and assumes $\phi = 0.052$.
Curve D is the same calculation, assuming
$\phi = 0.037$.

substantial improvement over earlier predictions made
by Doherty and Benedek [19] on the basis of Stephen's
theory [23]. The latter considers the Coulombic
interaction between a macroparticle and the mobile
counterions of the surrounding solvent. It predicts
an enhancement in the diffusivity that is at least
four times too large at the lowest measured ionic
strengths. However, as seen in Figs. 1-3, there
remains a substantial discrepancy between curve A and
the experimental data, especially at high BSA charge.
    Curve B represents the predictions of the same
Schor-Serrallach theory, but assuming a smaller value

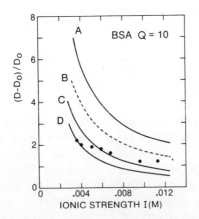

FIGURE 2.   Same as Fig. 1, except that the BSA charge
Q is 10.

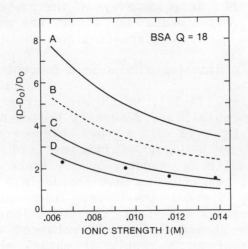

FIGURE 3.   Same as Fig. 1, except that the BSA charge
Q is 18.

for the volume fraction, ϕ = 0.052, corresponding to
the estimated prolate volume of BSA (discussed
above).   We believe that curve B may present a more

reliable estimate of the correction to the diffusivity due to the concentration dependence of the osmotic pressure.

Curves C and D show the results of our calculations, which include the hydrodynamic correction to D, equal to $(K_t + K_h)\phi$. The only difference in the two cases is the choice of ϕ. For curve C, we take $\phi = 0.052$, which assumes the prolate volume; for curve D, $\phi = 0.037$, which should be a reliable lower limit estimate. Considering the obvious scatter in the original data points; we find that the theory as presented is in good agreement with the experimental results due to Doherty and Benedek. It clearly is a substantial improvement over earlier approaches. Shortcomings in the theoretical model are related to such issues as nonspherical particle shape, exclusion of van der Waals attractions, and the lack of uniformity of charge density at the protein surface. Additional experiments on other charged globular proteins at several concentrations would be useful.

B.   Ionic Micelles at Low Salt Concentrations

Micelles [24-26], which are aggregates of surfactant molecules, are generally believed to contain a liquidlike core of monomer hydrocarbon chains, surrounded by a layer of charged head groups plus some neutralizing counterions. Since a sizable fraction of the monomers that reside in a micelle are effectively ionized (that is, fully dissociated), these aggregates are highly charged. Their resulting diffusivity should be greatly enhanced at low salt concentrations as a result of mutual electrostatic repulsions, according to the theory given above.

However, application of the linear interaction theory to micellar solutions may not be straightforward since the micelles are self-assembled aggregates. That is, they exist in rapid dynamic equilibrium with monomer surfactant molecules and other micellar aggregates, so that the distribution of micellar sizes should generally depend on the overall surfactant concentration [27,28]. Thus a straight-forward dilution of the micellar solution may yield a nonlinear plot of D versus ϕ because the intrinsic particle (micelle) size is not constant.

In Fig. 4 we show representative dynamic light-scattering results for micelles of cetyltrimethyl-ammonium bromide (CTAB) plus added NaBr for a range of surfactant and salt concentrations at fixed temperature ($T = 40°C$) [29]. Clearly, there is no simple function that describes the observed D versus [CTAB] (that is, CTAB concentration) behavior over the entire experimental domain. However, the behavior at low NaBr concentrations (viz., [NaBr] $\lesssim$ 0.08 M at 40°C) is particularly simple: D increases linearly with increasing [CTAB], with a slope that decreases with increasing [NaBr] [30]. Furthermore, the straightline plots of D versus [CTAB] all appear to converge to a common intercept D_0 in the limit of zero surfactant concentration (actually at the criti-cal micelle concentration or CMC ($< 10^{-3}$ M CTAB), below which there should be no micelles).

The "positive-fan" behavior of D versus [CTAB] found at low NaBr concentrations is consistent with Eq. (4). The simplest interpretation of this region is that the solution consists of "minimum-sphere" micelles [27] of approximately constant size; an increase in surfactant concentration results in a larger number of micelles of substantially the same size, which results in a larger perturbation of the diffusivity. The mean hydrodynamic radius a is then obtained by inverting Eq. (5): at 40°C, a = 29.2 Å. It must be stressed that the micellar radius is not an adjustable parameter; it emerges directly from the limiting diffusivity D_0 and hence is independent of the choice of a pair interaction potential. The value obtained above for radius a is consistent with rough estimates of the length of an extended CTAB monomer molecule (including some ambiguity due to the surrounding hydration layer) [26]. If the size of the micelle were to change significantly with either surfactant or salt concentration in this region, one

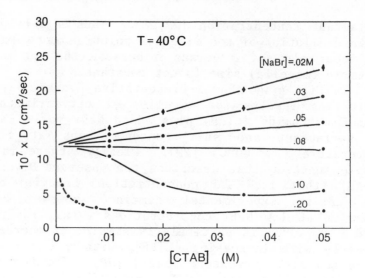

FIGURE 4.   Measured diffusion coefficient D for CTAB
as a function of surfactant concentration
[CTAB] and salt concentration [NaBr] at
T = 40° [29]. The solid curves smoothly
connect data points for a given value of
[NaBr].

would in general expect a nonlinear D versus [CTAB]
dependence. Hence it would appear that the observed
changes in D in the positive-slope region are due
mainly to intermicellar interactions and may be
accounted for by the DLVO-linear interaction theory
given above.

To obtain theoretical fits to our experimental
diffusivity results, we need to assign values to the
following parameters: the effective spherical radius
a; the micellar charge Q; the micellar aggregation
number N; and the Hamaker coefficient A. As stated
above, radius a is determined from D_0 and is not a
free parameter for these fits. The micellar charge
Q, however, is unknown. It is convenient to express
Q as the product αN, where α is the effective fractional ionization of the surfactant monomers that

reside in the micellar aggregate. A substantial fraction of the Br⁻ counterions must reside in the immediate vicinity of the micellar surface (that is, in the Stern layer) to be able to significantly screen the Coulombic repulsion between neighboring head groups and thereby yield a favorable free energy for micellization. Typically, α is believed to lie in the range 0.1-0.4 for most ionic surfactants; its determination by a variety of experimental methods has thus far proved to be quite unreliable (that is, to contain large uncertainties). Therefore, we treat α, rather than Q, as a free parameter in the theory. We see that α is determined relatively unambiguously by this theoretical fitting procedure.

To fix the micellar charge Q in the expression for $W_R(x)$ [Eqs. (10)-(12)], given a choice for α, we need to know the aggregation number N. A reasonable estimate [29] in the case of CTAB is $N = 120$ (corresponding to "minimum-sphere" micelles). To calculate $W_R(x)$, we also need to know the value of the solution ionic strength I, which affects the inverse screening length κ [Eq. (9)]. Both the added NaBr and the Br⁻ ions dissociated from the surfactant in the micelles (ignoring the small concentration of monomers well above the CMC) contribute to I: $I = 6 \times 10^{20}$ ([NaBr] + $\alpha/2$[CTAB])(cm^{-3}). The micellar contribution to I is completely negligible except at the largest CTAB concentration (0.05 M) and the lowest NaBr concentration 0.02-0.03 M). Hence κ is roughly constant for a given added salt concentration.

The remaining free parameter is the Hamaker coefficient A, which determines the strength of the van der Waals attractions. Since $W_A(x)$ diverges at $x = 0$, the integrals for K_t and K_h [Eqs. (7) and (8)] both require a lower cutoff $x = x_L$. We choose the value adopted by Corti and Degiorgio [11], which corresponds to a minimum separation of 2.5 Å, roughly equal to the Stern layer thickness, or $x_L = 0.04$. Parameters A and x_L are obviously coupled in the fitting procedure; different pairs of values can produce the same overall attractive contribution to K.

Effectively, we are left with the free parameters α and A. At the lowest values of [NaBr], the theoretical slope of D versus [CTAB] is very sensitive to α since the long-range repulsions are only slightly screened. At [NaBr] = 0.08 M, on the other hand, there is sufficient screening of the Coulombic repulsions that the fit is quite sensitive to the choice of A and relatively insensitive to α. For the best fits to all four D versus [CTAB] plots, we find $\alpha = 0.22$ and $A = 15k_BT$ ($T = 40°C$). The comparisons between theory and experiment are shown in Fig. 5b; clearly, the agreement is very good. We then test the fits at three additional temperatures, where all D versus [CTAB] lines are observed to systematically shift to smaller slopes with decreasing temperature. We keep the same values for parameters α and A but must change radius a somewhat at each temperature because of the observed shifts in D_0. Radius a is seen to increase monotonically with decreasing T: 26.8 Å at 55°C, 29.2 Å at 40°C, 32.2 Å at 25°C; and 34.9 Å at 15°C. The resulting comparisons of the theoretical fits with experiment are shown in Figs. 5a-d for 55°C, 40°C, 25°C, and 15°C, respectively [29]. The agreements are equally good for all temperatures (albeit with fewer positive-slope comparisons possible as T decreases).

To obtain the theoretical fits above, we arbitrarily fixed the micellar aggregation number for CTAB at $N = 120$. In fact, the quality of the fits turns out to be insensitive to the choice of N over a wide, physically relevant range (N = 100-200). The reason for this is that the linear correction coefficient K ($=K_t + K_h$) [Eqs. (7) and (8)] is nearly linear in N for a wide range of choices of α, A, and N. Since the micellar volume fraction is inversely proportional to N, $\phi = (6 \times 10^{20})(4/3)\pi a^3[(CTAB)/N]$, the product $(K_t + K_h)\phi$ in Eq. (4) is independent of N. Hence we cannot determine N (or, for that matter, Q) from this procedure.

The theoretically predicted slope of D versus [CTAB] at the lowest NaBr concentrations studied

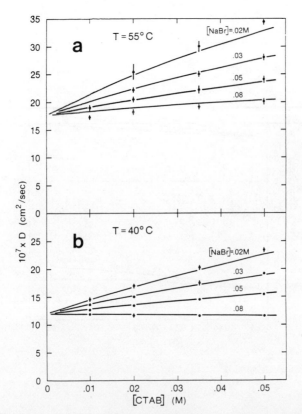

FIGURE 5a,b. D versus [CTAB]: theoretical fits versus experimental data, with the fitting parameters α = 0.22, N = 120, A = $15k_BT$ (T = 40°C, x_L = 0.04): (a) T = 55°C (R = 26.8 Å);   (b) T = 40°C (R = 29.2 Å).

([NaBr] < 0.03 M) depends almost entirely on the micellar charge and very little on the strength of the van der Waals attractions. Hence dynamic light scattering plus DLVO/linear-interaction theory provides a relatively unambiguous method for obtaining the micellar ionization fraction α at low ionic strengths. Of course, our simplified model assumes that α is approximately independent of NaBr and CTAB

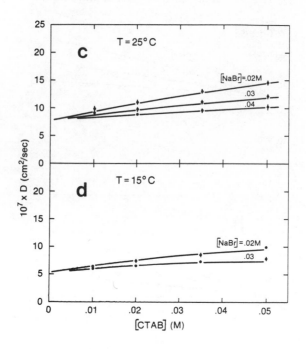

FIGURE 5c,d. D versus [CTAB]: theoretical fits versus experimental data, with the fitting parameters $\alpha = 0.22$, $N = 120$, $A = 15k_BT$ ($T = 40°C$, $x_L = 0.04$): (c) $T = 25°C$ ($R = 32.2$ Å); (d) $T = 15°C$ ($R = 34.9$ Å).

concentrations in the positive-slope region. Our value for CTAB, $\alpha = 0.22$, is consistent with other estimates in the absence of NaBr based on electrochemical techniques such as Br⁻ electrode and conductivity [31,32]. Using the recently reported data of Corti and Degiorgio for the system SDS + NaCl, we have obtained from our fitting procedure the fractional ionization $\alpha = 0.36$ at low NaCl concentrations ([NaCl] < 0.4 M at 25°C). The fact that α is substantially larger for SDS than for CTAB agrees with earlier findings with the use of other techniques.

Although we have confined our analysis to the positive-slope region of D versus [surfactant], where the net intermicellar interaction is clearly repulsive, the attractive potential $W_A(x)$ cannot be dispensed with. If we set $A = 0$ [that is, $W_A(x) = 0$] and adjust α to yield the best theoretical fit for the highest slope line, we again obtain $\alpha = 0.22$. However, the predicted D versus [CTAB] fits become progressively worse with increasing [NaBr]; the theoretical D values are too high at larger NaBr concentrations. Hence screening of the electrostatic repulsions by added NaBr cannot alone account for the observed decrease in slope of D versus [CTAB] with increasing [NaBr]. An additional attractive potential is needed, which we have represented by Eq. (13).

C. Fluorocarbon Emulsion

Because fluorocarbon oils are excellent solvents for oxygen [33,34], they are currently being investigated for use as artificial blood substitutes [35-38]. These oils (that is, the neat liquids) are insoluble in water; hence their successful use in intravascular applications requires that they be emulsified through the use of a nonionic surfactant. The resulting preparations vary greatly in stability, depending on the specific chemical structures of the perfluorinated compound and associated surfactant. Determination of the emulsion particle (droplet) size is important. Particles significantly larger than 0.1 μm tend to aggregate over time, leading to potential toxicity; those that are much smaller confer undesirable osmotic properties to the emulsion and also retard its elimination by the reticuloendothelial system.

522   D. F. Nicoli and R. B. Dorshow

A typical fluorocarbon emulsion currently under investigation [39] is designated "FC-47." It is prepared by mixing 10 ml of perfluorotributylamine (the oil) with 90 ml of 10% (w/v) Pluronic F68 (a nonionic surfactant). The latter compound is a long-chain polyoxyethylene polymer. This mixture does not spontaneously emulsify; rather, it must be sonicated to form a single-phase solution. The resulting preparation is not truly thermodynamically stable; given a sufficiently long period of time, it will phase separate.

The results of dynamic light scattering on FC-47 are summarized [39] in Fig. 6. The diffusivity D of the emulsion particles (presumably surfactant-coated oil droplets) is plotted against their volume fraction ϕ. The volume fraction of the starting emulsion was 0.19. This "stock" solution was subsequently diluted with water several times to yield a series of smaller volume fractions, decreasing ϕ to less than 0.005.

As seen in Fig. 6, D decreases markedly with increasing ϕ, which indicates the presence of significant interparticle attractions. The behavior of D versus ϕ can be divided into roughly two regions. For $\phi \lesssim 0.05$, the measured values of D fall nearly on a straight line of negative slope, which extrapolates to $D_0 = 4.07 \times 10^{-8}$ cm^2/sec at $\phi = 0$. For $\phi \gtrsim 0.05$, the diffusivity values also decrease with increasing ϕ, but with a smaller (negative) slope. We attribute this change in slope to the existence of multiple scattering at the higher solute concentrations; at these dilutions the solutions were noticeably cloudy and scattered light very strongly. Hence smaller values of ϕ were required to obtain a linear dependence of D with ϕ and a reliable value of D_0. From Eq. (5) we estimate the emulsion particle hydrodynamic radius to be $a = 0.063$ μm (63 nm), which agrees with measurements using electron microscopy.

Application of linear interaction theory to this system is relatively straightforward since the dispersed phase of this emulsion consists of uncharged

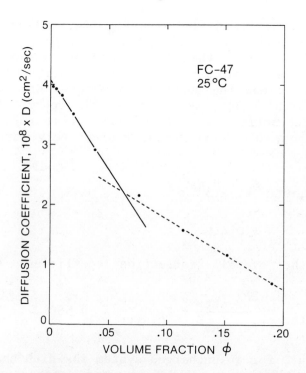

FIGURE 6.   Diffusion   coefficient D versus   volume
fraction ϕ for the fluorocarbon emulsion
FC-47 at $T = 25^{\circ}$C.

particles (neutral oil plus a nonionic surfactant).
Hence there are no electrostatic repulsions ($Q = 0$),
only hard-sphere repulsions and van der Waals attractions. We could fit the theory to the data in Fig. 6
and thereby obtain an estimate of the strength of the
Hamaker coefficient. However, such a determination
would not be very revealing from a physical point of
view since it would be difficult to relate the value
of A to the detailed physicochemical structure of an
emulsion droplet.

A more interesting way to characterize the
attractive perturbation was provided by Batchelor [7]
(recently reviewed by Finsy et al. [12]). He accounts

for the attractions by assuming that within a distance 2a from the center of a given particle there are more particles than would exist on average. This excess number of particles can be written as $\beta\phi$, where β is now the unknown parameter. From Batchelor's calculations, we find that the contribution of these attractions to the second virial coefficient and hydrodynamic correction, respectively, are given by

$$B_{2,attr} = -\frac{1}{2}\beta\frac{4}{3}\pi a^3 ,$$ (14a)

$$K_{h,attr} = 0.44\beta .$$ (14b)

Hence the overall correction coefficient becomes

$$K_{attr} = \frac{2B_{2,attr}}{(4/3)\pi a^3} + K_{h,attr} = -0.56\beta .$$ (15)

Thus, for an uncharged system in which there are only hard-sphere repulsions and attractions as described by parameter β above, we obtain, with Batchelor,

$$D = D_0[1 + (1.45 - 0.56\beta)\phi + \cdots]$$ (16)

[According to Felderhof [9], the hard-sphere constant is 1.56 (= 8 - 6.44).]

Applying this formalism to the D versus ϕ data of Fig. 6, we obtain $\beta = 15.7$. Hence, for example, at $\phi = 0.05$ there are on average 0.79 excess emulsion particles located within a range of 2a surrounding a given particle. This value for β is quantitatively similar to that found by Finsy et al. [12] for a neutral W/O microemulsion consisting of much smaller particles (a = 80 Å): $\beta = 13$. The challenge remains to try to calculate the strength of the attractive interaction for this and other systems from first principles.

REFERENCES

1. G. B. Benedek, in *Polarization, Matter and Radiation* (Jubilee volume in honor of Alfred Kastler), Universitaire de France, Paris, 1969.

2. B. Chu, *Laser Light Scattering*, Academic Press, New York, 1974.

3. B. J. Berne and R. Pecora, *Dynamic Light Scattering*, Wiley, New York, 1976.

4. T. Olson, M. J. Fournier, K. H. Langley, and N. C. Ford, J. Molec. Biol. *102*, 193 (1976).

5. A. Rohde and E. Sackmann, J. Phys. Chem. *84*, 1598 (1980).

6. C. Tanford, *Physical Chemistry of Macromolecules*, Wiley, New York, 1961.

7. G. K. Batchelor, J. Fluid Mech. *52*, 245 (1972).

8. G. K. Batchelor, J. Fluid Mech. *74*, 1 (1976).

9. B. U. Felderhof, J. Phys. A: Math. Nucl. Gen. *11*, 929 (1978).

10. M. Corti and V. Degiorgio, in *Light Scattering in Liquids and Macromolecular Solutions*, V. Degiorgio, M. Corti, and M. Giglio, Eds., Plenum Press, New York, 1980.

11. M. Corti and V. Degiorgio, J. Phys. Chem. *85*, 711 (1981).

12. R. Finsy, A. Devriese, and H. Lekkerkerker, J. Chem. Soc. Faraday II *76*, 767 (1980).

13. G. D. J. Phillies, J. Chem. Phys. *60*, 976 (1974).

14.   T. L. Hill, *An Introduction to Statistical Thermodynamics*, Addison-Wesley, Reading, Mass., 1960.

15.   M. M. Kops-Werkhoven and H. M. Fijnaut, J. Chem. Phys. *74*, 1618 (1981).

16.   E. J. W. Verwey and J. T. G. Overbeek, *Theory of the Stability of Lyophobic Colloids*, Elsevier, New York, 1948.

17.   P. C. Hiemenz, *Principles of Colloid and Surface Chemistry*, Dekker, New York, 1977.

18.   H. C. Hamaker, Physica *4*, 1058 (1937).

19.   P. Doherty and G. B. Benedek, J. Chem. Phys. *61*, 5426 (1974).

20.   R. Dorshow and D. F. Nicoli, J. Chem. Phys. *75*, 5853 (1981).

21.   F. Perrin, J. Phys. Radium *5*, 497 (1934); J. Phys. Radium 7, 1 (1936).

22.   R. Schor and E. W. Serrallach, J. Chem. Phys. *70*, 3012 (1979).

23.   M. J. Stephen, J. Chem. Phys. *55*, 3878 (1971).

24.   H. Wennerstrom and B. Lindman, Phys. Rep. *52*, 1 (1979).

25.   B. Lindman and H. Wennerstrom, Top. Curr. Chem. *87*, 1 (1980).

26.   C. Tanford, *The Hydrophobic Effect: Formation of Micelles and Biological Membranes*, Wiley, New York, 1980.

27. N. A. Mazer, M. C. Carey, and G. B. Benedek, in *Micellization, Solubilization and Microemulsions*, Vol. I, K. L. Mittal, Ed., Plenum Press, New York, 1977.

28. P. J. Missel, N. A. Mazer, G. B. Benedek, and C. Y. Young, J. Phys. Chem. *84*, 1044 (1980).

29. R. Dorshow, J. Briggs, C. A. Bunton, and D. F. Nicoli, J. Phys. Chem. *86*, 2388 (1982).

30. J. Briggs, R. B Dorshow, C. A. Bunton, and D. F. Nicoli, J. Chem. Phys. *76*, 775 (1982).

31. R. Zana, J. Colloid Interface Sci. *78*, 330 (1980).

32. J. W. Larsen and L. B. Tepley, J. Colloid Interface Sci. *49*, 113 (1974).

33. L. C. Clark, Jr. and F. Gollan, Science *152*, 1755 (1966).

34. E. P. Wesseler, R. Iltis, and L. C. Clark, Jr., Fluorine Chem. *9*, 137 (1977).

35. T. Suyama, in *Fifth International Conference on Red Cell Metabolism and Function*, Alan R. Liss, New York, 1980.

36. J. G. Reiss and M. LeBlanc, Angewandte Chemie *17*, 621 (1978).

37. D. H. Glogar, R. A. Kloner, J. Muller, L. W. V. DeBoer, E. Braunwald, and L. C. Clark, Jr., Science *211*, 1439 (1981).

38. T. H. Maugh, Science *206*, 205 (1979).

39. R. Ciccolella, L. C. Clark, Jr., and D. F. Nicoli, J. Colloid Interface Sci. (in press).

CHAPTER 20

Effects of Critical-Type
Behavior on Droplet Sizing
in Microemulsions

R. B. DORSHOW and D. F. NICOLI
University of California at Santa Barbara

CONTENTS

I.   INTRODUCTION

Two immiscible fluids, such as water and oil, will generally form a dispersion under agitation; the

529

relative quantities of each component will determine which is the dispersed phase and which is the continuous one. If the dispersion persists over an extended period of time (seconds to years), it is called an *emulsion*; however, emulsions are unstable and will eventually phase separate. The dispersed phase usually consists of droplets of radius greater than 1000 Å, resulting typically in strong light scattering. The large particle sizes that occur in emulsions are the result of the high interfacial tension between the two components, which favors a small oil-water contact area (droplet surface area).

The interfacial tension, and hence the particle size, can be reduced by the addition of a surfactant to the dispersion. A surfactant, or detergent, molecule contains a hydrophobic moiety (usually a long hydrocarbon chain) and a hydrophilic moiety (typically a polar head group) that can be cationic, anionic, zwitterionic, or nonionic. The surfactant will locate itself at the oil-water interface, with its head group in the water and its hydrocarbon tail in the oil, which lowers the water-oil interfacial tension and thus permits the formation of smaller particles. Such an oil-water-surfactant system, which consists of a single-phase solution, is known as a *microemulsion* [1,2].

In general, microemulsions need a fourth component to form thermodynamically stable solutions: a cosurfactant (usually a medium-chain alcohol). The function of the cosurfactant is not entirely clear; like the surfactant, it is believed to be located at the oil-water interface. Microemulsions are stable, transparent solutions with particle (droplet) sizes in the range 50-1000 Å. They exist in two general categories, depending on the relative proportions of oil and water: (1) water in oil (W/O), where the water is dispersed in the oil (the continuous phase) and (2) oil in water (O/W), where the oil is dispersed in the water. A plausible physical representation of an O/W microemulsion is shown in Fig. 1. Microemulsion systems are found in a variety

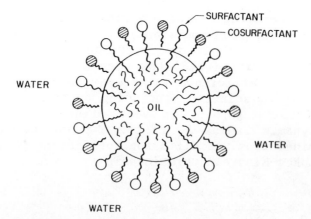

FIGURE 1. Idealized physical model of a droplet
particle in an O/W microemulsion.

of commercial products, such as polishes and waxes,
dry cleaning fluids, cosmetics, and lotions. A
potentially very important application of microemulsions exists in the area of tertiary oil recovery,
which attempts to recover residual petroleum that
cannot be obtained by secondary recovery (that is, by
water flooding the oil field at high pressures) [3].
One of several techniques employed is to flood the
oil field with solutions of micelles or microemulsions. Knowledge of the temperature and pressure
dependence of these systems is of prime importance in
determining the most efficient use of surfactants in
this application.
    Microemulsions that consist of an oxygen-
solubilizing fluorocarbon and a nonionic surfactant
of low toxicity are currently being studied for use
as artificial blood substitutes [4]. Here knowledge
of the emulsion droplet size is crucial; too large a
particle size may result in blockage of the capillary
bed, whereas too small a droplet size results in
retention of the fluorocarbon compounds by the
body for unreasonably long times. Hence there is a

critical need for reliable sizing of the particles in these fluorocarbon emulsions.

In recent years the technique of quasi-elastic light scattering (QELS) has been used to measure the diffusion coefficients of a wide variety of macromolecules in solution, including proteins, polymers, and micelles [5,6]. In very dilute solutions the hydrodynamic radius R_h of the macromolecules can be determined from the diffusion coefficient D by use of the Stokes-Einstein relationship

$$R_h = \frac{k_B T}{6\pi\eta D} , \tag{1}$$

where k_B is Boltzmann's constant, η is the solvent viscosity, and T is the absolute temperature.

In concentrated solutions, however, the measured diffusion coefficients can be greatly affected by interparticle interactions [7], thus rendering the deduced R_h values [from Eq. (1)] inaccurate. Nevertheless, reliable determinations of particle size can be made in semidilute solutions by correcting for the influence of these interactions on the diffusivity. At low solute concentrations the diffusivity would generally be expected to be linear in the volume fraction ϕ of the scattering particles:

$$D_\phi = D_0(1 + K\phi) , \tag{2}$$

where D_ϕ is the diffusivity at volume fraction ϕ and D_0 is the infinite-dilution value (that is, in the limit $\phi = 0$). Thus, from D_ϕ measurements at several volume fractions an extrapolation to $\phi = 0$ can be obtained, thereby yielding D_0 and a reliable value of R_h for the particles [Eq. (1) with $D = D_0$]. Coefficient K depends on the nature of the interparticle interactions, which in general include hard-sphere exclusion, electrostatic repulsions, van der Waals attractions, and hydrodynamic effects. This procedure has been extensively applied to polystyrene latexes, silica particles, charged proteins, and micellar systems at low ionic strengths [8,9].

Application of the extrapolation procedure described above is not straightforward in the case of microemulsions since the latter are multicomponent, self-assembled systems. Consequently, alteration of the sample solution may change not only the volume fraction of the emulsion droplets, but also their intrinsic size. Dilution must be accomplished by increasing the percentage of the continuous phase only while not changing the components of the dispersed (droplet) phase. (This problem is particularly vexing when the system contains a cosurfactant, as it is often ambiguous how the latter is divided between the dispersed and continuous phases.) Graciaa et al. [10] performed a straightforward dilution in studying a W/O microemulsion by light scattering. They found that the measured D_ϕ initially decreased, reached a minimum value, and then increased with increasing ϕ. They concluded that D_ϕ was perturbed by particle interactions and simply extrapolated their results to $\phi = 0$ to estimate the microemulsion droplet radius R_h. Finsy et al. [11] also studied a W/O microemulsion. They observed a linear decrease in the diffusion coefficient with increasing volume fraction and modeled the interactions, using hard-sphere repulsion, van der Waals attractions, and hydrodynamic effects.

Recently Zulauf and Eicke [12] and Gulari et al. [13] reported measurements on microemulsion systems that exhibit a liquid-liquid phase transition, going from a single solution to two separated liquids with increasing temperature. Both studies note that as the phase separation temperature is approached, the R_h values (uncorrected for interactions) increase dramatically; they both note the similarity of this behavior to critical phenomena previously observed in binary fluid mixtures. In this chapter we review the results of our light-scattering studies of an O/W microemulsion. We find that this microemulsion exhibits critical-like behavior [14,15].

II.   MATERIALS AND METHODS

A.   Microemulsion Chemistry

The microemulsion under investigation consists of an oil, *n*-octane (Aldrich, 99%); a cosurfactant, 1-butanol (Aldrich, spectrophotometric grade); a surfactant, cetyltrimethylammonium bromide (CTAB) (M. C. B.); sodium bromide (Mallinkrodt, A. R.); and water. The sodium bromide was dried before use and the water was deionized, distilled, and filtered through a Milli-Q (Millipore) water purification system with a 0.22 μm output filter. The CTAB was purified as reported [16].

In general, microemulsions form a single, stable solution over only a relatively narrow range of component concentrations. We fixed our system at 2.30% CTAB, 2.30% octane, and 4.60% butanol (percentages by weight); the sodium bromide concentration was initially varied in the range 1.3-1.7%, with the remaining weight percentage water.

B.   QELS Apparatus

Dynamic light scattering was performed on 0.5 ml sample solutions contained in 6 mm diameter, flame-sealed cylindrical glass tubes. The sample tubes were centrifuged at 5000 *g* for 10 min immediately prior to use to sediment dirt-impurity particles. The cylindrical glass sample tube was fixed at the center of a toluene-filled fluorimeter cuvette to provide index matching against stray light reflections. The cuvette was housed in a black-anodized aluminum cell block, whose temperature was regulated to $\pm$ 0.05°C by a Peltier thermoelectric element. The light source was a Spectra Physics Model 164 argon-ion laser operating at 5145 Å (power ~ 100 mW). Light scattered at 90° was collected from approximately one coherence area and imaged onto the slit of an EMI type 9789 photomultiplier tube (PMT). A Nicomp Model 6864 computing autocorrelator was used to calculate and display the diffusion coefficient and associated derived parameters from second- and third-order least-squares fits (method of cumulants) [17] to the intensity autocorrelation function.

C. Angular Dissymmetry Apparatus

A multipurpose light-scattering apparatus was used to measure the angular dissymmetry $d(\theta)$ ($\equiv I_\theta / I_{\pi-\theta}$) and diffusivity D_q as a function of angle θ (or wavevector q) and temperature T (Fig. 2). The output from a 5 mW He-Ne laser (Spectra Physics Model 120) is split into two beams of roughly equal intensities that are directed along two paths that intersect the sample volume from opposite directions. With a rotatable detector arm placed at angle θ with respect to beam 1, scattered intensity I_θ is collected from beam 1 and $I_{\pi-\theta}$ is collected from beam 2. A timing circuit controls a shutter mechanism that allows only one beam at a time to illuminate the sample, with a frequency of approximately 1 Hz. Use of a fixed detector and opposing beams, with measurements made at two supplementary angles, eliminates the necessity of solid-angle corrections. A microprocessor-controlled photon counting system measures the intensities at 0.5-1 sec intervals and displays the ratio [dissymmetry $d(\theta)$] plus the average and standard deviation. The detector arm may be rotated in the range $60° \lesssim \theta \lesssim 120°$. The temperature control system of the sample holder block is the same as that described in Section B. A rectangular glass cuvette (1 × 2 cm) is used for the scattering cell. The 2 cm width helps to minimize the effects of stray reflections at the detected volume due to the wide separation of the beam entrance and exit spots at the cuvette walls. A Teflon cap with Teflon inlet and outlet tubes is used to seal the cuvette, thus yielding a closed system that can be conveniently cleaned and filled.

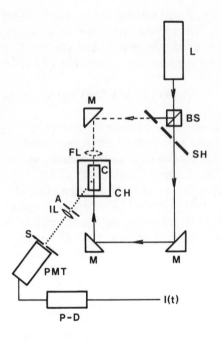

FIGURE 2.   Schematic diagram of the angular dissymmetry apparatus: L, laser (Spectra
Physics Model 120 He-Ne); BS, beam splitter; SH, reciprocating shutter; M, front-
surface mirror; FL, focusing lens (used
for dynamic light scattering, but not
angular dissymmetry); C, sample cell; CH,
cell holder; A, aperture; IL, imaging
lens; S, slit; PMT, photomultiplier tube
(RCA 7265); P-D, preamplifier-discrimina-
tor; $I(t)$, intensity output (TTL-logic
pulses).

Diffusivity measurements are made by blocking
beam 2 and using only beam 1 to illuminate the
sample. A focusing lens is inserted in the beam path
and light from approximately one coherence area is
collected at the photomultiplier tube (RCA 7265). A
turbidity correction is obtained by measuring the

laser beam intensity before and after it enters the cuvette by use of a photodiode. This information is used to correct the scattered intensities measured at 90°. Corrections to the measured dissymmetries were made to account for reflections of the incident beam at the sample-glass-air interfaces (effectively ~ 4% of the incident light impinges on the scattering volume from the opposite direction). Small turbidity corrections due to differences in path lengths of the opposing beams were also included in the data analysis. Finally, since the beam splitter did not yield two beams of equal intensity, we normalized all dissymmetries to the value at 90°, $d(\pi/2)$. The system was tested by using polystyrene latex spheres (Dow Diagnostics) of nominal radius 455 ± 29 Å. Angular dissymmetry and diffusion coefficient measurements were made over a wide range of angles. The former values were analyzed according to Rayleigh-Gans theory, yielding a radius of ~ 440 Å. The diffusion coefficient measurements were found to be angularly independent, with a D corresponding to R_h ~ 455 Å.

III. RESULTS

A. Diffusion Coefficients

The diffusion coefficients D measured at 90° scattering angle as a function of temperature for microemulsions containing 1.3, 1.5, 1.6, and 1.7% sodium bromide are shown in Fig. 3. We observe a minimum in D in the vicinity of 30-40°C that deepens with increasing salt concentration. A more suggestive result is obtained by plotting the apparent hydrodynamic radius R_h^{app} obtained from Eq. (1), thereby removing the temperature dependence of the solvent viscosity. The resulting curves are shown in Fig. 4. The striking feature of Fig. 4 is that R_h^{app} versus T for each salt concentration is sharply peaked at approximately 35°C. Far from the peaks

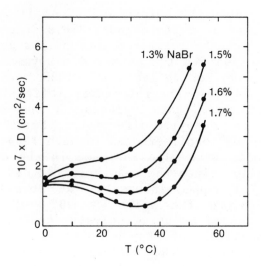

FIGURE 3.   Diffusion coefficient D versus temperature T for four microemulsions that contain differing NaBr concentrations (1.3-1.7%). The remaining constituents are (wt.%): 2.30% CTAB, 2.30% octane, and 4.60% butanol, with the remaining weight percentage water.

(for example, $\sim 0°C$), all the curves appear to converge to the same R_h^{app} value, ~ 70 Å.

The symmetrical peak structure of R_h^{app} versus T seen in Fig. 4 is reminiscent of critical-type behavior, in which the path being traversed (fixed [NaBr], variable T) brings the system close to a phase boundary. According to this interpretation, the higher the NaBr concentration, the closer the system is to phase separation. In that case, we expect a critical "slowing down" of the density fluctuations of the microemulsion near its critical point as the mean correlation length of the fluctuations diverges [18]. The R_h^{app} versus T behavior shown in Fig. 4 would thus not represent a changing hydrodynamic radius of an individual microemulsion

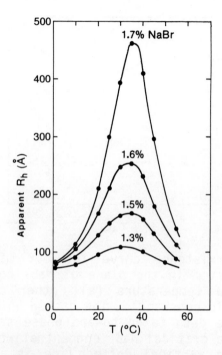

FIGURE 4.   Apparent   hydrodynamic   radius R_h (Å)
(obtained from D by use of the Stokes-
Einstein relationship) versus temperature
T for the four microemulsions whose diffusivities are plotted in Fig. 3.

droplet   radius   per   se;   rather,   it   represents   a
temperature-dependent mean *correlation* radius for the
system.
      Qualitative support for this interpretation is
shown in Fig. 5.   Here we have mapped out a very
small portion of the coexistence curve of the microemulsion, that which lies in the plane of NaBr concentration and temperature with all other components
held   constant.   For   this   multicomponent   system,
Fig. 5 obviously represents merely a small part of a
complex phase diagram. We obtained Fig. 5 by measuring the scattered intensity at 90° as a function of

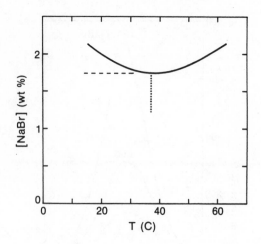

FIGURE 5.   Coexistence curve for our microemulsion
system in the plane of NaBr concentration
and temperature (all other constituents
held constant). The dashed line indicates
one path taken toward phase separation at
the critical NaBr concentration ($\cong 1.77\%$);
the vertical dotted line is another path
at the critical temperature T_c ($\cong 37.5°C$),
analogous to the critical isochore in
binary fluid mixtures.

both increasing and decreasing T for NaBr concentrations of 1.8, 1.9, and 2.0%; the temperatures at
which the intensity diverged (the solution became
opaque) located the coexistence curve. Its minimum
is located at approximately [NaBr] = 1.77% and T =
37°C. From the shape and location of the coexistence
curve, we can tentatively conclude that the maxima in
R_h^{app} are the result of the proximity of the system to
the phase separation boundary. It is important,
however, to note that the coexistence curve of Fig. 5
differs from those normally found in critical systems
(for example, binary fluid mixtures) in that it
possesses an extremum with respect to chemical composition (that is, NaBr concentration) rather than

temperature.   In effect, it has been rotated by 90°
with respect to the axes of composition and temperature.

From the behavior of the diffusivity when the
system is furthest from phase separation, we can
estimate the intrinsic microemulsion droplet size.
We observe (Fig. 4) that R_h^{app} is approximately independent of NaBr concentration (1.3-1.7%) at 0°C. It
is reasonable to assume that the Coulombic interparticle interactions are effectively screened at
these relatively high NaBr concentrations. If we
assume that the attractive interactions between
droplets can be ignored to first order, we are left
with only hard-sphere repulsions. Hence the linear
interaction coefficient K in Eq. (2), which is the
sum of the thermodynamic correction (+8) and the
hydrodynamic correction [7] (-6.44 according to
Felderhof [19] and -6.55 according to Batchelor
[20]), is approximately equal to 1.5. Since the
volume fraction ϕ of the microemulsion dispersed
phase is approximately 0.10 (depending on the location of the cosurfactant), we obtain roughly a 15%
increase in the measured diffusivity over the "bare"
value due to the excluded volume correction. Our
estimate of the true droplet hydrodynamic radius
[using Eq. (1) with $D = D_0$] is thus 75-80 Å.

This estimate is in good agreement with the
stoichiometry of the system. We model the microemulsion droplet as a sphere of radius R that contains all the octane in a smaller spherical core of
radius R_0; the latter is surrounded by an outer layer
of surfactant, of thickness approximately equal to
the length of an extended surfactant molecule, ~ 25 Å
in the case of CTAB. Thus for an overall droplet
radius R of 75 Å, we arrive at a ratio of surfactant
layer volume to oil core volume, V_{layer}/V_{core}, equal
to $(R/R_0)^3 - 1$, or ~ 2.4; for $R = 80$ Å, this ratio is
2.0. We are uncertain as to how the butanol dis-
tributes itself between the core, the surfactant
layer, and the continuous aqueous phase. Assuming
that it resides totally in the outer layer, we arrive

at $V_{layer}/V_{core} \sim 3$ [$\cong$ ([CTAB] + [butanol])/[octane] = $(2.3\% + 4.6\%)/2.3\%$]. If the butanol is divided equally between the surfactant layer and the continuous aqueous phase, we arrive at a ratio of 2. Considering the simplicity of this treatment, the agreement is good. These rough calculations have additional value — they virtually exclude the possibility that the dramatic increases observed in R_h^{app} (Fig. 4) correspond to actual growth of the intrinsic droplet particles. That is, for such large values of droplet radius, the ratio V_{layer}/V_{core} is much too small to be reconciled with the relative concentrations of oil, surfactant, and butanol available in the system.

B.   Critical-Type Behavior

In the case of binary fluid critical systems [21,22], if we assume an Ornstein-Zernike density correlation function for the density fluctuations, we obtain for the scattered intensity I_θ at scattering angle θ

$$I_\theta = A \; \frac{T\kappa_T}{1 + q_\theta^2 \xi^2} \; , \tag{3}$$

where A is an instrumental constant, κ_T is the isothermal compressibility, ξ is the correlation length, and q_θ is the scattering wavevector [$q_\theta = (4\pi n/\lambda)\sin(\theta/2)$, where n is the index of refraction of the medium and λ is the incident wavelength]. Correlation length ξ is a measure of the characteristic range of the spatially correlated regions. As the critical temperature T_c is approached, ξ increases dramatically, diverging at $T = T_c$. Similar behavior is seen for I_θ and thus for κ_T.

The typical coexistence curve for gases and binary fluids is described by the power law;

$$\frac{\rho - \rho_c}{\rho_c} \propto t^\beta , \qquad (4)$$

where the reduced temperature $t \equiv (T - T_c)/T_c$ and ρ_c is the critical concentration. Both ξ and κ_T obey a power law dependence in the reduced temperature and, when measured along the so-called critical isochore $(\rho = \rho_c)$, are given by

$$\xi(t) = \xi_0 t^{-\nu} , \qquad (5)$$

$$\kappa_T(t) = \kappa_0 t^{-\gamma} . \qquad (6)$$

These exponents obey the scaling law $\gamma = 2\nu$. Typical values of the critical exponents for binary fluid mixtures are $\nu \cong 0.63$, $\gamma \cong 1.22$, and $\beta \cong 0.35$.

The diffusivity in simple binary critical systems is found to be wavevector dependent as the critical temperature is approached. According to the theory of Kawasaki [23,24]:

$$D(q) = D(q = 0) \frac{H(x)}{x^2} , \qquad (7a)$$

where

$$H(x) = \frac{3}{4}[1 + x^2 + (x^3 - x^{-1})\tan^{-1} x] \quad (x = q\xi) . \qquad (7b)$$

To first order, $D(q = 0)$ and ξ obey a Stokes-Einstein relationship; therefore $D(q = 0) \propto t^\nu$.

Given this background, we proceed to examine the experimental results for our system. First, we investigated a microemulsion that contained the critical NaBr concentration, $\sim 1.77\%$. That is, we chose a path of constant NaBr concentration, which just intersects the coexistence curve at its extremum (at $T \sim 37°C$). This path is indicated by the horizontal dashed line in Fig. 5. We again stress that this path does not correspond to the usual critical isochore because the coexistence curve in Fig. 5 is rotated by 90° in the concentration-temperature plane with respect to conventional binary fluid systems. As we vary the temperature, we move along a path that is tangential, rather than normal, to the minimum in the coexistence curve.

Intensity angular dissymmetry data were taken over an angular range of 68-119° and a temperature range of 5-32°C. We assume that the Ornstein-Zernike expression [Eq. (3)] for the scattered intensity can be applied to our system. In that case we can express the dissymmetry $d(\theta)$ as

$$d(\theta) = \frac{I_\theta}{I_{\pi-\theta}} = \frac{1 + q_{\pi-\theta}^2 \xi^2}{1 + q_\theta^2 \xi^2} . \tag{8}$$

From a least-squares fit of Eq. (8) to the $d(\theta)$ results, we determined the best value of ξ at each temperature. In Fig. 6 we show the resulting fits along with the corrected data. [We note that for small $q\xi$, $d(\theta)$ is linear in cos θ.] The ξ values are listed in Table 1. Figure 7 illustrates the divergence of the correlation length as we approach phase separation.

From the scattered intensities at 90° (corrected for turbidity) and the ξ values derived above, we determined the "isothermal compressibility," given by Eq. (3): $\kappa_T \propto (I_{\theta=\pi/2})(1+q_{\pi/2}^2 \xi^2)/T$. We have used quotation marks to emphasize that ours is a multi-component system. In Fig. 8 we see that κ_T diverges as we approach the coexistence curve along the path of constant [NaBr].

Finally, QELS measurements were performed to determine the diffusion coefficient $D(q)$ as a function of scattering wavevector q. Below approximately 27°C (that is, relatively far from phase separation), $D(q)$ exhibited no appreciable wavevector dependence. We confined our measurements to temperatures below 33°C because of the onset of significant turbidity

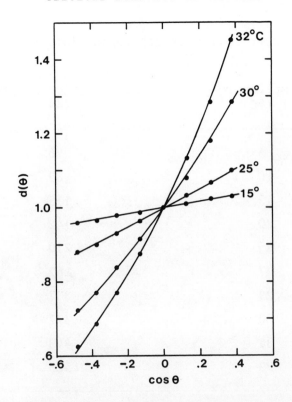

FIGURE 6.   Intensity angular dissymmetry $d(\theta)$ versus cos θ. The data points are plotted along with the theoretical curves obtained by using the values ξ derived from Eq. (8).

and multiple scattering at temperatures closer to phase separation. We analyzed our results by using Kawasaki's theory [Eq. (7)], even though our multi-component microemulsion may not be isomorphic to a simple binary fluid system, for which Eq. (7) has been shown to be valid. The values of D_0 obtained from a least-squares fit to Eq. (7) (using the ξ values obtained from angular dissymmetry) are listed in Table 2.

TABLE 1. Correlation Length versus Temperature

Temperature T (°C)	Correlation Length ξ (Å)
5	76
10	89
15	115
20	142
25	210
27	262
29	337
30	377
31	443
32	523

We have modeled the temperature dependence of the correlation length by use of the power law expression that successfully describes pure fluids and simple binary mixtures [Eq. (5)]:

$$\xi(t) = \xi_0 t^{-\tilde{\nu}} . \tag{9}$$

For the "critical exponent," we use the symbol $\tilde{\nu}$ rather than ν to emphasize that the thermodynamic path taken toward the coexistence curve does not coincide with the conventional critical isochore. A least-squares fit to Eq. (9), using the data in Table 1, yields $T_c = 310.7 \pm 0.5$ K, $\xi_0 = 6.0 \pm 0.6$ Å, and $\tilde{\nu} = 1.1 \pm 0.1$. The data points and best fit are shown in Fig. 9.

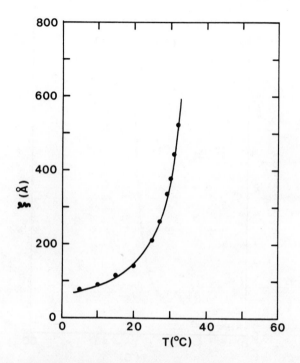

FIGURE 7. Experimental values of the correlation length ξ [obtained from $d(\theta)$, using Eq. (8)] versus temperature T. The smooth curve is a guide to the eye only.

Similarly, we model the temperature dependence of the "isothermal compressibility" κ_T to an equation of the same form as Eq. (6):

$$\kappa_T = \kappa_0 t^{-\tilde{\gamma}}, \qquad (10)$$

where $\tilde{\gamma}$ denotes the "critical exponent" obtained for our constant-[NaBr] path. A least-squares fit to Eq. (10) yields $T_c = 310.5 \pm 0.5$ K and $\tilde{\gamma} = 2.2 \pm 0.1$. The data points and corresponding best-fit curve are shown in Fig. 10.

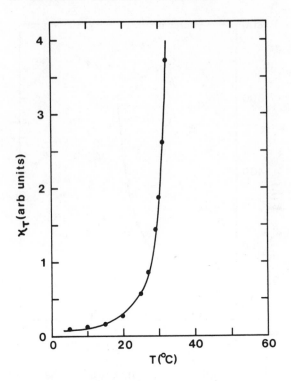

FIGURE 8.   Experimental values of the isothermal compressibility κ_T [obtained from $I_{\theta=\pi/2}$ and the ξ values from Fig. 7, using Eq. (3)] versus temperature T. The smooth curve is a guide to the eye only.

We emphasize that our multicomponent microemulsion differs markedly from simpler critical systems in that ξ changes significantly at temperatures normally considered to be far from T_c. Fully 10°C below T_c, the solution exhibits strong scattering and substantial turbidity, presumably caused by the increasing spatial correlation of relatively large particles (surfactant-coated oil droplets). Because of the turbidity and multiple scattering effects, we were unable to make reliable measurements of $d(\theta)$

TABLE 2.   Diffusivity versus Temperature

Temperature T (°C)	Diffusion Coefficient (Extrapolated to $q = 0$) $D(q = 0) \times 10^8$ (cm²/sec)
27	4.0
28	3.6
29	3.0
30	2.7
31	2.3
32	1.9
33	1.5

above 32°C. Thus our smallest values for the reduced temperature t are more than an order of magnitude larger than those usually reported for fluid critical systems; the relatively large uncertainties reported for T_c, ξ_0, and $\tilde{\nu}$ merely reflect this limitation. Nevertheless, the T_c values obtained from independent least-squares fits to the correlation length and isothermal compressibility data are in agreement, to within experimental uncertainites.

Several results are of interest. First, the exponent $\tilde{\nu}$ is almost twice the value of 0.63 typically found for the exponent ν for binary fluid critical systems. However, a direct comparison of $\tilde{\nu}$ and ν is not meaningful since the approach to the coexistence curve is qualitatively different in the two cases. We see below that we obtain a much smaller value if we adopt a path that is analogous to the usual critical isochore. Nevertheless, we find for our path that, to within experimental uncertainties,

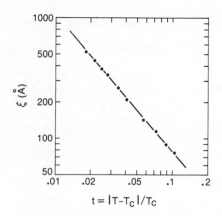

FIGURE 9.   Correlation length $\xi(\text{Å})$ versus reduced
temperature t. The solid dots represent
the experimental data and the straight
line the least-squares fit to power-law
behavior [Eq. (9)]. The path taken was
the dashed line ([NaBr] = 1.77%) in Fig.
5.

the "scaling law" $\tilde{\gamma} = 2\tilde{\nu}$ is satisfied, analogous to
the relationship that is obtained on a critical
isochore ($\gamma = 2\nu$).

Second, our prefactor ξ_0 is substantially larger
than that found for pure fluids or binary mixtures.
Although its size probably depends on the path taken
to criticality, the larger value is expected; the
basic structural unit in the microemulsion is a
droplet in the size range 75-100 Å, which is much
larger than the individual molecules that make up
conventional fluid systems. Hence far from phase
separation our system scatters much more light than
does a typical binary mixture.

The temperature dependence of the diffusion
coefficient in the limit $q = 0$ was also assumed to
follow a power law behavior:

$$D(q = 0) = D't^{\tilde{\nu}'} \ .$$

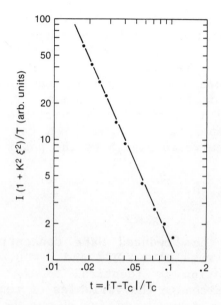

FIGURE 10.   Isothermal compressibility κ_T versus reduced temperature t. The solid dots represent the experimental data and the straight line the least-squares fit to power-law behavior [Eq. (10)]. The path taken was the dashed line ([NaBr] = 1.77%) in Fig. 5.

However, the data could not be analyzed as rigorously as in the case of the correlation length and isothermal compressibility; the small dependence of $D(q)$ on the scattering wavevector q resulted in relatively large errors in the extrapolated $D(q)$ values. Consequently, we fixed the critical temperature at T_c = 310.5 K and, by graphical analysis of the data (Table 2), we obtained the exponent $\tilde{\nu}' = 1.1 \pm 0.1$. Hence the dynamic "critical exponent" $\tilde{\nu}'$ is in agreement with its static counterpart $\tilde{\nu}$.

We now examine the critical behavior of our microemulsion along a path that more nearly resembles the critical isochore in a binary critical system.

As pointed out earlier, the coexistence curve in Fig. 5 is rotated by 90° with respect to its usual orientation in the plane of composition ([NaBr]) and temperature. The conventional critical isochore defines a path that intersects the phase boundary along its normal at the extremum. Hence it would appear useful to transpose the variables of composition and temperature. Indeed, we can describe the coexistence curve in Fig. 5 by the power law:

$$\frac{T - T_c}{T_c} \propto \varepsilon^\beta \tag{11}$$

where ε is the "reduced NaBr concentration," $\varepsilon \equiv (c - c^*)/c^*$, with $c^* \cong 1.77\%$ and $\beta = 0.42$. The form of this equation is identical to that of Eq. (4), provided we exchange the variables of temperature and concentration. Our value for β can be compared to the value found for typical binary systems, $\beta \cong 0.35$. By contrast, mean field theory predicts $\beta = 0.5$.

The path that is analogous to the critical isochore, therefore, which we might call the *critical isotherm*, is shown as a vertical dotted line in Fig. 5. Holding T constant at approximately 37.5°C, we vary [NaBr] (= c) and approach the critical concentration c^* ($\cong 1.77\%$) from below. Although detailed measurements of ξ and κ_T have not yet been made along this path, we can nevertheless gain useful insight into the critical behavior along this path by examining the diffusion coefficient data shown in Fig. 3. We model the D values at $T = T_c \cong 37.5$°C for the four NaBr concentrations studied (1.3, 1.5, 1.6, and 1.7%) by the power law: $D \propto \varepsilon^{\nu''}$. The data and best-fit line are shown in Fig. 11.

By this means we obtain a preliminary estimate of the critical exponent ν'' along the critical isotherm: $\nu'' = 0.76$. This value compares well with the value of 0.73 ± 0.04 previously obtained near the plait point of a ternary liquid mixture [25]. (The exponent of that ternary system was related to the

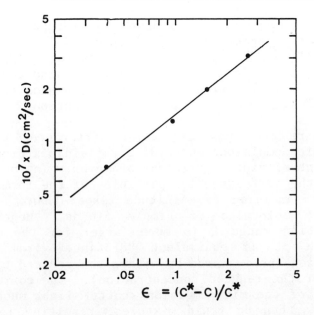

FIGURE 11.   Diffusion coefficient D (cm^2/sec) versus reduced NaBr concentration ($c^* = 1.77\%$). The solid dots represent the experimental data and the straight line, the least-squares fit to $D \sim \varepsilon^\nu$ power-law behavior. The path taken was the dotted line ($T = 37.5°C$) in Fig. 5, analogous to the critical isochore in binary fluid critical systems.

critical exponent of a binary fluid mixture by renormalization.) Hence, given an exchange of the roles of temperature and concentration, our microemulsion bears both a qualitative and quantitative resemblance to simple fluid critical systems.

From our analysis of both static and dynamic light-scattering data obtained for a family of multi-component microemulsions, we conclude that this system exhibits critical behavior that qualitatively resembles that found for simple fluid systems such as binary mixtures.   According to this interpretation, as the microemulsion approaches phase separation the intrinsic droplet particles (presumably oil surrounded by surfactant) become increasingly spatially correlated, while their basic size remains constant. There is a correlation range ξ that measures the mean size of regions of cooperative diffusion of these particles.

There are, however, major differences between these microemulsions and ordinary fluid mixtures from the point of view of critical behavior. In the first place, the self-assembled particles of the former are at least an order of magnitude larger than the constituent molecules of simple fluids. Hence the correlation range ξ is much larger for the microemulsion at an equivalent "distance" from phase separation (as measured by either the reduced temperature or the reduced concentration). The correlated regions of the microemulsion scatter light much more strongly than do fluid mixtures, resulting in substantial turbidity and multiple scattering for the former when it is nominally far from phase separation. Second, we find that the coexistence curve (in that limited part of the phase diagram that we have investigated) for our microemulsion is qualitatively different from that of conventional critical systems.

In conclusion, it appears that in general the problem of determining the intrinsic particle size of a multicomponent O/W microemulsion is nontrivial. If the system is located at a point in the phase diagram that is far from a phase boundary, the diffusion coefficient may require corrections for interparticle interactions if it is to yield an accurate Stokes-Einstein hydrodynamic radius. If one is not far from phase separation, the diffusivity may change substantially with temperature and/or composition as a result of changes in the spatial correlation range, rather than alterations in the intrinsic droplet size per se.

REFERENCES

1. L. M. Prince, Ed., *Microemulsions, Theory and Practice*, Academic Press, New York, 1977.

2. K. L. Mittal, Ed., *Micellization, Solubilization and Microemulsions*, Plenum Press, New York, 1977.

3. D. O. Shah and R. S. Schechter, Eds., *Improved Oil Recovery by Surfactant and Polymer Flooding*, Academic Press, New York, 1977.

4. R. Ciccolella, D. F. Nicoli, and L. C. Clark, Jr., J. Colloid Interface Sci. (submitted).

5. B. Chu, *Laser Light Scattering*, Academic Press, New York, 1974.

6. B. J. Berne and R. Pecora, *Dynamic Light Scattering with Applications to Chemistry, Biology and Physics*, Wiley, New York, 1976.

7. D. F. Nicoli and R. B. Dorshow, Chapter 19 in this volume.

8. M. Corti and V. Degiorgio, J. Phys. Chem. *85*, 711 (1981).

9. R. B. Dorshow, J. Briggs, C. A. Bunton, and D. F. Nicoli, J. Phys. Chem. *86*, 2388 (1982).

10. A. Graciaa, J. LaChaise, P. Chabrat, L. Letamendia, J. Rouch, C. Vaucamps, M. Bourrel, and C. Chambu, J. Phys. (Paris) *38*, L-253 (1977).

11. R. Finsy, A. Devriese, and H. Lekkerkerker, J. Chem. Soc. Faraday II 76, 767 (1980).

12. M. Zulauf and H. F. Eicke, J. Phys. Chem. *83*, 480 (1979).

13.   E. Gulari, B. Bedwell, and S. Alkhafaji, J. Colloid Interface Sci. 77, 202 (1980).

14.   D. F. Nicoli, F. de Buzzaccarini, L. S. Romsted, and C. A. Bunton, Chem. Phys. Lett. 80, 422 (1981).

15.   R. Dorshow, F. de Buzzaccarini, C. A. Bunton, and D. F. Nicoli, Phys. Rev. Lett. 47, 1336 (1981).

16.   C. A. Bunton, L. S. Romsted, and C. Thamavit, J. Am. Chem. Soc. 102, 3900 (1980).

17.   D. E. Koppel, J. Chem. Phys. 57, 4814 (1972).

18.   H. L. Swinney and D. L. Henry, Phys. Rev. A 8, 2586 (1973).

19.   B. U. Felderhof, J. Phys. A 11, 929 (1978).

20.   G. K. Batchelor, J. Fluid Mech. 52, 245 (1972); ibid. 74, 1 (1976).

21.   H. E. Stanley, *Introduction to Phase Transitions and Critical Phenomena*, Oxford Press, New York, 1971.

22.   J. M. H. Levelt Sengers, in *Experimental Thermodynamics*, Vol. II, B. LeNeindre and B. Vodar, Eds., Butterworths, London, 1975.

23.   K. Kawasaki, Phys. Lett. 30A, 325 (1969); Ann. Phys. (N. Y.) 61, 1 (1970); Phys. Rev. A 1, 1750 (1970).

24.   B. Chu, S. P. Lee, and W. Tscharnuter, Phys. Rev. A 7, 353 (1973).

25.   K. Ohbayashi and B. Chu, J. Chem. Phys. 68, 5066 (1978).

Index

1-MONTH